ELECTRICITY PRICING IN TRANSITION

Topics in Regulatory Economics and Policy Series

Michael A. Crew, Editor

Center for Research in Regulated Industries
Graduate School of Management, Rutgers University
Newark, New Jersey, U.S.A.

Previously published books in the series:

ELECTRICITY PRICING IN TRANSITION

edited by

Ahmad Faruqui
Charles River Associates
Oakland, California, U.S.A.
(formerly with EPRI)

and

B. Kelly Eakin
Laurits R. Christensen Associates, Inc.
Madison, Wisconsin, U.S.A.

Kluwer Academic Publishers
Boston/Dordecht/London

Distributors for North, Central and South America:
Kluwer Academic Publishers
101 Philip Drive
Assinippi Park
Norwell, Massachusetts 02061 USA
Telephone (781) 871-6600
Fax (781) 681-9045
E-Mail: kluwer@wkap.com

Distributors for all other countries:
Kluwer Academic Publishers Group
Post Office Box 322
3300 AH Dordrecht, THE NETHERLANDS
Telephone 31 786 576 000
Fax 31 786 576 474
E-Mail: services@wkap.nl

 Electronic Services <http://www.wkap.nl>

Library of Congress Cataloging-in-Publication Data

ELECTRICITY PRICING IN TRANSITION
Ahmad Faruqui and B. Kelly Eakin
ISBN 0-7923-7600-5

A C.I.P. Catalogue record for this book is available
from the Library of Congress.

Printed on acid-free paper.

Printed in the United States of America

CONTENTS

CONTENTS

CONTRIBUTORS

Massoud Amin leads research and development in Infrastructure Security and serves as lead of mathematics and information science at EPRI in Palo Alto, California. Prior to joining EPRI in January 1998, he held positions of associate professor of systems science and mathematics and associate director of the Center for Optimization & Semantic Control at Washington University in St. Louis, Missouri. Dr. Amin is the author or co-author of more than 85 research papers, editor of five collections of manuscripts, and serves on the editorial boards of four international journals. He is listed in various Who's Who volumes and is the recipient of several awards. Dr. Amin holds B.S. (cum laude) and M.S. degrees in electrical and computer engineering from the University of Massachusetts-Amherst, and M.S. and D.Sc. degrees in systems science and mathematics from Washington University.

Ziyad Awad is a partner in Awad & Singer. From 1981 through 1991 Ziyad worked for the Southern California Edison Company in Rosemead, California. During much of that time he conducted and later supervised a market research function. Later, he was given responsibility for managing strategic planning and regulatory interface for Edison's energy management efforts. Since 1991 he has been a partner in Awad & Singer providing a wide variety of management consulting services to electric and combination utilities with a focus on marketing and market organization and structure. He has worked with several companies in transition from traditional monopolies to competitive businesses. He received a Ph.D. in Psychology from the University of Southern California in 1983.

Seth A. Blumsack is a researcher at the Carnegie Mellon Electricity Industry Center, located at Carnegie Mellon University. He has been an energy economist at Economic Insight, Inc., where he analyzed energy and electricity market issues and authored reports for both public and private clients. He was also the editor of Pacific West Oil Data and an occasional editor and contributor to the Energy Market Report. He has also been an economist for the Oregon Department of Fish and Wildlife, working on fisheries management issues.

Steven Braithwait is a Vice President at Christensen Associates. He specializes in competitive retail pricing with an emphasis on measuring customer price response and accounting for the effect of demand response programs. He has assisted clients in developing pricing strategies by designing market-based service offerings, including real-time pricing and time-of-use programs. He is also an expert in load forecasting, and has been asked to defend utility load forecasting practices in the U.S. and internationally. He has provided expert testimony before public service commissions and arbitration panels in the areas of load forecasting and least-cost planning guidelines. Previously, Dr. Braithwait managed numerous projects in load forecasting, demand-side management and planning at EPRI. He has delivered papers at numerous industry conferences on the topics of load response, load forecasting, customer demand management, and competitive pricing. He received a Ph.D. in Economics from the University of California-Santa Barbara in 1976.

Romkaew P. Broehm specializes in economic analysis of the electric utility industry, including market power analysis, evaluation of demand-side management, service unbundling and cost structure, market price forecasts, optimal rate design, and pricing and costing of ancillary services, particularly in the hydro-electric power market. Prior to joining The Brattle Group in 1996, Dr. Broehm was an economic consultant at Christensen Associates where she participated in the development and implementation of a "cutting edge" time- and spatial-differentiated marginal costs in real-time pricing program, interruptible service, and transmission pricing. She helped design a demand model used for optimizing rate designs, and has evaluated customer response to real-time prices. She also taught economics and statistics at the University of Wisconsin-Milwaukee and Cardinal Stritch College. Dr. Broehm holds a Ph.D. in economics from the University of Wisconsin-Milwaukee.

Robert J. Camfield is a Senior Economist at Christensen Associates. His electricity industry experience includes numerous projects involving market organization, pricing strategy, cost assessment, and price forecasting. He has recently managed a large restructuring project for power markets in Central Europe. With a focus on newly emerging market structures, he has developed a two-part tariff approach to price transmission services. This work incorporates location-based usage charges and risk hedging. Mr. Camfield has also developed an innovative approach to pricing unbundled distribution services utilizing marginal cost methods, and is currently Program Director of a national transmission pricing school. In addition, he has testified on numerous occasions, most recently on transmission interface congestion. Before joining Christensen Associates, he worked with numerous organizations within The Southern Company, including Pricing and Economic Analysis, Economic Evaluation, and Cost Analysis, as well as The Southern Company's Strategic Planning group. Prior to that, Mr. Camfield was chief economist for the New Hampshire Public Utilities Commission. Mr. Camfield received his M.A. from Western Michigan University in 1975.

Allan Chung is a Product & Pricing Specialist in BC Hydro's Rates & Pricing Department, where he specializes in the development of market-based pricing programs. His expertise has been utilized in the development of RTP and TOU programs for industrial and large commercial customers. He has also worked extensively in the area of demand-side management. Prior to joining BC Hydro, he co-authored several published papers on economic productivity and growth while working at the University of British Columbia. He received a B.A. in economics from the University of British Columbia and an M.A. in economics from Queen's University, at Kingston, Ontario.

Carl R. Danner is a Director in the San Francisco office of LECG, LLC, where his consulting practice emphasizes public policy affecting energy, telecommunications, and other regulated industries. Previously, he served as advisor and Chief of Staff to former California Public Utilities Commission President G. Mitchell Wilk. Dr. Danner holds a Ph.D. in public policy from Harvard University.

B. Kelly Eakin is a Vice President with Christensen Associates. Dr. Eakin is a specialist in the economic and financial aspects of competitive product pricing. In addition, he has experience with the organization and regulation of industries, and environmental economics. His major projects in the energy industry have included the development of innovative pricing and service designs, the assessment of customer price responsiveness and product choice, and analyses of competitive impacts of restructuring proposals. Dr. Eakin joined Christensen Associates from the U.S. Department of Agriculture (USDA) where he developed expertise in environmental and resource economics. Prior to his service with USDA, Dr. Eakin was on the faculty at the University of Oregon for seven years. At the University of Oregon, Dr. Eakin was active in the graduate programs for the Economics Department and the College of Business. His scholarly writings have been published in a number of prestigious journals including *The Review of Economics and Statistics*, *Journal of Human Resources*, and *Southern Economic Journal*. Dr. Eakin also co-edited the book *Pricing in Competitive Electricity Markets*. He received his Ph.D. from the University of North Carolina-Chapel Hill in 1986.

Ahmad Faruqui is a Vice President with Charles River Associates in the Energy and Environment practice. He leads consulting engagements focusing on pricing strategy for retail and distribution companies. He has also been involved in projects dealing with market forecasting, energy efficiency planning and risk management. In his career, he has held senior management positions with EPRI, Barakat & Chamberlin, EDS, Hagler Bailly, and Battelle. Dr. Faruqui has taught economics at the University of California - Davis, San Jose State, and Karachi University. He has co-edited three books on energy planning and marketing and has written more than 100 papers on energy economics. He holds B. A. and M. A. degrees from the University of Karachi, both First Class First, and a Ph. D. in economics from the University of California, Davis, where he was a Regents Fellow. He is listed in Who's Who in America, and is a Fellow of the American Institute for International Studies.

J. David Glyer is a Senior Economist at Christensen Associates. He specializes in applied microeconomics, program design, and statistics. He has managed projects on, and is an expert in, the design, modeling, and evaluation of competitive pricing products, including real-time pricing, as well as interruptible and standby services. Dr. Glyer has testified in regulatory cases in Wisconsin and Missouri. He has also developed commercial grade PC-based software that simulates customers' hourly demand, customer choice among competing products, and utility profitability from offering specific product lines. He has held faculty appointments at Oregon State University and the University of Colorado-Denver. He received a Ph.D. from Claremont Graduate School in 1990.

Richard L. Gordon is Professor emeritus of Mineral Economics and MICASU faculty endowed fellow emeritus, College of Earth and Mineral Sciences at Pennsylvania State University. Starting with his Ph. D. thesis, Dr. Gordon has engaged in research on many aspects of energy economics. While his initial stress was on coal, the importance of electric power as a market for coal inspired close attention to electric-utility problems. This has produced several books and many articles. He has been consultant to numerous government agencies, consulting firms, and industrial corporations. He holds an A.B. Dartmouth and a Ph.D. from MIT.

Hillard G. Huntington is the Executive Director of the Energy Modeling Forum at Stanford University, where he conducts studies to improve the use of energy and environmental policy modeling. His research publications span a range of interests, including most recently the modeling of restructured electricity markets. He edited several electricity publications including a special issue on transmission pricing in different countries and a book, Designing Competitive Electricity Markets. He is a Senior Fellow and past President of the United States Association for Energy Economics. He has also been Vice President for Publications for the International Association for Energy Economics and a member of a joint US-Russian National Academy of Sciences Panel on Energy Conservation Research and Development and the American Statistical Association's Committee on Energy Statistics. He received a B.S. from Cornell University and a Ph.D. in economics from the State University of

CONTRIBUTORS

New York at Binghamton. Prior to coming to Stanford in 1980, he held positions at Data Resources and the U.S. Federal Energy Administration.

John L. Jurewitz is the Director of Regulatory Policy for the Southern California Edison Company. He holds a Bachelor's degree from the University of San Francisco and a Doctorate in Economics from the University of Wisconsin. Before joining Edison, Dr. Jurewitz was an assistant professor of Economics at Williams College and Pomona College. Dr. Jurewitz joined Edison in 1978. He has testified extensively on a wide range of electric utility regulatory policy issues before the California Public Utilities Commission, the California Energy Commission, the California State Legislature, and the Federal Energy Regulatory Commission. Today Dr. Jurewitz continues to teach courses in Energy Policy, and Environmental and Natural Resource Economics at Pomona College and Claremont Graduate University.

John H. Kalfayan is an independent consultant in Madison, Wisconsin. Previously he was a Senior Economist at Christensen Associates. He specializes in the development of quantitative methods to provide rigorous support for structuring competitive pricing strategies to increase supplier profit and manage risk. He has been responsible for the development of software used industry-wide in the evaluation of a range of market-based products, including interruptible service. Mr. Kalfayan has taught economics at Ripon College and his recent publications include work in econometrics and development economics. He received his M.S. from the University of Wisconsin-Madison in 1988.

Laurence D. Kirsch is a Senior Economist with Christensen Associates in Madison, Wisconsin. He has spent two decades specializing in economic analyses of the electric power industry. He has studied bulk power markets, power pool operations, and utility cost structures. He has developed and applied methods for pricing generation and transmission services, including ancillary services, at the wholesale and retail levels. Dr Kirsch has evaluated the merits of various schemes for auctioning wholesale power, participated in the development and implementation of pricing policies for independent power producers, and assessed a wide variety of retail electricity pricing practices. He is becoming increasingly involved in measuring the potential or actual exercise of market power in wholesale electricity markets. Dr. Kirsch holds a Ph.D. in economics from the University of Wisconsin-Madison.

Jeff Lam is the Manager of Market Access, at Powerex Corp. He facilitates Powerex's access to wholesale power markets in California and the Southwest

United States. Prior to joining Powerex, Mr. Lam managed the retail pricing team at BC Hydro where he introduced several innovative pricing programs. He has over 10 years of experience in the fields of electrical distribution, marketing, demand side management, retail energy pricing, and energy trading. He has a B.A.Sc. in Electrical Engineering from the University of British Columbia.

Karl McDermott is a Vice President with National Economic Research Associates, Inc., specializing in public utility regulation. Since Dr. McDermott joined NERA he has directed and participated in numerous projects in both the energy and telecommunications areas. His main focus has been the development of performance-based regulation mechanisms and advising clients on strategic regulatory options. Dr. McDermott also advises clients on competitive electric and gas markets including regulatory policy, generation location decisions, unbundling, tariff design and corporate reorganization. Prior to joining NERA, Dr. McDermott served as Commissioner on the Illinois Commerce Commission during the negotiation of the Illinois electricity restructuring law. As a Commissioner, Dr. McDermott also lectured extensively in Eastern Europe and South America on regulatory reform and restructuring. He has also assisted the country of Poland since 1994 with their efforts to privatize and restructure their electric supply industry. Dr. McDermott has taught economics at Illinois State University, University of Illinois at Urbana-Champaign and Parkland College and has written on issues related to the electric, gas and telephone industries. Dr. McDermott received a B.A. in economics from Indiana University of Pennsylvania, a M.A. in public utility economics at the University of Wyoming, and a Ph.D. in economics from the University of Illinois at Urbana-Champaign.

Robert J. Michaels is Professor of Economics at California State University, Fullerton and a consultant affiliated with Tabors, Caramanis & Associates of Cambridge, Massachusetts. He holds an A.B. from the University of Chicago and a Ph.D from the University of California, Los Angeles. His writings on deregulation and competition in the electricity and natural gas industries regularly appear in academic and industry publications. His current interests center on the regulatory treatment of market power in emerging electricity exchanges. He has served as an expert in state electricity restructurings, and testified before the Federal Energy Regulatory Commission and the U.S. House Commerce Committee.

Michael T. O'Sheasy is a Vice President at Christensen Associates. Previously, Mr. O'Sheasy directed real-time pricing and other innovative pricing programs at Georgia Power Company, the largest operating company in the Southern

Company system, from 1980 – April 2001. He was responsible for retail and wholesale rate filings and other regulatory requirements, and routinely testified before various commissions on both costing and pricing. He has published numerous articles on pricing in several journals including the TAPPI Journal, Public Utilities Fortnightly, Electric Perspectives, EPRI Journal, and the Electricity Journal. On a national media level, Mr. O'Sheasy has been interviewed in USA Today, Newsweek, and The Wall Street Journal, and on CNN FN and National Public Radio. He received an MBA from Georgia State University in 1974.

Robert H. Patrick is an Associate Professor in the Rutgers Business School, at Rutgers University. His current research includes designing market structure and rules to introduce competition in network industries (particularly electricity and natural gas); incentive regulation, finance and nonlinear pricing options in these industries; and testing the impacts of such on consumer and producer welfare. Prior to joining Rutgers' faculty, he was a Manager at EPRI and held academic positions at Purdue University, the Colorado School of Mines, and Stanford University. He has contributed numerous articles on pricing, regulation, energy, and environmental economics to professional journals and books; served on editorial, governmental, and private advisory boards, nationally and internationally; and is a charter member of the New Jersey Council of Academic Policy Advisors. He earned his Ph.D. in Economics from the University of New Mexico and his B.A., magna cum laude, from Blackburn College.

Peter Fox-Penner is the Chairman of the Brattle Group, specializing in economic, regulatory, and strategic issues in network industries. He advises managements on regulatory issues and corporate strategy in response to restructuring. His practice centers on energy and environmental policies and electric utility deregulation, where he is the author of several books, including the highly acclaimed Electric Utility Restructuring: A Guide to the Competitive Era (Public Utility Reports, Vienna, VA, 1997) and is a frequent expert witness. He is also experienced in the natural gas, communications, transportation and environmental industries. Dr. Fox-Penner serves on the firm's management committee and board of directors. As founding director of the firm's Washington office, Dr. Fox-Penner built the office from one professional to over 40 total staff in three years. He has been named to the Advisory Boards of the Aspen Fund and the Center for National Policy. He has also been a policy advisor in the last three President campaigns and is active in the National Democratic Committee. Prior to joining The Brattle Group, Dr. Fox-Penner was a Principal Deputy Assistant Secretary in the U. S. Department of Energy. In this role he served as COO for a unit of the Department with a budget of $1 billion, 690 full time employees and over 6,000 contractors. The unit, Energy Efficiency and Renewable Energy, funds and oversees most of the Federal Government's research on distributed generation, transportation technologies, renewable energy, and utility policy. He also served briefly as a Senior Advisor in the White House Office of Science and Technology Policy, where he monitored budgets and policymaking for the entire U.S. government's $40 billion civilian R&D program. Prior to federal service, Dr. Fox-Penner was a Vice President at Charles River Associates, a Boston-based economic and energy-consulting firm. Dr. Fox-Penner has a Ph.D. in economics from the business school at the University of Chicago, and M.S. and B.S. degrees in engineering from the University of Illinois, Urbana-Champaign. He lives in Falls Church, Virginia with his wife Susan Vitka, and daughter Emily.

George R. Pleat is the Regulatory Affairs Manager with Constellation Energy Source Company. Prior to this position, Mr. Pleat was a Principal Pricing Analyst with Baltimore Gas and Electric Company (BGE) where he directed the Company's efforts in unbundling electric retail prices and designing fees for supplier services. Over his 17-year tenure with BGE, Mr. Pleat provided the lead for the Company on strategic cost studies and subsequent price design applications. From 1981 to 1984 Mr. Pleat worked with the Minnesota Public Service Department as a utility rates analyst testifying on electric and gas rate design and cost of service issues before the Minnesota Public Utilities Commission. Prior to 1981, Mr. Pleat was a market specialist for the U.S. Postal Rate Commission. Mr. Pleat has a M.A. degree in Economics from George Washington University (1981) and a B.S. degree in Business Administration from Duquesne University (1975). George is currently the Chairman of the Southeastern Electric Exchange Rate Section.

Carl Peterson is a Consultant with National Economic Research Associates, Inc., specializing in energy and public utility regulation. At NERA Mr. Peterson provides research and analysis on a broad range of policy and analytical issues in the electric, gas and telecommunications industries. In particular, Mr. Peterson has worked on performance-based regulation for gas and electric utilities, rate design, energy efficiency and electric restructuring both in the US and internationally. Mr. Peterson has held senior positions with the Illinois Commerce Commission where he advised the Commission on specific energy issues including electric restructuring and performance-based regulation. He has taught economics at Illinois Central College and Eureka College and has written on energy pricing, performance-based regulation, access pricing and electric restructuring. He holds B. S. and M. S. degrees in economics from Illinois State

University, and is currently completing his Ph. D. in economics at the University of Illinois at Chicago.

Anjali Sheffrin is the Director of Market Analysis at the California Independent System Operator (CAISO). In this position she is responsible for managing market monitoring activities for California's deregulated energy markets to ensure an open and efficient market for energy and transmission services. She has testified before the Federal Energy Regulatory Commission, U.S. Congressional Committees, and the California Public Utilities Commission on market power issues. She has 22 years of managerial and technical experience in the electric utility industry working on utility deregulation, market design, strategic business plans, generation and transmission planning, and load and market research. She holds a Ph.D. in Economics from the University of California, Davis. Prior to joining the CAISO, Dr. Sheffrin was Manager of the Power Systems Planning Department at the Sacramento Municipal Utility District. She directed a staff of 40 engineers, economists, and financial analysts. Dr. Sheffrin was the chief economist at SMUD through the closure of the Rancho Seco nuclear power plant and the evaluation of bids for replacement power. She started her professional career as Senior Economist in the Load Forecasting Department of the Potomac Electric Power Company in Washington D.C.

Fereidoon Sioshansi is the President of Menlo Energy Economics, a consulting firm based in Menlo Park California, specializing in market restructuring issues. Dr. Sioshansi is also the editor and publisher of monthly newsletter, EEnergy Informer, now in its 11th year of publication. He has previously been with Southern California Edison Company (SCE), Electric Power Research Institute (EPRI), and National Economic Research Associates (NERA). He is the author of several books and reports and has published widely in professional journals.

Jerry Taylor is the Director of Natural Resource Studies at the Cato Institute and an adjunct scholar at the Institute for Energy Research. Mr. Taylor has written extensively on energy conservation, trade and the environment, sustainable development, electric utility restructuring, and waste management. He has also contributed to several anthologies, including Market Liberalism: A New Paradigm for the 21st Century (1993), The Cato Handbooks for the 104th, 105th, and 106th Congress (1995, 1997, 1999), China as a Global Economic Power: Market Reforms and the New Millennium (1998) and Earth Report 2000 (2000). Mr. Taylor is a graduate of the University of Iowa, where he received a BA in political science.

David J. Teece is the Mitsubishi Bank Professor at the Haas School of Business at the University of California, Berkeley, where hc also directs the Institute of Management, Innovation and Organization, and is Chairman of the Consortium for Research on Telecommunications Policy. Dr. Teece has been Chairman of LECG since 1988. He organized the conference of energy experts that authored the Manifesto.

Samuel A. Van Vactor is the President of Economic Insight, Inc. and a researcher at the University of Cambridge. He has been economist at the U.S. Treasury and the International Energy Agency. He has taught economics for the University of Maryland, European Division and Portland State University and has lectured on energy economics on six continents. He is the author of over two-dozen books and articles concerning energy economics and the market structure of the energy industries. He has appeared frequently as an expert witness in the valuation of energy commodities and associated regulatory issues. Economic Insight, Inc. publishes the Energy Market Report and Pacific West Oil Data.

Peter VanDoren is the Editor of *Regulation* magazine, published by the Cato Institute, and Associate Professor of Public Policy at the University of North Carolina at Chapel Hill. Dr. VanDoren has written extensively on the political regulation of banking, housing, land, energy, transportation, health, environment, and labor markets. He is the author of Politics, Markets, and Congressional Policy Choices (1991) and Cancer, Chemicals, and Choices: Risk Reduction Through Markets (1999). Dr. VanDoren received his Bachelor of Science degree from the Massachusetts Institute of Technology and his masters and Ph.D. degrees from Yale. He has taught at the Woodrow Wilson School of Public and International Affairs at Princeton University, at Yale University's School of Organization and Management, and in the Public Administration Program at the University of North Carolina at Chapel Hill.

Daniel Violette is a Principal in the strategy practice at Summit Blue Consulting. Prior to helping found Summit Blue, Dr. Violette co-managed the retail strategy practice at PHB Hagler Bailly. At Hagler Bailly, he was one of the founders of the firm and he led the firm's utility practice for over a decade. He held corporate positions of Sr. Vice President in Law and Economics and as a Director with the firm. Dr. Violette is recognized in the utility industry for his leadership in the development of effective market strategies and use of quantitative methods to support utility decisions. Dr. Violette currently serves as the Vice Chair of the Peak Load Management Alliance (PLMA) comprised of energy providers and large consumers and he served three elected terms as

President of the Association of Energy Services Professionals (AESP). Dr. Violette previously held Sr. Vice President positions with Electronic Data Systems (EDS Management Consultants) and XENERGY, Inc. Dr. Violette received his Ph.D. in economics from the University of Colorado.

Michael V. Williams is the Director of Strategy Planning for XENERGY Inc. He has experience developing market and asset strategies for large investor owned, as well as publicly owned utilities, His experience covers market strategy development and modeling, asset acquisition strategy, integrated planning and real and strategic options analysis. He has worked in the U.S. and internationally and has extensive experience in California. Dr. Williams began his energy career 25 years ago as a manager for Southern California Edison and has worked on many issues within California, first at the utility and subsequently as a senior consultant to several California utilities. Dr. Williams has provided high level strategic consulting to utility executives as well as technical model development and data analysis.

FOREWORD

One of the popular premises for the deregulation of electricity was that customers want choice and we all thought that meant choice of suppliers.

Maybe that case can be made for large commercial and industrial customers, but the vast majority of other customers have little interest in having a choice of supplier. It turns out that all customers, regardless of whether they are in a state that has deregulated or one that hasn't, and also regardless of whether they are big or small, want to have choice-- but not necessarily the choice of suppliers. What they want to choose is whether or not to purchase electricity, and how much electricity to purchase. To make that choice, they need to know the price of electricity at the time they are making the buying decision, and they need to know what their usage habits are likely be over the time frame they will be paying that price.

The public policy benefits of dynamic pricing are becoming better understood and well known. Overall cost reduction for all customers and the associated environmental benefits have been identified and several thoughtful papers have now been written, but the customer choice benefits are just now being evaluated and understood.

Contrary to popular belief, it is not just the large commercial and industrial customers who can and will respond to dynamic pricing signals. Residential and small commercial customers are demonstrating a willingness and a desire to have the information that allows them to manage and control their energy usage and their bills. So while they may not be interested in choosing their supplier, they are interested in the choice of how much electricity they purchase based on how much they have to pay for it.

In a large scale pilot program involving 300,000 residential and small commercial customers, Puget Sound Energy has shown conclusively that customers do shift usage away from high priced periods towards lower priced periods, even with modest differentials in electricity prices. Peak usage has gone down by about 5% and off-peak usage has gone up by about 5%, consistently over a period of ten months beginning with May 2001. This program has demonstrated that residential customers have an overwhelmingly positive reaction to having timely usage and price information to manage and control their usage. Ninety percent of surveyed participants said they had actually taken actions to both shift and reduce their usage while over 90% said they would recommend the program to their friends and neighbors.

For the electricity industry this is very good news. It means that for customers to have choice it doesn't matter if their state has deregulated its electricity business or not. It means that provided with timely usage and price information, individual customers can attain the benefits of lower electricity bills through there own actions. Most importantly, what it really means is we can achieve many if not most of the benefits associated with deregulation without having to restructure the entire industry.

Dynamic pricing, with its associated benefits, has the potential to help reshape the electric business in a way that is better for all. The essays in this book deal with various aspects of pricing electricity, as the industry makes a transition to a competitive environment. Written by practitioners, academics and researchers at the cutting edge of change, they should be of great interest to utilities and regulators alike.

Gary B. Swofford
Senior Vice President &
Chief Operating Officer
Puget Sound Energy
Bellevue, Washington

ELECTRICITY PRICING IN TRANSITION
PREFACE

In the winter of the year 2000, we co-edited a book in this series on *Pricing in Competitive Electricity Markets*. At that time, the California market had been operating satisfactorily for two years, and many other markets were on their way to being opened. Analysts envisaged a future in which:

- More than half of US population would reside in states with restructured power markets

- Most customers would have a choice of retail energy service provider

- There would a veritable cornucopia of pricing products in the energy marketplace.

We anticipated that full retail and wholesale competition would become the norm in a few years time. In our opening chapter, we spoke optimistically of innovations in electricity pricing that would involve the provision of a wide range of risk-management services, and of creation of new opportunities for creating value by bundling the core energy product with several value-added services.

We recognize that our anticipation of retail competition has not been realized. To our chagrin, we are retreating and titling this book *Electricity Pricing in Transition*. Nevertheless, we maintain our optimism about the potential for competitive electricity markets and believe competitive retail electricity markets ultimately will come to be.

Regrettably, few of the pricing and product innovations envisioned in the earlier book have emerged in the marketplace. We believe this is due, in large part, to the California energy crisis. The California market exploded in May 2000, and the subsequent events resulted in the bankruptcy of one of the nation's largest utilities.[1] Then came an inevitable rate increase in California in the spring of 2001, the entry of the state of California as a power supplier during

---

[1] For an overview of what wrong in California, see Ahmad Faruqui, "California's Travails: Guest Editorial," *The Electricity Journal*, August/September 2001, pp. 65-67. A more detailed analysis is contained in Ahmad Faruqui, Hung-po Chao, Vic Niemeyer, Jeremy Platt and Karl Stahlkopf, "Analyzing California's Power Crisis," *The Energy Journal*, 22 (4), pp. 29-52, 2001.

the summer of 2001, and as this book goes to press, the state of California is planning to issue revenue bonds to pay for the price of electricity. In order to enable these bonds to sell at reasonable prices, the state of California was forced to annul the provision of customer choice that had been a key feature of the state's restructuring plan.

Ironically, in a paper published in the spring of 2001, Steve Braithwait and Ahmad Faruqui have shown that a way to solve the crisis in California's electricity market would be to implement real time pricing for large commercial and industrial customers. Simulations with EPRI's models and databases pertaining to real time pricing showed that voluntary implementation would reduce peak demands by 2.5%, ultimately reducing wholesale prices by 24% during peak hours. That is, the presence of real time pricing would help reduce wholesale power costs and price volatility, benefiting not only the customers who are on RTP but all customers.[2] Furthermore, with two-part real time pricing, participating customers would become "virtual generators" and market power concerns would be significantly mitigated.

The California Energy Commission (CEC) has embraced real-time pricing for its ability to alleviate the crisis by introducing demand response into the retail market. In June 2001, the CEC filed a voluntary two-part real time pricing tariff with the California Public Utilities Commission (CPUC). The CEC proposal was modeled after the very successful program pioneered by the Georgia Power Company in the 1990s. Citing a variety of concerns, the CPUC did not adopt the CEC proposal. While the CPUC remains on the record in favor of a real time pricing program, as of this writing no program has been approved. California's inability to implement real time pricing in the summer of 2001 shows that, in the real world, implementation can never be expected to flow automatically.

California's crisis sent shock waves throughout the world of electricity. Its most immediate impact was to slow down the move toward full competition in the United States. Not just in California, but also in Pennsylvania, many customers who had switched providers were returned to their local distribution utility, which was required to serve them at a default price that does not fully reflect market risks created by wholesale price volatility. This happened in many cases because the energy service providers were unable to procure wholesale power at a lower cost than the retail rates they were contractually

---

[2] See Steve Braithwait and Ahmad Faruqui, "The Choice Not to Buy," *Public Utilities Fortnightly*, March 15, 2001.

PREFACE

locked into. In other cases, it happened because the retail providers went bankrupt. Finally, in some cases it happened because the customers had not locked in attractive fixed prices with their retailers, and found it advantageous to return to their utilities and obtain power under fixed default service rates.

In today's chaotic and rapidly changing environment, a whole host of new issues have emerged. Utilities need to deal with surviving and ultimately prospering in an environment where retail and wholesale markets are often disconnected. Retailers need to demonstrate capability to prospective customers, and deliver on it. Regulators need to provide a meaningful set of rules for market participants, and ensure that the benefits of competition flow through to all customers. Customers need to make well-informed choices that improve their economic well being. The new issues facing utilities, retailers, regulators and customers include:

- Defining and executing a sustainable value proposition for customers that builds upon – but doesn't rely exclusively on – price

- Developing a cost-effective acquisition strategy for new customers, in which marketing and sales costs are balanced with expected customer profitability

- Investing in world-class back-office systems for customer relationship management and billing, so that the company can deliver on its promises embedded in its market offerings

- Developing a pricing strategy during this transition period and (perhaps) a different strategy for the steady-state competitive market

- Understanding the role that market-based demand response programs play in bridging the gap between retail and wholesale markets[3]

- Developing a portfolio of short, medium-term and long-term supply contracts in a retailer's asset portfolio, in order to manage price and volume risk

---

[3] Ahmad Faruqui, Hung-po Chao, Victor Niemeyer, Jeremy Platt and Karl Stahlkopf, "Getting out of the dark: market based pricing could prevent future crises," *Regulation*, Fall 2001, pp. 58-62.

This book is written to address many of these new issues, which characterize the industry in transition. It is organized into five sections. Section I deals with the new restructured organization that has emerged from yesterday's vertically integrated, regulated monopoly company. Mike O'Sheasy and Robert Camfield lay out the critical role that pricing will need to play in the restructured companies of tomorrow. George Pleat discusses how to price generation supplier coordination obligations, and argues that this is a new service that investor-owned utility distribution service companies can provide. Massoud Amin lays out a vision of restructured electric enterprise, and Richard Gordon surveys the practice of industry restructuring.

Section II deals with issues in competitive pricing. John Jurewitz discusses the specific challenges that underlie designing default service. Robert Michaels reviews different approaches to pricing electric capacity. Sam Van Vactor and Seth Blumsack review approaches to making competition succeed in electricity markets. The two of us, Kelly Eakin and Ahmad Faruqui, examine whether market-based pricing is a form of price discrimination. And Mike Williams and Ziyad Awad discuss the role of market research in a competitive environment.

Section III reviews the role of demand response and product design in today's chaotic marketplace. Romkaew Broehm and Peter Fox-Penner argue that price responsive demand is a national necessity, and not just an option anymore. Dan Violette assesses price-responsive load among mass-market customers. Steve Braithwait and Mike O'Sheasy survey the empirical evidence on customer response to real time pricing. Perry Sioshansi asserts that the hour for "negawatt pricing" has finally struck. Allan Chung, Jeff Lam and William Hamilton describe a case study of innovative retail pricing in the Pacific Northwest. Robert Camfield, David Glyer and John Kalfayan bring forth the notion that electricity products can be designed by customers, rather than being developed by utilities. Finally, Robert Patrick and Frank Wolak discuss the development of demand-side bids, using information on customer response.

Given the signal importance of California's energy crisis, and the fact that it will be studied for years to come, we have devoted Section IV to studying the lessons learned from this crisis. Carl Danner, James Ratliff and David Teece discuss the California Manifesto that was developed to provide guidance to the state in responding to the crisis. Jerry Taylor and Peter VanDoren analyze what went wrong in California; who is to blame; and what to do to fix the crisis. Anjali Sheffrin provides empirical evidence on strategic bidding behavior in the ISO Real-Time Market. Elizabeth Farrow summarizes the results of a Stanford University Energy Modeling Forum conference on retail participation in competitive power markets.

PREFACE

The final section of the book deals with markets and regulation. Hill Huntington assesses the outlook for electricity prices in the US, using a multi-model evaluation. Karl McDermott and Carl Peterson discuss the essential role of earnings sharing mechanisms in performance-based regulation programs. Finally, Laurence Kirsch discusses how transmission affects market power in reserve services.

The summer of 2001 was mild in much of the country. Loads stayed under their historical peaks, and electricity prices were substantially lower than in the year 2000. California did not experience a single hour of an electric outage, let alone the hundreds of hours that were widely forecast to occur. Peak demand was lower by 7,000 MW, compared to a year ago, after normalizing for the effects of weather and economic activity. Of this, about 5,000 MW was attributed by analysts to the effects of the state's aggressive conservation programs and 2,000 MW to customer "belt tightening" that was induced by the Governor's appeals, unprecedented daily media coverage, and customer's desires to avoid triggering a power blackout.

As we enter the year 2002, some are calling the California crisis a one-off instance off a "perfect storm." Common sense tells us that this not the case. Much of what happened in California can occur elsewhere. The good news, however, is that crises like California's can be managed. This requires the use of risk management instruments to hedge asset portfolios, and by providing the necessary incentives to suppliers to add generation capacity and to buyers to use energy efficiently. However, there is no way to prevent electricity prices from displaying volatility, given that electricity cannot be stored economically in large quantities. Low prices this year don't mean that prices will always be low. Years of alternating high and low prices are a confirmation of price volatility. It is our view that price volatility will continue to be the hallmark of a restructured market, and that there will be no move back to a vertically-integrated industry structure. Congress will not repeal the Energy Policy Act of 1992. However, volatility is not an evil to be avoided at all costs. It creates opportunities for creating value, by risk arbitrage.

A key aspect of successful market design is demand response. As noted by Commissioner William Massey of the Federal Electricity Regulatory Commission (FERC), this will require action from both federal and state regulators:

> Load serving entities buy in FERC jurisdictional wholesale markets but sell in state jurisdictional retail markets. The Commission can facilitate demand responsiveness by requiring that RTOs allow the

demand side to bid so called "negawatts" into the market along with supply side resources, and we can facilitate the resale of consumption rights. However, it is up to the states to facilitate both the right price signals reaching customers and customers ability to react to those signals.[4]

It is our hope that the essays in this book will provide practitioners with guidance on how to avoid the major pitfalls to in pricing electricity while the market is in transition. We hope they will draw upon the insights and lessons learned from the experience of others that are documented in this book. We hope that our work will also inspire theoreticians at the cutting edge of methodology development, by exposing them to some of the industry's really hard problems that no one has yet solved.

A.F.
Danville, California

K.E.
Madison, Wisconsin

---

[4]   Remarks in New York City, New York, November 15, 2001.

Section I

The Restructured Organization

Chapter #1

Profiling the Critical Role of the New Pricing Organization

Mike O'Sheasy and Robert Camfield
Laurits R. Christensen Associates, Inc.

Abstract: This paper describes the changing role of the electric rates group within evolving power markets. In the world of franchised utilities, the rates group was focused on gaining approval from regulators and implementing tariffs that satisfied revenue needs. The Energy Policy Act of 1992 provided the framework for competition in wholesale generation markets, which have since flourished under open access transmission. Since 1992, competition has also nibbled away at the periphery of retail markets, and retail access is being implemented in several states. No one is surprised about the enlarged role of the rates group – or perhaps more appropriately, the pricing organization – as the design of retail services is now strategically important. This paper describes how the rates group is changing, and presents a vision of the pricing organization of the future

Key words: Pricing organization; rates department; electric restructuring; reorganization; transitioning to competition.

1. INTRODUCTION

Faced with competition from all corners, electric utilities are redefining products and services, with increased emphasis on pricing. As ratepayers have become customers and rates have become prices, the rates organization is evolving into the pricing organization within the incumbent providers.

Price attributes are fast becoming the features that distinguish service providers, and the pricing organization is being asked to play a pivotal role in strategically repositioning the host supplier within incumbent markets.

A. Faruqui and B.K. Eakin (eds.), Electricity Pricing in Transition, 3-17.
@ 2002 *Kluwer Academic Publishers.*

This enlarged purview is the subject of our discussion. We explore the questions: "What is the emerging role of the pricing organization, and what are the processes and functional activities necessary to fulfill this enlarged responsibility?" We begin with a brief review of market restructuring and the expected reorganization initiatives by incumbent utilities. We then discussion the modus operandi of the rates organization under the traditional mode of setting prices, and present a vision of how pricing fits into the dynamics of competitive market arrangements. The end state of openly competitive power markets will be reached only gradually, and thus we conclude with a detailed review of the emerging pricing organization within a process referred to as transition mode.

2.   RESTRUCTURING

Restructuring of the electric services industry (ESI) is accelerating worldwide, and the implications are both predictive and instructive. The consensus view holds that integrated services will be unbundled into essential components. Generation services including energy and reserves will be competitive, while transport services including network transmission and distribution wires services will remain largely regulated services.[1,2] The functions that have become known as customer services including metering and billing may also be rendered competitively. Merchants will procure and rebundle electric service components into retail products and, given perceived preferences for risk, employ the principles of modern finance to tailor such products to the needs of customers. Electric services may be bundled with other energy and non-electron services. The table below describes the key pricing characteristics of the restructured ESI in terms of the traditional functions of the electricity business.

---

[1] Transmission and distribution services can be further broken into transportation – i.e., the transport of energy and reserve services from the locations where they are produced to the locations where they are consumed – and connection services.

[2] Transmission can generally be distinguished from distribution by voltage level. Within the U.S., transmission service is often defined as transport services rendered at voltages equal to or greater than either 69 kV or 115 kV, although numerous exceptions can be cited. Also, the technology used to supply transmission service is usually a meshed network topology, whereas distribution service is usually supplied with a radial or loop configuration.

Table 1: Function

	Generation Services	Transmission	Distribution	Merchant Services
Price and Service Drivers	Competitive regional marketplace	Market design features	PBR and Standards Service Quality	Competitive marketplace Economies of scale and scope
Product Innovation Opportunities	Largely unencumbered bilateral arrangements	Limited to − PBR and − LMP	Moderate − Distributed resources − Cable TV, data transmission	Numerous: augmented by supporting information
Profit Risk Considerations	Significant, though hedging strategies and asset diversity can mitigate risks	Moderate, though risks may be larger under PBR price setting regime	Moderate, though profit risks may be significant under service quality regulation	Significant, though vehicles to hedge risks will proliferate

The unbundled markets paradigm implies choice: retail customers will have numerous service options as products and suppliers proliferate – at least initially. Motivated to sustain profits, suppliers will be largely unencumbered in the design of service offerings. We need not worry about price gouging as we entrust competition to discipline the behavior of suppliers and maintain prices close to economic costs. In the world of unbundled services and open access, the rates organization finds itself at the nexus of power markets. Price is the quintessential attribute of retail products, and *pricing* is the primary instrument to beget profits and sustain market share. The pricing function now assumes a vastly expanded and strategic role that contributes immeasurably to the success of the firm.

3.   REORGANIZATION

Market restructuring and service unbundling affects *all* facets of our industry. The direction of reorganization will be determined by regulatory requirements,[3] new market arrangements, and the perceived profit potential within various niche markets.[4] [5] The implied changes are manifold, and many

---

[3]   For example, regulatory concerns about market power in generation services may give rise to the prescriptive sale of incumbent providers' generation assets.

[4]   Profit potential and other advantages to expanding these functional areas are inherently determined by internal capabilities, which can generally be described as core competency and economies of scale and scope.

incumbent providers will redefine themselves in ways that better align capabilities to markets.

Major market areas will be exited as assets – in particular, generation assets – are auctioned to large, specialized firms. Reorganization can produce essentially new firms as internal activities are abandoned, curtailed, or expanded. Redefined incumbents will integrate and merge complementary resources into expanding functions and activities. The drive for increased efficiency and effectiveness will cause internal hierarchies and departmental structures to evolve toward highly focused processes. The internal processes and activities of the firm will be much more tightly aligned with products and the markets in which those products are offered. The focus will be on performance and the contribution of various internal processes to the success of these products, and thus the firm will be assessed.

Within newly reorganized service providers, some functions and activities that were heretofore essential to the achievement of the firm's objectives may become less influential; other organizations and functions will flourish. Routine activities that lend themselves to economies of scale and scope may be outsourced to specialized firms. To better enable the firm to manage risk and revenue within new market conditions, new activities will arise.

4.   TRADITIONAL MODE

In the old order of franchised monopoly, tariffs and standards of service define the relationship between the utility and its customers. The governance structure of agency regulation establishes these tariffs and standards through a longstanding system of administrative procedure. In large part, this institutional arrangement also determines the role of the rates organization, which is deeply integrated within the procedural process.

In traditional mode, overall prices are determined under the auspices of the *compensatory rates standard*, which essentially affirms that the service provider is entitled to recover the costs of resources devoted to the convenience and necessity of the public. Embedded cost-of-service (COS) is the core paradigm to determine prices. In principle, average prices are set at

---

[5]   Recently, the industry media are replete with a host of announcements by incumbent providers about plans to enter various market niches. Examples include distribution operations, distribution design and construction, transmission planning and construction, generation plant operations, nuclear relicensing, meter maintenance, customer billing, and nationwide retail power supply under the auspices of highly valued incumbent brand names.

a level sufficient to recover prudently incurred costs, including operating expenses and capital charges on investment-valued at vintaged cost. While details across regulatory jurisdictions vary, the first step in setting rates is for the incumbent service provider – often an integrated electric system – to file for a rate change.[6] Preparations for the incumbent's rate filing can be described in three broadly defined steps:

1. Determine revenue requirements based upon total embedded costs.[7] [8]
2. Perform a COS separation study. This study allocates shares of the total revenue requirements to customer groups or classes – i.e., fully-distributed costing.
3. Develop class rates, which are designed to obtain a flow of revenue across classes equivalent to total costs.

While COS separation studies are rarely used to actually determine class revenues, they influence the *direction of the allocation* of revenue requirements among classes.

In traditional mode, the rates group generally resides within the utility's financial organization and is at the center of the process to determine regulated prices. Setting prices involves individuals at all levels, and the rates group often coordinates the internal rates process, which involves strategy, analysis, and decisions within the utility. Using as input historical and projected revenues, sales, and total costs provided by others, the rates group typically performs cost allocation studies and develops average prices for classes and rates for the various tariffs. Within the regulatory proceeding, the rates group will present and defend rate proposals. The process of administrative procedure can be a rather open yet contentious proceeding, with considerable disagreement among parties about pricing philosophy and the assumptions that underlie the overall revenue level and its allocation to customer groups or classes.

---

[6] The highly detailed scrutiny given rate change filings over the past two decades has vastly expanded the amount of formally prepared information and data that accompanies a filing. Many regulatory agencies have established standardized filing requirements and other procedural mechanisms to manage the enlarged information flow and streamline the process of setting prices for utility services.

[7] Revenue requirements are set according to a notion of total costs. Generally speaking, revenue requirements are based upon either the observed historic costs within a recent consecutive 12-month period, adjusted for the effects of weather on revenues and costs, or estimated total costs for a forward period.

[8] The traditional mode of setting prices is complicated by the numerous accounting procedures used to compute embedded costs, the forecasts of costs and sales levels, and the influence of stakeholders.

Let's examine the rate setting process of the traditional mode more closely with the following example: a TOU rate option.

4.1   Rate Design Problem:

In view of new rate options available to customers, the utility faces the issue of customer migration to the new rates from existing tariffs. Customers can be expected to select the least expensive rate option facing them, and migration from one tariff to another can affect realized revenues and lead to revenue erosion.   Traditionally, the utility might estimate customer migration and revenue impacts on the basis of differences in bills, given the billing determinants of eligible customers.

Assumptions:

a) Incumbent Electricity Company (IEC) files a rate case requesting an overall revenue requirement of $1.2 billion.[9] A class cost-of-service study reveals class revenue responsibilities of $400 million, $500 million, $250 million, and $50 million, respectively, for residential, commercial, industrial, and street-lighting customer groups.

b) A new time-of-use (TOU) rate option for commercial customers whose load is equal to or greater than 1 MW is desired. The regulatory agency prescribes that the rate design for the rate option will include demand charges. Since IEC is capacity constrained during summer peak loads, the commission's objective is to discourage summer load, thus avoiding additional generating plants. The technical staff of the regulatory agency indicates that the rate option should be sufficiently attractive that 10 percent of the commercial customers will select the TOU option.

c) Prices of current tariffs plus new rate options will be set (adjusted) to levels that match the commission-approved overall and class revenue requirements.

4.2   TOU Design Process:

Let's now examine the manner in which the TOU rate is traditionally developed. This process begins with key input assumptions and COS study results:

---

[9]   The *filed for* revenues can be factored down significantly by regulatory agencies. Indeed, the experience of recent years suggests that the regulatory agency rarely allows the incumbent utility to implement an overall rate level equivalent to the level requested in the filing.

1. A *migration target* for the TOU rate option is set at 10 percent, as mentioned above.
2. The embedded cost study suggests that, for the commercial class:
   - the customer-related costs are $50/customer/month;
   - the demand-related costs per monthly coincident peak kW is $20/kW;
   - the energy-rated costs are 3¢/kWh.

These inputs are used to obtain initial prices, as follows:

First Iteration Prices:

Eligibility	=	_____
Customer Charge	=	$25/customer/month
Demand Charge	=	$17/summer month/customer maximum monthly peak kW; $10/non-summer month/customer maximum monthly peak kW
Energy Charge	=	8¢/kWh during the summer on-peak hours of 10 am – 10 pm; 5¢/kWh during other summer hours; 4¢/kWh during non-summer hours

A comparative bills study suggests that 8 percent of the eligible commercial customers will select the TOU option, short of the 10 percent target. That is, the comparison study reveals the option to be the lowest cost alternative for 8 percent of the commercial customers. The study presumes that those customers that realize bill savings under the option will choose it. The study outcome is based upon rather heroic assumptions: *1) historically observed billing determinants do not change under the option, 2) customers have equivalent information to make comparisons and decisions, and 3) customers that experience bill declines choose the option.* Through a second iteration analysis, the target participation shortfall gives rise to revised prices, as follows:

Second Iteration Prices:

Eligibility	= No change
Customer Charge	= No change
Demand Charge	= No change for summer months; = $9/non-summer month/customer maximum monthly peak kW
Energy Charge	= No change for any hour type

The bill comparison study now suggests that 10.25 percent of the commercial customers will select the TOU option, closely matching the target migration level.

The final step is to adjust the commercial class billing determinants to reflect participation on the TOU option, and adjust the allowed class revenue requirement accordingly. For our example, billing determinants for the existing commercial tariff now represent only those commercial customers that are expected to remain on that tariff, or 89.75 percent of the commercial class.[10] Also, revenue erosion estimates due to migration to the TOU option imply adjustments elsewhere to maintain overall revenue neutrality.

The traditional mode of determining prices has serious limitations, and is quite distinct from the dynamic interaction inherent to competitive markets consisting of numerous buyers and sellers. We can distinguish the traditional mode in at least four ways. First, tariffs and pricing serve as revenue collection devices rather than tools to create customer value and improve the competitive position of suppliers. Second, while the traditional mode contains considerable give and take, the process of setting prices is a rather mechanical, computational task *once assumptions and data inputs are defined*. Third, the process focuses on embedded costs rather than economic costs.[11] Fourth, the quantitative analysis that underlies the determination of prices ignores customer behavior, including the selection of rate options and load response to energy and demand price

5.   COMPETITIVE MARKETS MODE

It is sometimes stated that the unbundling of electric markets causes electric service to be revealed as a commodity. In view of the inherent complexity of the technology of electric supply, we find this view to be a simplified interpretation, and in stark contrast to the proliferation of choices that is beginning to appear in the marketplace. In addition, the preferences of customers as reflected in their selection behavior when confronted with rate choices, can vary greatly. If adequately understood, customer preferences for product attributes can be exploited for the development of differentiated products. Knowledge of customer preferences thus provides a powerful vehicle to maintain market share and profits. Arguably, efficient

---

[10] It is likely that the TOU price terms would also be fine tuned in order to account for the affects of the adjusted existing rates. This step is ignored here for the sake of simplicity.

[11] Economic cost refers to the value of resources as determined by competitive market processes. Alternatively, economic cost can be described as opportunity cost, or the highest-valued alternative employment of a resource.

pricing and innovative product design become the only remaining components in retail markets for improving margins and achieving strategic market goals.

As so often expressed, electric services will be driven *by the needs of the marketplace.* More precisely, we expect that the design and packaging of retail services will be a highly focused process geared to the perceived attribute preferences of customers. What does this mean? We can start by defining electric service as the delivery of power to the premises and facilities of customers, which are specific locations within power networks. When one looks more closely, however, the provision of electric service at individual locations has at least three basic characteristics or dimensions, including the quality of the electricity flow and signal, the continuity of service (*i.e.,* reliability), and the amount of money paid for it.[12] The latter two dimensions, in particular, have significant dimensions of risk. That is, service reliability and price – as with many services – are inherently stochastic. Indeed, because of the characteristics of non-storability and locational externalities of power networks, electric service can exhibit exceptionally high price variation, as recent wholesale price experience suggests. Attributes of risk – or, attribute combinations and sets – become the basis upon which retail electric service offerings can be differentiated.[13]

Pricing, then, is a matter of determining the price for various packages of electric service, differentiated by the risks that attend service attributes. Packages of service options are packages of risky services; customers can select the attribute bundle that allows them to manage risk consistent with their needs and preferences. Customer selection of an option reveals the customer's preference for risk.

Because customers are generally risk averse, they are willing to pay more for low-risk product bundles (or packages) than high-risk bundles. Conversely, service options with low service risks are more costly to provide than high-risk options. The service provider is the arbiter of service risks by *bundling attributes (as insurance) within electric product offerings.* Thus, the key to developing and pricing optimal service design packages is to understand customer preferences for the attributes of services, differentiated by risk, and the implied costs of offering them. Customer behavior and economic costs are the foundations for pricing and electric product design

---

[12] The method of payment can be interpreted as a fourth dimension.

[13] At the most general level, price risk attributes can be differentiated according to, for example, the variation in prices, the frequency of price changes, and the notice of price changes. Similarly, the reliability of power supply can be differentiated in terms of frequency of power outages and notice of power outages. Moreover, reliability attributes provide a basis for a price-differentiated menu of service, from which customers can self-select the level of service reliability that they prefer.

within fully competitive retail power markets. Product design will become a strategic core competency of the firm; it is what the firm must do best.

Acquiring knowledge of the behavior of customers and economic costs necessary is a highly technical undertaking, and a plausible road map is available. Surveys, conjoint analysis, and other market research tools can be used to estimate customer preference surfaces for product attributes. These surfaces are a quantitative interpretation of customer preferences, and they provide a basis to estimate customer selection of product options from a menu of choices. Along with econometric estimates of customer response to the price terms of service options and knowledge of economic costs necessary, optimal product packages can be developed.

Because pricing is the salient dimension differentiating electric products, we anticipate that the pricing organization will be central to carrying out much of the research and analysis that underlies the development of profitable service packages. This purview reflects the newfound importance of pricing and service design, and implies changes for the pricing organization, including functions, tasks, skill sets, methods, and tools.

Whereas the rates group of the traditional mode is often staffed by individuals with general business backgrounds, we observe that these core staff are fast being augmented by a cadre of individuals with backgrounds in economics, engineering, finance and risk management, statistics, and market research. The modern pricing organization of the competitive mode will be staffed by individuals facile with computer modeling, information systems, and data management. We anticipate that the pricing group will become more deeply integrated into marketing, as service providers become *process oriented*. Pricing is inseparable from marketing. Because the process of product design reaches into many areas of the incumbent supplier, product development and implementation will involve several organizations and many individuals. We anticipate that the pricing organization can assume a coordinating responsibility for the design process and also contribute to implementation.

It is perhaps useful to describe a vision of the service design process of the future. This on-going process will likely center on monitoring and strategy, and will heavily utilize information systems to integrate data regarding sales, market share, revenue, and prices of wholesale services.[14] We expect that successful merchant firms will view retail service as a portfolio of contracts, and that these data, gathered through the monitoring function, will be used to value the portfolio in terms of economic profits (or

---

[14] Within the context of the end state of competitive markets, wholesale services include generation energy and ancillary services, transmission services, and distribution wires services.

margin) and risk in on-going fashion. In fact, the retail book can be valued daily. The monitoring function can also track the attitudes and views of customers regarding the delivered service quality provided by the host supplier via survey methods and other means.[15] It seems natural to place these monitoring activities within the financial and marketing groups of the firm.

The pricing organization will rely upon the monitoring activity and long-term strategy to guide product development initiatives. The development process will likely be on-going, and proposed new product offerings will undergo an initial screen prior to broader review within the firm. The screen will search for feasible candidates, and is likely to focus on product concepts. The initial screen is likely to be largely qualitative, and will rely upon on-going market monitoring and research, management's sense of power markets, and feedback realized from direct contact with customers. Nevertheless, the screen should delve into certain analysis issues, such as market targets, market acceptance, expected net profitability, response by competitors, and product development costs. Feasibility issues such as implementation and administration costs, billing, and metering complications are important, and thus marketing, customer information, and billing organizations need to be involved early on. As both coordinator and leader of the product design process, the pricing organization can also assume responsibility for ensuring the adequacy and completeness of the screening process. Once approved, the development process commences.

At its core, formal development is based upon quantitative evaluation aimed at determining the attribute set and price terms for candidate product concepts that survive the screen. As alluded to above, the process can begin with market research focused on customer preferences for product attributes and risk features. Conjoint analysis and other formal statistical techniques can be applied to market data. These data, gathered through survey methods, can be used to describe customer response to and selection of products *with respect to a change in an attribute of the design.* In competitive markets, we can expect that market research and analysis will become truly integrated into the design process.

We anticipate stark contrasts in product pricing between the traditional and competitive market environments, as shown below:

---

[15] Several electric systems are currently monitoring customer satisfaction.

Table 2: Contrasts in Market Environments

	Traditional/Regulated Market	Competitive/Less Regulated Market
The Role of Price in Markets	• Collect revenue requirement • Induce load shape changes via DSM	• Maximize profitable contribution • Enhance customer value • Achieve market share
Considerations of Price Setting	• Parity effects • Peak load impacts • Revenue requirements • Political consequences • Little consideration of marginal cost	• Load response and customer selection • Marginal cost floor • Price of wholesale services • Market share • Time to market
Resultant Price Setting Characteristics	• Regulatory commission approval • 6 months – 1 year implementation timeframe • Long product life cycle • Fixed price • Emphasis on load research	• Marketplace acceptance • Quick concept to market implementation time frame • Emphasis on market research

6.   TRANSITION MODE

6.1   The Context and Problem of Transitioning

Retail markets in several regions are currently evolving toward open access market regimes. While the end state of retail markets may not be present currently, it is clearly anticipated and retail access regimes are being selectively implemented in several states.[16] [17] We refer to this state as *transition mode*, a market state that contains pressures to deviate from longstanding class-wide tariffs based upon cost-of-service principles. We might say that transitioning has elements of the competitive markets mode.

In transition mode, competition is nibbling at the periphery of franchised monopoly and is manifest in various forms of competitive entry, including but not limited to:

- competitive choice retail pilot programs;

- requests for special arrangements between independent generators and large industrial customers;

---

[16] These states include Illinois, Pennsylvania, New York, California, Massachusetts, New Jersey, Maryland, and Texas.

[17] In addition, it should be mentioned that since 1976, large customers within the state of Georgia have had a one-time choice of power supplier.

- requests for special pricing arrangements;

- competitive bidding for new service extensions;

- gradual opening of retail markets, with large customers gaining access first.

Customers are more informed and increasingly sensitive to prices. Even in those regions where progression toward retail access is cautious, customers are demanding rate choices from their host utility.

In transition mode, tariffs are no longer viewed as devices to collect revenue only. With an eye toward an increasingly competitive future, electric companies have a renewed focus on customer service and are generally committed to improving customer satisfaction and maintaining market position. Expanding retail products is viewed as an essential vehicle to achieve these ends, and incumbent utilities understand price structure as the key product attribute.

Nationally, many incumbent electric systems are in transition mode, and rates groups are being deployed as *pricing organizations* with close alignment to product development. This change in title often suggests a considerably enlarged responsibility and an explicit mandate driven by strategic marketing objectives. Indeed, offering pricing options is a demonstrable and observable action that can increase customer value and profits immediately.

The changing purview of the pricing group has both organizational and regulatory aspects. First, we find that the new pricing organization is often teamed closely with the marketing organization. Experts from the pricing organization are now involved with field marketing and have direct contact with customers regarding pricing issues. Implicitly, pricing is geared toward developing products tailored to customer needs and thus the *pricing process* is more closely knitted to marketing in transition mode.

Second, because transition mode remains well within the world of price regulation, new services and pricing options continue to fall within the overall umbrella of the revenue requirement constraint. Pricing remains constrained to the development of *product options* for customers currently served under broad tariff categories.

The development of product options aimed to improve customer value and market position requires an understanding of customer behavior and the impact on the service provider. The problem of product development begs the question, "what is the impact or effect of implementing candidate options?" The answer is founded in the process of how value is created and the role of price within markets, and is manifest in two dimensions of customer behavior including 1) *load response* and 2) *selection*. The idea of economic demand suggests that customers will consume electric service at a level where the incremental value of consumption is equal to the price paid for it; the consumption decisions of customers can be referred to as *load-response behavior*. *Selection behavior* is reflected in product choice; when confronted with electric service options, customers will choose the option that is expected to provide the greatest value.

The notion of developing product options that increase value and profits represents a fundamental departure from the *rates as revenue collection devices* view that is inherent to the traditional mode of determining prices. That is, "rates" are no longer assessed according to the somewhat inflexible cost allocation rules implicit within embedded COS principles. Rather, pricing options employ a cost-benefit analysis framework that explicitly accounts for the economic profits of the firm.

7.   WHERE ARE WE NOW?

Many incumbent utilities are in transition mode and have formed pricing organizations that serve as a forum for the development of pricing strategies. This enlarged role is emerging because, in the world of customer choice, pricing decisions and retail product innovations have significant impacts on profits and the market position of the incumbent supplier. Moreover, the longstanding rate administration function is being supplemented by a product development activity based upon a much more rigorous quantitative approach to the development of pricing.

Market research is being utilized in various ways, including strategic market information systems to support innovations in product offerings. The pricing organization is directing market research activities including, for example, customer surveys geared to understanding customer preferences. Survey results can be integrated into analysis processes to evaluate proposed new product offerings.

In some companies, a host of new necessary product options are being developed, including time-of-use (TOU), real-time pricing (RTP), curtailable service menus, price risk protection, economic development rates, fixed bill rate options, two-part tariffs, and cross-product bundling. As suggested above, each product option implies demand- and supply-side impacts. Each must be assessed *quantitatively within the context of the overall service design package or portfolio*.

8.   CONCLUSION

Rate choices can be expected to proliferate under restructuring, and pricing is the biggest game in town. The task at hand is to develop and implement profitable service packages. This is challenging, and the pricing organization is emerging as a strategically important force within power suppliers.   While the expanding role of the pricing organization is proceeding faster in some companies than in others, all incumbent providers ultimately will either build or acquire these enlarged capabilities.

The capabilities of an effective pricing organization cannot be built overnight, and two key elements are needed.   First, pricing strategy should be crafted under the auspices of an overarching objective, such as customer retention, market share, or profitability. The objective should be supported by a conceptual framework and clearly defined principles. The framework serves to guide pricing strategy.

Second, the technical nature of this rather daunting task requires the integration of technical skills, cost and market information, and computer models.   The overall effectiveness of the pricing organization will be severely muted without information systems that effectively integrate these elements.   In particular, data and information must be integrated within an evaluation paradigm that provides the capability to estimate the effects of new pricing products, including customer selection of – and load response to – product options.   The evaluation paradigm is the core of the pricing process.

Chapter #2

Pricing Generation Supplier Coordination Obligations
The New Service of the Investor Owned Electric Delivery Service Company

George R. Pleat
Baltimore Gas & Electric Company

Abstract: Investor owned electric utility delivery service companies are assuming the
              arduous permanent responsibility of coordinating the retail supplier generation
              market. This chapter surveys the challenges facing such companies, and
              suggests pricing approaches that they may use based on incremental costs. To
              maximize expected fee revenue, these companies should apply a diversified
              menu of pricing strategies.

Key words: Delivery service; pricing of electricity; coordination of generation supplier
              services.

1.   PREFACE

As the dust settles on retail generation choice in many states, investor
owned electric utility delivery service (DS) companies are assuming the
arduous permanent responsibility of coordinating the retail supplier
generation market. The coordination services consist of (1) supplier account
management; (2) load profiling and settlement; (3) switching customers
from one supplier to another; (4) Electronic Data Interface (EDI) pre-
enrollment and post-enrollment testing; (5) DS consolidated billing; (6)
supplier requests for monthly consumption and interval data history; and (7)
supplier requests for off-cycle meter reading, meter testing, and meter
removal.

As DS companies prepare for providing value-added coordination
services to retail suppliers, corresponding cost centers are developing in

A. Faruqui and B.K. Eakin (eds.), Electricity Pricing in Transition, 19-26.
@ 2002 *Kluwer Academic Publishers.*

parallel. Therefore DS stockholders will be expecting to receive a full recovery of variable expenses and a reasonable return on new investment. Consequently DS companies will be pressured to seek rate relief from the local public service commissions (PSC) in the form of fee assessments on retail generation suppliers. The utility must demonstrate to the regulators and suppliers that the fees are cost based, avoid double cost recovery and prevent cross subsidies from the electric retail delivery service customer. In addition fees must be fair and equitable to the retail generation supplier and should be easily understood and easily administered by the DS company.

One pricing approach that ensures that the above goals are met by the DS company is to design supplier coordination fees on incremental costs. The incremental costs should measure both short and long run decision making in providing coordination services. To maximize expected fee revenue, the DS company should apply a diversified menu of pricing strategies ranging from monthly average daily MW charges assessed retail suppliers to having expenses deferred into a regulatory asset account and surcharged all retail customers at some later date.

2.   NEW PERMANENT SERVICE OBLIGATIONS

With the establishment of generation retail choice in many regions in the U.S., local public service commissions are mandating that the host DS company provides retail suppliers with the necessary coordination services to make generation choice viable.[1] These services do not encompass electric supply purchasing or trading but instead are administrative in nature. The administrative aspect of the service by no means diminishes the importance - - in contrast inadequate supplier coordination services could cause major obstacles to providing generation choice to customers. Moreover, by taking on this awesome responsibility of providing supplier coordination activities, the DS company is adding value to retail suppliers process of selling electric generation to customers. Since generation choice is likely to stay in many regions of the U.S, (not withstanding the halt in California),[2] coordination

---

[1]   The DS companies are likely to be selected by the regulators to perform these services because of experience in meeting generation and delivery service reliability standards. In addition, utilities are likely candidates because they typically earn good bond ratings and have solid earnings to finance the coordination infrastructure requirements.

[2]   As of today it appears that the California State government has assumed the role as a monopoly retail supplier for large service territories. The government is arranging for long-term wholesale power purchases and selling directly to retail customers, consequently establishing itself as the dominating retail power supplier. Though retail

services will be permanent just like any traditionally regulated retail wires and customer service.[3]

The new value added service provided to the retail supplier by the DS company consists of several coordination functions. These functions are usually defined by a consortium of stakeholders comprising of the DS company, retail suppliers, regulators, and customer associations -- with subsequent review and approval by the local PSC. The functions (or service obligations) offered by the host utility consist of many facets.[4]

Supplier Account Management – This function involves the DS company organizing through a central area all direct contact with retail suppliers. The coordination duties consist of educating suppliers on the market rules and regulations; establishing the supplier in EDI; performing credit analysis on entrant suppliers; processing customer enrollment; implementing rescind letters to customers who are switching suppliers;[5] maintaining supplier information web site; communicating load profiling and settlement data to the supplier and power pool, and administrating all supplier fee billing.

Load Profiling and Settlement – In this function lies the greatest responsibility for the DS company when dealing with supplier coordination services. Each day the DS company must estimate, through load software applications, for each customer enrolled with a retail supplier the daily supplier capacity obligation and transmission peak load contribution along with day after hourly load shapes for settlement among suppliers. The data is forwarded to the power pool where power capabilities are matched with expected demands.   As can easily be seen, any deterioration of this service could lead to shortages and excess supplies in the regional power pool -- to the detriment of suppliers and sellers of electricity.

DS Consolidated Billing – It is likely the DS company will be asked by the local PSC to be the backstop service in the competitive billing arena. If the customer chooses the local DS company to do the full supplier billing, the entity will be obligated to serve. Given this premise, the DS company will piggyback on its own bill the retail supplier billed information. This service may require the DS company to expand its current customer

---

deregulation has broken down in California, it is assumed that supplier coordination responsibilities will remain with the DS company.

[3] For a discussion on pricing traditional DS wires and customer services, see "Unbundling Retail Prices: An Electric Utility Prepares for Life as a DISCO," Pleat, George R., Public Utilities Fortnightly, May 15, 1999.

[4] Typical supplier coordination services are illustrated in the BGE Direct Testimony of Sheldon Switzer, Schedule 1 and Rider 9 – Supplier, Third Party and Customer Fees and Charges, May 22, 2000, Before the MD PSC Case No. 8794/8804.

[5] When a customer switches suppliers, the DS company will forward a letter to the customer allowing several days to rescind a decision to go to a new supplier.

information system to accommodate extra bill information, could cause extra postage, and require EDI transactions along with appropriate pre and post EDI testing.

Consumption Data Requests – One of the more important coordination services necessary for stimulating retail generation choice is the DS company meeting the continuous demand by retail supplier and third party entities[6] for customer consumption data profiles. The DS company provides these entities with monthly data histories or interval data histories, and if complicated needs arise special load data studies are performed. One efficient way of providing interval load data is to set up a web based system where suppliers can retrieve, at will, the data through a password. The consumption data service is necessary for compiling supplier billing charges or to provide "just look see" profiles to the retail supplier prior to market entry on a prospective customer.

Off-Cycle Meter Reading, Meter Testing, and Meter Removal - At the request of the retail supplier, the DS might be asked to perform certain metering services. If a supplier is not content with the DS company billing cycle, having of-cycle meter readings might enable them to establish a tailored billing cycle. If competitive metering is allowed, suppliers may want to install their own advanced meters prompting the DS company to make a special trip to remove the existing standard meter. Also, the supplier may suspect a faulty meter and could request the DS company to perform a field meter test. The DS company probably does not consider these metering services as required for retail generation choice and will be very hesitant to perform them without fees.

3.   STOCKHOLDERS MUST BE COMPENSATED FOR COORDINATION RISK

By taking on the value added supplier coordination obligations, the DS company is exposed to financial risk that warrants compensation from the group causing the cost. Not fulfilling these service obligations correctly could mean severe financial impacts on the power pool, suppliers and the enrolled retail customers. By assuming these responsibilities, the stockholder of the DS company should be entitled to recover the labor and material expenses, capital costs and a fair rate of return. Any compensation received short of these cost burdens will impact earnings and jeopardized the ability of the DS company to provide long–run reliable wires and customer

---

[6]   Third party entities could include licensed suppliers not enrolled with customers, retail aggregators, energy consultants etc.

service to its core retail customers. Also an unfavorable cost recovery outcome could effect the stockholder's eagerness to invest in the DS company. If the DS company fails to get compensation from the retail supplier, eventually the utility will petition the local PSC to recover all costs from the retail customer without regard to cost causality pricing[7].

The Double Dipping Dilemma - In order to protect the stockholder, the DS company must make its case before the local PSC that coordination supplier obligations are new and the associated costs are not already recovered in current retail tariffs. This is referred to as the so-called "double dipping" issue. There is no question that the stakeholders[8] impacted by proposed supplier coordination fees will be looking with a careful eye the possibility that the DS company has already recovered these expenses in current tariffs.

One common argument from the stakeholders involves the "zero expense" game. Though the new coordination services are incremental in nature, the DS company can shed other responsibilities that are left over from the vertically integrated electric service days, like centralized marketing. The argument concludes that total expenses are not changed and therefore no new fees should be allowed. This argument can be easily discounted however by the DS company by focusing on that fact that supplier coordination obligations are new and if avoided could result in operating cost savings i.e., supplier coordination expenses are incremental by definition and warrant cost recovery just like any traditional service offered by the utility. In addition, it is likely that the current retail tariffs were developed in a test year prior to retail choice and would exclude any coordination obligation expenses in them.

Retail Choice Adds Complexity in Fee Request - When proposing new fees for new services under utility regulation, the burden is on the DS company to show that service is adding value to the customer and the fee is cost justified. Cost justification is especially important given that the service is an obligation that the DS company must meet as prescribed by the local regulator. In other words there is one supplier of the service, which forces the "reasonable" rates review. These criteria alone make it imperative that the DS company present a sound proposal when asking for fee assessment on retail suppliers.

Another factor that complicates the situation for the DS company when asking for cost recovery of supplier coordination costs is the fact that these

---

[7] By shifting coordination costs to the general retail customer, allocation issues among customer groups will exist.

[8] Stakeholders would consist of PSC Staff, retail suppliers, and customer group representatives.

new obligations are coincident with the start of retail generation choice. The political locomotive for having retail choice rolls at day one. The retail choice locomotion is mighty and strong because the legislators and regulators have thrown the dice that generation competition, in lieu of regulation, is good for retail customers and ultimately voters. Any obstacles to this mission will be perceived as anti-competitive and a reduction in the general welfare of retail customers.

There is no doubt that coordination service fees assessed on retail suppliers are going to eat into the supplier's margins. This is usually the front line of defense for suppliers in roundtable discussion -- that coordination services sans fees is the only way that suppliers will participate in the marketplace. In response the DS company should demonstrate that these fees are a small proportion of the retail supplier's acquisition costs especially when compared with the commodity cost of electricity[9]. Also the DS company should argue that supplier coordination service are value added to making a generation market run smoothly and thus enabling suppliers to compete fairly and easily. The fee assessments are normal operating expenses of the supplier and should not be borne by indirect benefactors.

4.   INCREMENTAL COST PRICING ENSURES EQUITY AMONG ALL STAKEHOLDERS

To ensure that stockholders are adequately compensated for risk associated with investing in supplier coordination services; to ensure that retail suppliers are fairly treated and not unduly billed burdened, and to avoid cross subsidization and "double dipping, the DS company should design fees for supplier coordination based on short and long-run incremental costs. Moreover, incremental cost based pricing is especially important when the coordination service is optional for the retail supplier (e.g., competitive billing). Once incremental costs are developed by coordination service, the DS company should pursue a package of rate design proposals to maximize fee revenue collection.

Measuring the Cost - The question is simple for calculating these costs, what are the expenses associated with making the short and long decisions for providing obligated coordination services on a permanent basis? Or on the flip side, what are the savings realized if the DS company does not have to perform these obligations? For many of the new coordination services,

---

[9]   Based on one major utility's coordination cost profile, the fee assessment is about .4% of the supplier's revenue collected form a 5 MW sale to a C&I customer.

the corresponding incremental costs consist of primarily labor[10] and some software support. Other services require capital expansion to accommodate the coordination -- like the DS consolidated billing service. The incremental costing approach divorcees itself completely from embedded accounting costs. This shelters the fees from recovering costs that are related to past economic decisions, which are most likely captured in current tariffs.

After estimating the incremental supplier coordination costs, it will be tempting and prudent for the DS company to compare the fee levels with other utility supplier fees. When making the comparison, however, it behooves the DS analyst to be cognizant of the fact that other utility supplier fees have most likely been negotiated down by the stakeholders. Consequently, "fresh" incremental cost estimates might appear high in the comparison. This dilemma should be highlighted to management in the review process.

Basis for the Price to Beat - Setting the appropriate incremental costing level is key to setting benchmark market prices to beat. For example, billing services may be open to semi-competitive status where the DS company must provide consolidated billing if requested by the retail customer. The customer, however, can also ask their supplier to do the full billing instead -- but the supplier can deny providing the service. This forces the DS company into a backstop consolidated billing service entity. This status requires the PSC to regulate the DS company fee charged the supplier.

If the supplier feels the fee is "too high," it can provide its own consolidated billing service to the customer, if the customer agrees. In effect, the DS consolidated billing fee becomes the price to beat. If the incremental costs are "too high", then the supplier will have incentives to do their own billing so long as their costs are less then the fees savings when DS billing is avoided. If the DS consolidated billing fee is "too low"; the DS company can be accused of "cornering" the billing market through a subsidized rate. It is in the interest of the DS company to do a good job in estimating the true incremental cost, because in theory, competitive market pricing should equal these expenses. If a risk adverse retail supplier realizes small savings by doing their own billing, they will more than be delighted to have the DS company do the billing for them.

Maximizing Expected Fee Revenues – The DS company can minimize its risk of getting no fees by applying a diverse array of pricing strategies that will attract stakeholders. For instance, the supplier account management account function is directly related to alternative retail suppliers needs and the corresponding costs could be assessed the supplier on a monthly average

---

[10] Including fringe benefits, department and corporate overhead.

daily capacity obligation MW charge. Other coordination costs would be excluded to minimize the direct billing impact on retail suppliers.

Other supplier coordination costs could be deferred and set aside into a regulatory asset account. The load profiling and settlement function would be a good candidate for this type of price making. The annual costs would be accumulated and when the next delivery service rate case arrives, a surcharge would be designed to recover these expenses from all retail customers The prospective costs would part of the test year revenue requirement and bundled into wires delivery service charges or assessed on retail suppliers [11]. This price design approach might alleviate some of the political concerns of throwing too many fees at day one retail choice.[12]

The consumption data services provided by the DS company requires different pricing approaches that will help the DS company save on expenses in data collection. For example, in post enrollment, suppliers will be needing on-going web site access to interval load data. Certainly the DS company does not want to monitor individual web site hits by suppliers for fee assessment calculations. One way to reduce monitoring costs is to charge a monthly rental fee to suppliers based on number of customer accounts per supplier and offering unrestricted web site access each month.

Other coordination services can be more traditionally designed. For the DS consolidated billing, a charge per month bill per customer account by supplier can be assessed.

For metering services, pay as you go charges to the supplier are the most reasonable since they involve extraordinary trips out to customer sites. Metering supplier fees will act as a deterrent to potential abuse of these optional services – thus reducing metering congestion costs for the DS company.

Finally, the Delivery Service Company needs to decide what coordination fees are important and what fees are not for the future. In the spirit of good faith negotiating with stakeholders, it may be necessary to waive some fees.

---

[11] Load profiling and settlement costs directly assigned to retail suppliers are a cloudy issue if standard offer generation service is offered by the DS company. In most cases this function is supporting both standard offer service and alternative retail suppliers. Consequently, one might argue that all retail customers should pay the expenses. However, once standard offer service is expired, it is entirely appropriate to recover these expenses directly from retail suppliers.

[12] The DS strategy aimed at recovering regulatory asset supplier coordination costs can awaken stakeholders, representing customer groups, to the realization that cost allocation wars can be avoided by applying fees directly to the suppliers who cause the cost. These stakeholders may drop their initial opposition to supplier coordination fees in general in exchange for eliminating any kind of regulatory asset accumulation recovery issue in the future.

Chapter #3

Restructuring the Electric Enterprise
Simulating the Evolution of the Electric Power Industry with Intelligent Adaptive Agents[*]

Massoud Amin
Electric Power Research Institute (EPRI)

Abstract: A model and simulation of the "Electric Enterprise" (taken in the broadest possible sense) have been developed. The model uses autonomous, adaptive agents to represent both the possible industrial components, and the corporate entities that own these components. An open access transmission application and real-time pricing has been implemented. Objectives are: 1) To develop a high-fidelity scenario-free modeling and optimization tool to use for gaining strategic insight into the operation of the deregulated power industry; 2) to show how networks of communicating and cooperating intelligent software agents can be used to adaptively manage complex distributed systems; 3) to investigate how collections of agents (agencies) can be used to buy and sell electricity and participate in the electronic marketplace; and ultimately to create self-optimizing and self-healing capabilities for the electric power grid and the interconnected critical infrastructures.

Key words: Restructuring; intelligent adaptive agents; complex adaptive systems; enterprise restructuring; real time pricing.

[*]   I express my gratitude to the editor of this volume, Dr. Ahmad Faruqui, for his encouragement and continued interest in this subject. I also express my gratitude to Dr. Tariq Samad, Dr. Steve Harp, and Dr. Martin Wildberger for many earlier discussions and insightful suggestions during 1998-2000.

A. Faruqui and B.K. Eakin (eds.), Electricity Pricing in Transition, 27-50.
@ 2002 *Kluwer Academic Publishers.*

1.   INTRODUCTION

1.1   The Electricity Enterprise: Today and Tomorrow

The North American power network may realistically be considered to be the largest machine in the world since its transmission lines connect all the electric generation and distribution on the continent. Through this network, every user, producer, distributor and broker of electricity buys and sells, competes and cooperates in an "Electric Enterprise." Every industry, every business, every store and every home is a participant, active or passive, in this continent-scale conglomerate. Over the next few years, the Electric Enterprise will undergo dramatic transformation as its key participants – the traditional electric utilities – respond to deregulation, competition, tightening environmental/land-use restrictions, and other global trends.

While other, more populous, countries, such as China and India, have greater potential markets, the United States is presently the largest national market for electric power. Its electric utilities have been mostly privately owned, vertically integrated and locally regulated. National regulations in areas of safety, pollution and network reliability also constrain their operations to a degree, but local regulatory bodies, mostly at the State level, have set their prices and their return on investment, and have controlled their investment decisions while protecting them from outside competition. That situation is now rapidly changing. State regulators are moving toward permitting and encouraging a competitive market in electric power.

In this chapter we shall present a model and simulation of the "Electric Enterprise" (taken in the broadest possible sense) that has been developed. The model uses autonomous, adaptive agents to represent both the possible industrial components, and the corporate entities that own these components and are now engaged in free competition. The goal in building this tool is to help these corporations evolve new business strategies for internal reorganization, external partnerships and market penetration.

Development of this tool takes advantage of recent research in Complex Adaptive Systems (CAS) which has begun to produce an understanding of complexity in natural systems as a phenomenon that emerges from the interaction of multiple, simple, but adaptive, components. Agents are no strangers to the electronic marketplace, Internet versions of this software are commonly known as "softbots" or just "bots." Most common applications have involved accessing Website contents or search engines. In contrast to earlier software, these goal-seeking agents have been semi-autonomous in achieving their objectives.

From a computer programming point-of-view, agent-based modeling and simulation is a natural extension of the prevailing object-oriented paradigm.

Agents are simply *active objects* that have been defined to simulate parts of the model. Discrete event simulations with multiple quasi-autonomous agents (usually called actors or demons) have been used for at least twenty-five years to assist human decision-making in areas such as batch manufacturing, transportation, and logistics. The revolutionary new idea that comes from the computer experiments of CAS is to let the agents evolve, with each one changing in a way that adapts to its environment while that environment is modified by external forces and by the evolutionary changes in the other agents.

The agent community is allowed to evolve by causing innovative changes in the parameters of individual agents to be generated randomly and/or systematically. These parameter changes, in turn, produce changes in the agents' actions and decisions, so that the agents "tinker" with the rules and the structure of the system. Agents subjected to increased stress (resource shortages, environmental pressures, and financial losses) increase their level of tinkering until some develop strategies that relieve that stress. Some individual agents succeed (grow, reproduce, increase their profits) while others fail (shrink, die, are replaced, bought out).

Business enterprises, financial markets and the economy itself can all be viewed as complex adaptive systems and they give rise to practical problems that are often mathematically intractable. The methods developed to study CAS, as well as the insights derived from these studies, have been applied to all these areas with some success in the last decade. Practical market applications of more advanced agents represent buyers and sellers and carry out negotiations on their behalf. Agents are also used to represent stakeholders as they attempt to secure goods and services in an auction setting. Typically, the stakeholder is an individual user bidding for a good. However, auctioning may not be just for individuals. The Electric Power Research Institute, for example, has funded research into agent-based auctioning as a way to address the fierce competition for resources. Electric power marketers have emerged, and wholesale electric customers are learning to shop around for the best electric suppliers. This has peaked interest in bargaining agents that trade on behalf of various stakeholders. Like agents that represent individual human users, the bargaining agents decide how much to buy, who to buy it from, how much to pay, and how they will manage the exchange of goods and money. In a power market, however, there is also concern that the entire market not be harmed by the sale. Thus, looking at how agents complete their transactions and learn from them, provide insight into the dynamics of a complex supply and demand system.

Simulations of multiple, autonomous, intelligent agents, competing and cooperating in the context of the whole system's environment have had considerable success in providing better understanding of phenomena in biology and ecology, and, more recently, in financial markets. A CAS model is particularly appropriate for any industry made up of many, geographically dispersed components that can exhibit rapid global change as a result of local actions – a characteristic of telecommunications, transportation, banking and finance as well as gas, water and oil pipelines, and, especially, the electric power grid.

The first version of this tool treats several aspects of the operation of the electric power industry in a simplified manner. For instance, it uses a DC model. However, it includes base-classes for agents representing generation units, transmission system segments, loads, and corporate owners. Users may modify and interconnect these agents through a graphical interface. Simple adaptation strategies for the agents have also been implemented. More complex ones have been designed, and implemented. Scenarios have been prepared to illustrate open access and real-time pricing. This simulation tool can be further enhanced to provide greater physical and market realism by the inclusion of an AC model and futures trading, and to model co-generation, retail wheeling, and the effects of new technological developments, such as: storage, power electronics and superconductivity.

1.2   Deregulation, Competition, Re-Regulation and New Institutions

In 1978, the United States Federal Government began the movement toward deregulation by allowing competition in several strategic sectors of the economy, starting with the airlines and followed by railroads, trucking, shipping, telecommunications, natural gas and banking. Adam Smith succinctly stated the philosophy behind this movement in 1776: "Market competition is the only form of organization, which can afford a large measure of freedom to the individual. By pursuing his own interest, he frequently promotes that of society more effectively than when he really intends to promote it." More recently, Prof. Alfred Kahn of Cornell University, who guided the airline deregulation as the head of the Civil Aeronautics Board, expressed it in a different way: "Deregulation is an admission that no one is smart enough to create systems that can substitute for markets."

Throughout most of the history of electric power, the institutions that furnished it have tended to be vertically integrated monopolies, each within its own geographic area. They have taken the form of government departments, quasi-government corporations or privately owned companies subjected to detailed government regulation in exchange for their monopoly status. Selling or borrowing electric power among these entities has been carried out through bilateral agreements between two utilities (most often neighbors). Such agreements have been used both for economy and for emergency back up. The gradual growth of these agreements has had the effect that larger areas made up of many independent organizations have become physically connected for their own mutual support.

In recent years, some of the local monopolies have found it beneficial to be net buyers of power from less costly producers and the latter have found this to be a profitable addition to their operations. For instance, it is typical in the western United States and Canada for surplus hydroelectric power to be transmitted south for air conditioning in the summer; while less expensive nuclear power is transmitted northward in the winter when the reservoirs are low or frozen and only nighttime heating is needed in the south. These wide area sales and the wheeling of power through non-participant transmission systems are international in extent, especially in Europe and the Americas. There is evidence of a worldwide drive to use these interconnections intentionally:

- To create competition and choice, with the hope of decreasing prices,

- To get governments out of operating, subsidizing or setting the price of electric power, and

- To create market-oriented solutions in order to deliver increases in efficiency and reductions in prices.

In order to unbundle the monopoly structure of electric power generation in the United States, Congress passed the National Energy Policy Act of 1992. National monopolies in the United Kingdom, Norway and Sweden have been de-nationalized and unbundled into separate generation, transmission and distribution/delivery companies. In most approaches to deregulation, transmission is kept as a centrally managed entity, but generation is broken into multiple independent power producers (IPP), and delivery is left to local option. New IPP are encouraged or, at least, permitted, as are load aggregators and electric power brokers, both of whom own no equipment, but are deal-makers who operate on commissions paid by the actual producers and users.

The concept behind this arrangement is that electricity, much like oil and natural gas, is a *commodity* that can be sold in the cash or spot market. As a commodity, it is possible to buy and sell future options and more complex derivative contracts based on electricity prices. However, it is not clear that electricity meets all the necessary criteria for commodity trading. The original assumptions of NYMEX and its traders were based on the model of natural gas, which, unlike electricity, can be stored economically. Once a unit of electricity is produced it must be consumed almost immediately; however, a true commodity can be stored for some length of time and consumed when and how desired. Electricity storage devices are capable of handling only a small percentage of an area's electricity requirements. Storage limitations and capacity constraints on inter-regional transfer prevent all available suppliers across the continent from head-to-head competition.

An alternative, and more entrepreneurial, view is that furnishing electricity is a *service* to the end user. Electric service may be segmented into more specific markets such as heating, cooling, lighting, building security, etc., or combined with other consumer services such as telephone, cable TV, Internet connections, etc. Both views may be reconcilable by separating the *product*, handled by generation and transmission companies, from the *service*, performed by distribution companies.

1.3   Modeling the Future

The real issue, not yet being faced in United States (or in many other nations that are moving toward greater competition in electric power) is whether such an open, competitive market can be can fair and profitable to all participants, while continuing to guarantee to the ultimate consumer of power, at the best possible price, secure, reliable electric service, of whatever quality that consumer requires.

Some utilities are contending that sudden deregulation is unfair and are seeking government reimbursement for "stranded assets" – equipment that, for technical or financial reasons, cannot be made efficient enough to compete. In order to free the most profitable parts of their operations from regulation, other utilities are unbundling into separate and independent generation, transmission and delivery companies; or at least separate services, each optimizing its performance based on different criteria and all operating at arms length from each other. Still other utilities are merging with, buying or being bought by, companies that may not have been in the electric power business at all. Combinations are taking place, or proposed, in which parts of former electric power monopolies join with companies whose chief product or service has been natural gas, telecommunications, cable television, engineering or finance.

Current approaches to predicting the new business structure of the electric power industry are all driven by assumed scenarios. One such scenario, based on the experience of other industries and other nations, expects that in five years there will be only a few dozen companies engaged in the actual generation, transmission and distribution of electricity. The generation companies will be completely deregulated, except for some environmental constraints. The distribution companies will still be regulated, along the lines of today's local telephone companies, but major industrial/commercial customers, and cooperatives of individual residential customers, will generate their own power or buy it from the lowest bidder. The transmission companies will be partly regulated in an attempt to ensure open access and non-discriminatory pricing for "wheeling" power between any generator and any user or distributor, while maintaining some level of system security despite their lack of control of either generation or load. However, this is just one hypothetical future scenario and various other scenarios are emerging.

The topology of these alternative scenarios/business structures dictate features of the future power system infrastructure which, in turn, suggest the most profitable re-arrangements of capital assets and market segments for each company. Hence, the predictive accuracy of this "top-down" approach depends entirely on the actual occurrence of the scenario or family of scenarios postulated. As an alternative approach, EPRI is developing a model and simulation of the "Electric Enterprise" (taken in the broadest possible sense) that uses a "bottom-up" representation of the whole system without any preconceived scenarios. Its major endogenous constraints will be the laws of physics and the cost or availability of possible technological and economic solutions. Autonomous, adaptive agents represent both the possible industrial components, and the corporate entities that own these components and are now engaged in free competition with each other. Political accommodations and corporate restructuring will appear as global emergent behavior from these locally fixed agents cooperating and/or competing among themselves. As these artificial agents evolve in a series of experiments, the simulation should expose various possible configurations that the market and the industry could take, subject to different degrees and kinds of cooperation, competition and regulation. Possible results will be the development of conditions for equilibria, strategies or regulations that destabilize the market, mutually beneficial strategies, the implications of differential information, and the conditions under which chaotic behavior might develop. This view, of course, has considerable similarity to the mathematical theory of games of strategy, but, unlike the generalized games solved by von Neumann or Nash, these are repeated games with non-zero sum payoffs. Information theoretic considerations are pertinent and these may, in turn, be represented by entropy in the state or phase space in which the system operates.

The primary goal in building this tool is to help individual companies evolve new business strategies for internal reorganization, examine the potential of entering into new partnerships or attempting to exploit new market segments. Computer experiments with this model can also provide insight into the evolution of the entire electric power industry. Within this "scenario-free" testbed, all the global behaviors that are possible in the system can emerge from local agents cooperating and/or competing among themselves in response to "what if" studies and computer experiments hypothesizing various forms of exogenous constraints. In addition, the model will serve as a practical way to estimate the benefits of implementing any proposed new technology or making hypothetical changes to existing equipment and operating practices.

2.   COMPLEX ADAPTIVE SYSTEMS (CAS)

Development of this tool takes advantage of recent research in Complex Adaptive Systems (CAS) which has begun to produce an understanding of complexity in natural systems as a phenomenon that emerges from the interaction of multiple, simple, but adaptive, components. Researchers associated with the Santa Fe Institute have conducted much of this work. Simulations of multiple, autonomous, intelligent agents, competing and cooperating in the context of the whole system's environment have had considerable success in providing better understanding of phenomena in biology and ecology. Using computer experiments on CAS models that simulate biological phenomena has been called, somewhat extravagantly, "artificial life."

The attractiveness of these methods for general purpose modeling, design and analysis lies in their ability to produce complex emergent phenomena out of a small set of relatively simple rules, constraints and relationships couched in either quantitative or qualitative terms. Inventing the right set of the local rules to achieve the desired global behavior is not always easy, although it often seems obvious afterward. For instance, flocking behavior requires only two basic rules: (1) stay close to the nearest bird, (2) avoid colliding (either with another bird or any obstacle).

Business enterprises, financial markets and the economy itself can all be viewed as complex adaptive systems and they give rise to practical problems that are often mathematically intractable. The methods developed to study CAS, as well as the insights derived from these studies, have been applied to all these areas with some success. Other CAS simulation techniques such as spin glass models, sand piles and random Boolean networks have been, for some time, standard tools in certain relatively narrow areas such as condensed matter physics.

From a computer programming point-of-view, agent-based modeling and simulation is a natural extension of the prevailing object-oriented paradigm. Agents are simply *active objects* that have been defined to simulate parts of the model. Discrete event simulations with multiple quasi-autonomous agents (usually called actors or demons) have been used for at least twenty-five years to assist human decision-making in areas such as batch manufacturing, transportation, and logistics. The revolutionary new idea that comes from the computer experiments of CAS research is to let the agents evolve, with each one changing in a way that adapts to its environment while that environment is modified by external forces and by the evolutionary changes in the other agents. Several pertinent questions arise:

1) What is an agent? Agents have evolved in a variety of disciplines— artificial intelligence, robotics, information retrieval, and so on—making it hard to get consensus on what they are. Most researchers agree, however, that a truly intelligent agent has these attributes:

Reactivity. It can sense the environment and act accordingly

Autonomy. It does not need human intervention

Collaborative behavior. It can work with other agents toward a common goal

Inferential capability. It can infer various task-related issues from the environment.

Temporal continuity. Its identity and state persist over long periods.

Adaptivity. It can learn and improve with experience.

The more advanced agents may also have other attributes, such as mobility (it can migrate from one host platform to another, either by directing itself or following a pre-programmed schedule) and personality (manifesting some human qualities, such as cooperation for the "greater good," caution, and even greed).

2)What types of Agents are there? There are probably as many ways to classify intelligent agents as there are researchers in the field. Some classify agents according to the services they perform. System agents run as parts of operating systems or networks. They do not interact with end users, but instead help manage complex distributed computing environments, interpret network events, manage backup and storage devices, detect viruses, and so on.

Interface agents are intelligent interfaces that use speech and natural language recognition capabilities. Their main job is to reduce the complexity of information systems.

Filtering agents filter out data the user does not need. *Retrieval agents* search and retrieve information from various sources on the web and serve them to the user like an information aggregator. *Navigation agents* help users navigate through external and internal networks, remembering shortcuts, preloading caching information, and automatically bookmarking interesting sites, among other tasks. *Monitoring agents* provide users with information when a particular event occurs, such as a Web page being updated. Amazon.com customers, for example, get Eyes, agents that monitor catalogs and sales and notify customers when particular books are available.

Profiling agents gather information on Web site visitors, which the site uses to tailor presentations for that visitor.

A *heterogeneous agent system* contains two or more agents with different agent architectures.

3) How Adaptive Agents Work? An adaptive agent has a range of reasoning capabilities. It is capable of innovation—developing patterns that are new to it—as opposed to learning from experience (sorting through a set of predetermined patterns to find an optimal response). Adaptive agents can be passive—respond to environmental changes without attempting to change the environment—or active—exerting some influence on its environment to improve its ability to adapt. In effect, an active adaptive agent conducts experiments and learns from them.

Individual agents must be able to respond to environmental conditions and to other agents in a way that enhances their survival or meets other goals. To learn a strategy that increases its "fitness," the agent has to gather and store enough information to adequately forecast and deal with changes that occur within a single generation. The population then adapts through the diversity of its individuals. Some individuals will always survive, and their individual actions benefit the population goals. Thus, the population evolves over many generations, surviving as a recognizable organization.

The agent community is allowed to evolve by causing innovative changes in the parameters of individual agents to be generated randomly and/or systematically. These parameter changes, in turn, produce changes in the agents' actions and decisions, so that the agents "tinker" with the rules and the structure of the system. Agents subjected to increased stress (resource shortages, environmental pressures, and financial losses) increase their level of tinkering until some develop strategies that relieve that stress. Some individual agents succeed (grow, reproduce, increase their profits) while others fail (shrink, die, are replaced, bought out).

3.   UNDERSTANDING COMPLEX SYSTEMS

The demonstration market provides an interesting way to gain some insight into the many issues that affect the power market as it struggles with adapting to changes caused by deregulation. To gain insights into any large-scale network, however, we need some way to model the dynamics of the market. Current modeling techniques are unsuitable because they typically rely on top-down, scenario-driven methodologies, limited to a small set of preconceived scenarios. Agent systems offer an attractive alternative because they allow a bottom-up representation of the system that will not be restricted to preconceived or hypothetical scenarios. The North American power grid, for example, can be considered a complex adaptive system because it comprises many, geographically dispersed components and can exhibit global change almost instantaneously from actions taken in only one part of it.

EPRI is using CAS work to develop modeling, simulation, and analysis tools that may eventually make the power grid self-healing, in that grid components could actually reconfigure to respond to material failures, threats or other destabilizers. The first step is to build a multiple adaptive agent model of the grid and of the industrial organizations that own parts of it or are connected to it.

SEPIA (Simulator for Electrical Power Industry Agents) is an example of this adaptive agent model; it is a comprehensive, high fidelity, and scenario-free modeling and optimization tool developed with funding from EPRI by Honeywell Technology Center (HTC) in conjunction with the University of Minnesota. EPRI members, who sponsored the research, use SEPIA to conduct computational experiments for any kind of scenario, which gives them insights into the true dynamics of the power market, and assists in gaining strategic insights into the electricity marketplace.

SEPIA is an object-oriented, fully integrated Windows application with plug-and-play agent architecture. Users can readily adapt simulations to a parallel computing environment, including multiprocessor PCs and PC networks. SEPIA agents are autonomous modules that encapsulate specific domain behaviors. They are implemented as independent ActiveX applications, which communicate with each other by messages sent through the SEPIA agent bus. The user interface, which is modeled after the Windows GUI, lets users specify agents and agent relationships and modify agents, and provides mechanisms for to guide and monitor the simulation.

Within SEPIA, agents communicate through messages; the messaging mechanism is sufficiently flexible to handle the variety of communication needs necessary (this includes, for example, simulations of electric power transmission, of information flows between corporate agents, and of money transfers). Numerous agent classes have been designed and implemented: generating units, generating companies, loads, consuming companies, power exchanges, and transmission zones.

An open access transmission application has been implemented. Users can conduct simulations by defining scenarios through drag-and-drop operations on icons representing the agents, then interconnecting the agents, and pressing a "run" button. Simulation results are shown dynamically on graphs and reports, and the policies and parameters of agents can be modified dynamically as well.

This work has also resulted in the development of more sophisticated business scenarios for the operations of a deregulated power industry are articulated in some detail next.

The user interface, based on the familiar Windows GUI, allows users to specify agents and agent relationships, permits agent modification, and provides mechanisms for simulation steering and monitoring. SEPIA uses standard file input/output formats, such as the PSS/E data format for transmission networks, that are in common use today, so that EPRI members will be able to base their computer experiments with SEPIA on their own system data.

In Phase 1 of SEPIA, the agent model, the simulation engine and the graphical user interface (GUI) have been implemented. Simple adaptation strategies for the agents have also been implemented (Figure 1). More complex ones have been designed, and their implementation is underway.

The next phases will emphasize improvements to physical and market realism, such as power electronics devices, superconducting cables and various forms of storage, as well as the effects of trading in futures, options and various derivatives. Further enhancements will emphasize greater fidelity in modeling the implication for each transaction of the resulting power flow (stability, security, etc.) on the existing network. The physical realism will be enhanced with an AC model, models of Flexible AC Transmission (FACTS) devices, superconducting cables, and storage. These extensions will allow users to evaluate potential technological investments. Improvements to market realism will include a futures market, exchange and bilateral contracts, and exogenous inputs. This will permit the development of scenarios involving the revenue impact of load forecasting, and various control algorithms. Parallel processing, agent template libraries, and more readily customizable agents will enhance performance and flexibility of the tool itself.

Figure 1: Agent Architecture and Adaptation Agent design determines when and how Online Algorithms modify internal state based on experience

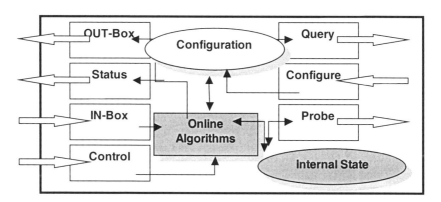

The first version of SEPIA includes base-classes for agents representing: generation unit agents, transmission system agents, load agents, and corporate agents, which may represent either a net power consuming company (LoadCo – Figure 2) or a net power generating company (GenCo). All agents continually make decisions, and these decisions affect the behavior of other agents. The design and implementation of these agents is sufficiently generic as not to limit how users may extend the system by specializing their classes or by defining new ones to allow for different kinds of generation, transmission, loads, and corporations. All agents consist of layered components, some specific to that particular agent and some applicable to other agents, so that different configurations can be assembled rapidly.

Figure 2: Load schedule function and Load Company Agent (LCA). LCA must decide: When to issue an RFQ; Hours and Amounts in the RFQ; Expiration date for RFQ; Whether to accept a quote; and When to accept a quote

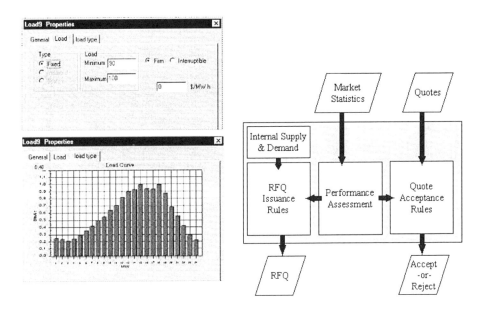

Agent adaptation in SEPIA means that the agent's online algorithms modify its internal state based on experience. Agent design determines when and how this occurs. All learned knowledge is stored in the internal states of agents, but it is also possible to have adaptation at multiple levels of organization, i.e.: distributed over a population of agents, or within a cohort of related agents, as well as internal to a single agent. The current version of SEPIA offers two reusable adaptation algorithms:

Learning Classifier System (LCS) using rule representation, with discovery via a genetic algorithm and blackboard architecture for reinforcement learning. The LCS is implemented as a generic modular LCS C++ class template. The rules, conditions and actions are separate classes of objects and may be reused for different kinds of agents' instantiated with different conditions and actions. The incremental genetic algorithm, triggered periodically, uses crossover, mutation, tournament selection and a simplified bucket brigade algorithm.

As customers pay hourly for energy delivered on contracts negotiated sometime earlier, the money is accumulated in bins tied to the given rate structure (and thereby to the action that produced the rates). If the agent must be "bailed out" by the generator of last resort, the associated debt is also accumulated. At the end of each epoch, these rewards are disbursed to the actions, updating the appropriate cells in the Q-Table. The rewards are normalized and discretized into four values, producing some immunity to small variations caused by load fluctuation.

It is not possible to accurately assess the profit from a contract until after generation takes place. Each agent's profit depends on the whole load on the grid, determined by all the agents. Hence, rewards affect actions two epochs later. The first two epochs in each run can only reflect whatever initial values were selected for rates, not any policy decision on the part of the learner.

When SEPIA is complete, there will be several more adaptation options from which users may select. There will also be hooks to incorporate new, custom algorithms. Continuing use will build up a library of adaptation algorithms. Users will have the freedom to mix multiple algorithms in a single simulation, or even in a single agent, and of course, users can also disable all adaptation.

4.   THE REAL TIME PRICING (RTP) SCENARIO

With Open Access to the continental grid and rapidly disseminated information about all bids and offers, Real Time Pricing of electric power is becoming possible, both at immediate spot market rates and at forward prices for various horizons. SEPIA is implementing a scenario that allows its adaptive agents to engage in this market under a wide variety of user-defined hypothetical arrangements.

4.1   Real Time Pricing (RTP) of Electricity

Real-Time Pricing (RTP) in the context of electric power seems to mean many different things. In the case of the recent RTP project, jointly pursued by EPRI, Consolidated Edison (ConEd), Honeywell and the Marriott Manhattan Hotel, ConEd posted an hourly schedule of prices for the next day. There was no negotiation and no alternative sources of power, but the prices were guaranteed, i.e.: ConEd would receive the gain or incur the loss if its estimates of its own costs for power differed from its actual real-time costs at the times when the Hotel drew that power. The Hotel was simply given the opportunity to plan its operations so as to reduce its use of electricity at times of high price. On the other hand, this was a part-contract. The Hotel could buy as much or as little power from ConEd as it wished during each hour.

In this case, RTP means that a utility announces a price schedule at which it will sell electricity at specified future times (and sometimes just to specified customers). The prices announced by the utility are a form of forward price although they do not meet its strict definition: i.e., the price to be paid <u>now</u> for a specified amount of a commodity to be delivered at a specified time in the future. In the case of the Marriott Hotel and ConEd, no payment was required until after delivery and the amount to be purchased was left open. With open access and retail wheeling, the Hotel could, in theory, buy its power from any producer in North America. In fact, it could buy from many different producers, switching, again in theory, every nanosecond!

While every State in the USA might actually impose a different structure on a free market in electricity, the most open arrangement would be to:

- Allow any kind of bilateral contract, between a single producer and a single consumer, or between aggregates of either, entered into at any time up to the actual consumption of the power.

- Allow the formation of multiple market pools with open bidding by both buyers and sellers. These pools would typically take the form of a double Dutch auction, with sellers gradually lowering their asking price and buyers raising their offers until a series of bilateral contracts clears the market.

- Establish an agreed formula for the imposition on each exchange of a fair price for transmission between seller and buyer. This formula would have to include ways to allocate <u>all</u> the costs of system stability, unintended flows, contingencies, back-up power (when not, itself, included in the bilateral contract), etc.

4.2   RTP in SEPIA

The RTP scenario being incorporated in SEPIA includes only a few of the possibilities mentioned above, and of the many other arrangements needed to make RTP practical. The basic principles behind the RTP scenario for SEPIA are:

- Future power may be traded on the power exchange.
- Contracts are for 100 kWh lots delivered in specified 1-hour time slots.
- A specified margin deposit is required for each open contract.
- Variable economic parameters affect both the demand for power and the cost of borrowing.
- Corporate agents must honor all contracts and stay within credit limits (or go bankrupt).
- Corporate agents try to maximize profits subject to constraints.

- Generic goals that affect all corporate decisions include:
- Maximize profit by optimizing power production/consumption schedules.
- Maintain liquidity by managing cash flow.
- Reduce market risk by hedging production /consumption in the Power Exchange.
- Reduce production risk by keeping adequate fuel/raw material inventories.
- Account for their Economic/Environmental risk using projections to make their budgets and schedules robust.

Agents for this scenario include the Power Exchange, Power Producing Companies (PPC), Power Consuming Companies (PCC), and the Economy/Environment.

The Power Exchange provides a market for buying and selling spot and future power, and acts as clearinghouse for all bids and offers. The Power Exchange has three major functions: Exchange Management, Finance, and Brokerage.

PPCs buy and consume fuel, produce and sell power; PCCs buy power and raw materials, produce and sell manufactured goods. Both classes of agents try to earn a profit, but are required to pay operating, finance and tax expenses and their cash flow is constrained by a limited line of credit.

The agent representing the Economy/ Environment in which the other agents operate provides the inputs they need for making business decisions, for instance:

1. Cost of fuel for electric generation.
2. Short-term interest rates (prime rate).
3. Weather.
4. Rate of economic growth.
5. Consumer demand for manufactured products.
6. Price of raw materials to manufacturers.

It generates the states of these variables at any time from stochastic differential equations representing Poisson processes. The Economy/Environment agent issues periodic market reports predicting the future behavior of these parameters as well as corrupted estimates of their steady-state values and standard deviations of each state separately. Future extensions to the current RTP scenario include:

- Allowing bilateral contract agreements (off exchange).
- Handling transmission issues – zones, independent system operator involvement, transmission costs.

- Expanding fuel-types (oil, gas, hydro and nuclear) as well as addressing long-term supply agreements.

- Possible plant outages, scheduled and unscheduled.

- More detail in financial accounting: i.e., adding structure to the liquid asset portfolio, paying preferred stock dividends, and income tax.

4.3   Scenarios and Examples

To further clarify the types of simulations that can be conducted with SEPIA, we illustrate in this section how the elements of the program discussed above can be combined to create and run scenarios. The first set of scenarios implemented in SEPIA model the wholesale world of open-access electric utility operations. These scenarios involve the types of agents discussed above: generation company agents, generator agents, consumer company agents, load agents, and a transmission network operator agent. In these scenarios, consumer company agents purchase all the energy they need from generation company agents through direct "bilateral" contracts. Periodically, each generation company determines the unmet hourly power needs of each of its loads for the next week and broadcasts a "request for quotes" (RFQ) to all generation company agents. Generation companies receive such broadcasts and determine whether to submit a quote for some or all of the power requested by the RFQ. Deciding whether to respond to an RFQ and determining the price to charge for the energy is a difficult problem that is further complicated by the limits of the transmission network.

All contracts that require power to be transmitted across zone boundaries must be checked against an available transmission capability (ATC) table for each hour; the ATC data are maintained by the transmission network operator agent. As transactions are agreed on by load company agents and generation company agents, the transactions are given to the transmission network operator agent and a new ATC table is calculated and posted. An important last component of these scenarios is the generator of last resort for each zone. These GLRs are always willing to sell energy for a very high, constant price to any consumer company. The purpose of specifying GLRs is to model the behavior of spot market prices and prevent unlimited price escalation due to tacit collusion among all power generation companies.

Figures 3 and 4 illustrate scenarios set up in SEPIA. In the first of these, for example, four zones are defined. Zones 2 and 4 contain one load and one ConCo each. Zones 1 and 3 contain one (nuclear) and two (hydro and fossil) generation plants, respectively, along with separate GenCo's for each.

Figure 3: Four-zone scenario with three generators and two loads

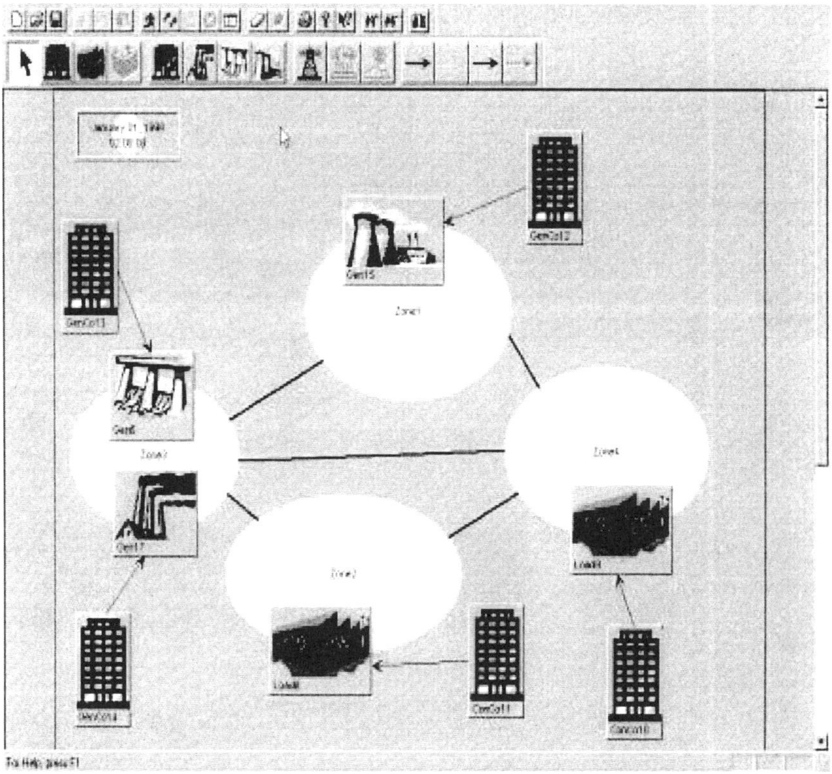

Consumer company agents are responsible for purchasing energy for each of their individual loads. Each consumer company agent operates as an independent business and will try to purchase power for the lowest possible price. Consumer companies can determine their hourly power need profile by aggregating the needs of each of their loads. Once the power need is determined, the consumer company can submit requests to all power generation agents, regardless of zone. Each RFQ will contain the power needs for every hour for the next week and will be valid for a short time period. Once the validity of the RFQ has expired, the consumer company can evaluate all the quotes it has received and accept one, many, or none of them. It is important to note that the burden for securing permission from the transmission network operator falls on the generation company. Therefore, once a consumer company has accepted a quote from a generation company, it can rely on receiving the energy promised by that quote.

Figure 4: Second example showing four loads and 15 generators

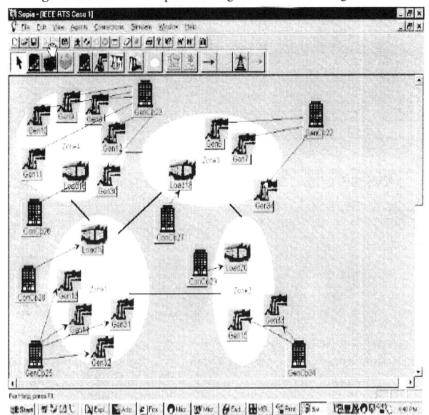

Each generation company controls one or more generators and will attempt to maximize its profit by selling its power for the highest possible price. Generation companies will attempt to establish attractive bilateral contracts with consumer company agents by responding to appropriate RFQs. Several factors must be considered when deciding how to respond to an RFQ. SEPIA currently takes into account the generation cost function and the megawatt capacities for each specific generator.

In these scenarios, generation company agents take on the risk and responsibility of delivering all energy they quoted to consumer agents. This means that before they submit a quote, generation company agents will check the public ATC table and will reserve transmission rights as soon as a quote is accepted. It is possible that a transmission that was permissible when a quote was submitted is no longer permitted when the quote is accepted. This is a risk generation company agents assume when doing interzone business. In these rare cases, the generation company agent is responsible for buying energy in the consumer's zone at inflated spot prices (the GLR rate).

The transmission network operator will calculate available transmission capability (ATC) and post ATC for all generation and consumer companies to access. The transmission network operator agent does not allow transactions if they violate transmission limits. The ATC and the accept/not accept decisions on specific transactions are based on security analysis checks (as noted earlier, base case limit checks and first contingency checks are undertaken).

4.4 Open-Access Scenario Example

In this section, we take a simple example scenario and walk through its configuration and simulation. This scenario looks at the competition between two power generation companies located in separate zones with a significant transmission bottleneck. Interested readers can "replay" the discussion below by downloading a self-running narrated demonstration (a Lotus ScreenCam executable for Windows PCs) from http://www.htc.honeywell.com/projects/sepia.

First, SEPIA is used to define the simple scenario. In this example, the user has defined two independent zones and then connected them with a tie line. Next the user adds a load to Zone 2 and a generator to each of the two zones. Two power generation companies and a power consumer company are also added and associated with the generators and the load.

For this simulation, each generator is capable of generating up to 100 MW and the load consumption per hour is a random quantity between 90 and 100 MW. Each of the two power generation companies is set to learning mode and given an initial price of $10/MW. The GLR in Zone 2 is given a fixed price of $60/MW.

Next we modify the electrical properties of the transmission network between the two zones to reduce the maximum transmission capacity. We can access the property sheet for the tie line by simply double clicking on it and effect desired changes.

As an example, we now set the simulation run time to 2000 days (through a property sheet) and start the simulation by selecting the run button. Once the simulation is complete, we can examine the results. SEPIA includes a third-party charting and visualization package through which a variety of simulation-generated data can be displayed in different forms (see Figure 5).

Figure 5: SEPIA chart depicting the competition between two generation
company agents

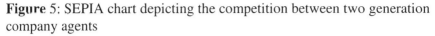

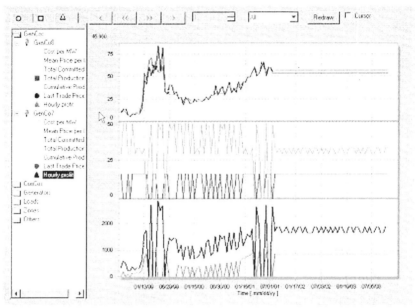

The top graph in Figure 5 displays the price each GenCo charges per
megawatt of power throughout the simulation. Both GenCos started raising
their prices together from the starting price of $10/MW. Once the price per
megawatt exceeded $60, the GenCo's started losing business to the GLR; the
price is then lowered and raised again until settling on a level just below
$60/MW.

The remote GenCo settles on a price slightly below that of the local
GenCo. The local GenCo is content with this arrangement since the
transmission operator limits the remote GenCo to a maximum of only around
18 MW. The middle graph displays how much power each GenCo sold at
each point in time. Note how the amount of power sold by a GenCo goes to
zero once its price goes over $60/MW. The final state reached by this
simulation has the remote company selling all the power it can transmit for a
price slightly lower than that of the local GenCo, which is selling all
remaining power for a price just slightly lower than that of the GLR.

The bottom graph in Figure 5 displays the profit generated at each hour.
This graph mirrors the production graph (middle) fairly well. Note that the
production and profit graphs for the local GenCo oscillate because the
demand of the load has a random component.

5.   SUMMARY

The U.S. electric power system developed over the last hundred years without a conscious awareness and analysis of the system-wide implications of its current evolution under the forces of deregulation. The possibility of power delivery beyond neighboring areas was a distant secondary consideration. Today, the North American power network may realistically be considered to be the largest machine in the world since its transmission lines connect all the electric generation and distribution on the continent. With the advent of deregulation, unbundling, and competition in the electric power industry, new ways are being sought to improve the efficiency of that network without seriously diminishing its reliability.

To address these and other emergent issues involving economic effects of deregulation on the "Electric Enterprise," EPRI is developing a "bottom-up," scenario-free model for exploring the evolution of the power industry, constrained only by the physics of the system components. This model and simulation of the "Electric Enterprise" which uses autonomous, adaptive agents to represent both the possible industrial components and the corporate entities who own these components. In this report, we have presented a brief summary of this model, its objectives, the background against which it is being developed, and the present state of its implementation as a computer simulation.

In many complex networks, for instance in the organization of a corporation, the human participants are both the most susceptible to failure and the most adaptable in the management of recovery. Modeling these networks, especially in the case of economic and financial market simulations will require modeling the bounded rationality of actual human thinking, unlike that of a hypothetical "expert" human as in most applications of artificial intelligence.

Although the focus of this chapter has been on the specific topic of restructuring and SEPIA, I have implied above that electric power systems are one example of a more general class of systems which we can refer to as complex interactive networks. A recent research program being conducted at more than 25 universities in the United States and supported by the U.S. Department of Defense and EPRI is emphasizing this broader perspective. Readers interested in more details on this program, the Complex Interactive Networks/Systems Initiative, are referred to http://www.epri.com/targetST.asp?program=83 and to references indicated below.

How to control a heterogeneous, widely dispersed, yet globally interconnected system is a serious technological problem in any case. It is even more complex and difficult to control it for optimal efficiency and maximum benefit to the ultimate consumers while still allowing all its business components to compete fairly and freely.

REFERENCES

Amin, M. 2001. "Toward Self-Healing Energy Infrastructure Systems," *IEEE Computer Applications in Power* 14 (no. 14, January): 20-96.

Amin, M. and Ballard, D. 2000. "Defining New Markets for Intelligent Agents." *IEEE IT Professional* 2 (No. 4, July/August): 29-35.

Amin, M. 2000. "National Infrastructures as Complex Interactive Networks," Chapter 14 in *Automation, Control, and Complexity: New Developments and Directions*, edited by Samad & Weyrauch. John Wiley and Sons: NY.

Bak, P. 1996. *How Nature Works: The Science of Self-Organized Criticality*. Springer-Verlag: New York, NY.

Commercial Practices Working Group (CPWG). 1997. Industry Report to the Federal Energy Regulatory Commission on the Future of OASIS, November 3.

Cowan, G.A., D. Pines, and D. Meltzer, (Eds). 1994. *Complexity: Metaphors, Models, and Reality*. Addison-Wesley: Reading, MA.

EPRI. 1999a. "Simulator for Electric Power Industry Agents (SEPIA): Complex Adaptive Strategies," November 1999, Palo Alto, CA, Report TR-112816

EPRI. 1999b. "Prototype Intelligent Software Agents for Trading Electricity: Competitive/ Cooperative Power Scheduling in an Electronic Marketplace," Dec. 1999, Palo Alto, CA, Report TR-113366

EPRI. 2000. "E-Commerce Applications and Issues for the Power Industry," May 2000, Palo Alto, CA, Report TP-114659

Friedman, D. 1991. "Evolutionary Games in Economics." *Econometrica* 59 (No. 3, May): 637-666.

Harp, Steven A., Sergio Brignone, Bruce Wollenberg, and Tariq Samad. 2000. "SEPIA: A Simulator for Electric Power Industry Agents." *IEEE Control Systems* 20 (No. 3, August): 53-96.

Holland, J. H. 1995. *Hidden Order: How Adaptation Builds Complexity*. Addison Wesley: Reading, MA.

Soros, G. 1995. *Soros on Soros: Staying Ahead of the Curve*. John Wiley & Sons, Inc.: NY, NY.

Wildberger, A.M. 1997. "Complex adaptive Systems: Concepts and Power Industry Applications." *IEEE Control Systems* 17 (No. 6, December): 77-88.

Chapter #4

Electricity Restructuring in Practice*

Richard L. Gordon
The Pennsylvania State University

Abstract: Severe flaws exist in the theoretic case for regulation and practice in electric
          power is worse than theory predicts. Theory suggests regulation was both
          unneeded and unlikely to succeed. New Deal initiatives in electric power were
          defective; energy interventions in the 1970s were totally unjustified. Crises
          arose because restructuring was implemented in disregard of these lessons. In
          self-defense, electric utilities are merging.

Key words: Electric-utility regulation; electric-utility restructuring; electric-utility mergers;
          New Deal reforms in electricity; national energy regulation in the 1970s;
          nonutility generation; the California energy crisis.

1.   INTRODUCTION

At a 1995 conference on then-emerging changes in electricity regulation,
I deemed the proposals ill conceived. The changes seemed only massive
alterations in how regulation was practiced. The removal of regulation was

---

*   Professor emeritus of Mineral Economics and MICASU faculty endowed fellow emeritus,
     College of Earth and Mineral Sciences, The Pennsylvania State University. This draws
     upon Gordon 2001. To cover restructuring and other industry developments, I examined
     reports from the U.S. Energy Information Administration, discussions by several think
     tanks, and the annual reports and 10ks of leading U.S. electric utilities.

A. Faruqui and B.K. Eakin (eds.), Electricity Pricing in Transition, 51-63.
@ 2002 *Kluwer Academic Publishers.*

preferable. The case was supported by a detailed review of theory and practice.[1]

Briefly, severe flaws exist in the theoretic case for regulation and practice is, as is typical for intervention, worse than theory predicts. This article summarizes my prior work and examines the subsequent practice of restructuring. It is argued that, as is usual with ill-advised intervention, implementation was worse than anticipated.

2.   THE THEORY RECONSIDERED

Two bases exist for opposing regulation of electricity as a natural monopoly. First, electricity may not be a natural monopoly or second, regulation may not be a desirable way to react to natural monopoly. While claims are made that monopoly does not prevail, handling that issue is one of empirical analysis, and efforts to date have not produced overwhelming support. A far stronger case exists for the second rationale, that the advocacy of regulation or government ownership is unjustified.

Posner (1969, 1999) stated the gist of the case in a massive, rambling pioneering article. The essence of his case was regulation was both unneeded and unlikely to succeed. He recognized that the industries that are regulated could and did employ precisely the forms of price discrimination that textbooks indicate will produce efficient output. Regulation, therefore, was justified only as an income-redistribution program. He doubted whether the redistribution effects were large enough to justify the effort.

In turning to implementation, he, in effect, applied two now classic elements of free-market attacks on intervention. First, he noted the problems of attaining efficiency because of the information-cost problems that Hayek often has raised. Second, in anticipation of subsequently developed economic theories of regulation, he noted that regulators were unlikely to pursue efficient outcomes.

In at least three ways, Posner understated the case. First, the depth and perversity of income-distribution problems are worse than he recognized. Second, he followed the bad tradition in natural-monopoly studies of ignoring joint production. Third, he ignored the economic defects of accounting practices used throughout the economy including in regulation.

In addition to being small, the redistribution effects are unlikely to be sensible. Regulators lack the resources and expertise to effect desirable

---

[1]   Due to various setbacks, the work was not published until 2001 when experience and well-publicized concern had emerged. That article, nevertheless, retained the original stress on examining both the theory and the historical record since the 1930s.

changes. In particular, regulators (in every realm) fall victim to the direct-impact fallacy. The effects on direct use of households are treated as so important that the harms to households of inefficient industrial prices are ignored.

Moreover, redistribution is a more pervasive influence than even my prior writing indicated. There, I stressed that the essence of public-utility regulation was transferring rents to users of the regulated industries' output. I recalled that no guidance existed for dividing that rent among those customers. That discussion failed to note that rents to both monopoly and inframarginal facilities are involved. Thus, I neglected my long-standing concern that too often rents are used unwisely to subsidize production and consumption.[2] The frenzy over problems with restructuring has inspired new efforts to employ rents to subsidize consumption.

As economists somehow eventually learn, the market allocates the costs of jointly produced goods.[3] Viability only requires that the collective revenues from all products cover costs. The strength of demand determines how much each product contributes. However, public-utility regulation employs a cost-of-service concept that tries to allocate costs independently of demand. This defect is ignored in the numerous treatises on natural-monopoly regulation.

Finally, a small but acute literature[4] deals systematically with the problems with standard accounting methods. Accountants like to base their numbers on historical experience. This violates fundamental economic principles. Among the key defects are failure to adjust for inflation and measuring depreciation by mechanical formulas rather than by market developments.

All this suggests that regulation does not merit its widespread support.

3. HISTORICAL PRACTICE RECONSIDERED

My prior review of history had two basic components. First, the four New Deal initiatives in electric power were defective. Second, the energy interventions in the 1970s were totally unjustified.

---

[2] That concern, in fact, dates back to my 1960 Ph.D. thesis on Western European coal policy that among other things used the rents of lower-cost mines to subsidize additional output.

[3] While many articles make this point, the only major textbook of which I am aware that treats the point is Stigler.

[4] Edwards and Bell (1961), while dated, is the best reference of which I am aware. It provides an extensive and valid development. A more recent but less thorough discussion appears in Fisher, McGowan, and Greenwood (1983).

The New Deal measures were creating a federal agency for regulating electric power (and later natural gas), trying under the Public Utility Holding Company Act (PUHCA) to restructure public utility holding companies, starting large federal electricity-generation project, and providing subsidies for rural electrification. The most germane new interventions of the 1970s were the pressure for non-utility generation (NUG) under the 1978 Public Utility Regulatory Policy Act (PURPA) and the pressures for conservation under the National Energy Conservation Policy Act (NECPA). In addition, existing public-utility regulation acted excessively to limit profitability and investment.

The complaints about three of the New Deal initiatives are straightforward. The main defect of federal-level regulation is that it extends the error of trying to regulate. Moreover, it is not clear that problems exist that only national controls can handle. Certainly, disputes over proper sharing of costs among different states cannot be resolved since this involves allocating joint costs. No *sound* economic basis exists for either government production of power or subsidy of rural or any other form of electricity supply.

The case against PUHCA is more complex (and of more immediate relevance). In even earlier work (Gordon 1992), I concluded that PUHCA was inadequately criticized. A mindless 1935 Federal Trade Commission report had charged holding companies with evils. The Act and many subsequent commentators then uncritically repeated the charges. However, these charges, in effect, were simply that the holding companies followed the still standard industrial practice of employing those "approved" accounting methods that produced the most favorable outcome. It was inadequately proved that the holding company had major advantages in altering accounts.

More critically, intensive examination of the resulting restructuring suggested that the outcome was questionable. Many of the surviving holding companies were and remain large interstate ventures; several became larger. However, most of the individual holdings of the old empires were made independent entities of widely different size and regional importance. Thus, whatever the optimal size might be, it clearly was not universally attained. Some states such as Michigan and Georgia still had only two significant electric utilities. Others, most notably Iowa, had a half dozen.[5]

Moreover, enforcement long bent and in the 1990s broke the Act's basic principles that a holding company must operate as a unified company in a single sector in a contiguous area. Several electric holding companies had gas operations or were, according to industry jargon, combination

---

[5]   Subsequent mergers reduced the Iowa six to two.

companies. Territories were not always fully contiguous.[6] Companies were not unified.[7]

A classic problem in public policy is the adoption and vehement defense of economically nonsensical options. Energy conservation is an example. The not-very-good best explanation is that the measures correct the effects of unwise price controls and subsidy. However, eliminating the distorting program is preferable. It is cheaper and, thus, less painful than forced conservation.

Economics, of course, stresses consumer sovereignty, that individuals are the best, if not the perfect, judges of what is preferable. This denies the possibility raised by proponents of imposed conservation of inefficient responses to price. At most, governments might efficiently improve information availability. Governments possibly could provide data on the costs of energy alternatives at a cost less than that of private provision. This too is a dubious proposition and, even if valid, does not justify any of the extent conservation programs. These theoretic doubts are reinforced by the shoddy quality of the "proofs" that enormous potential exists efficiently to reduce energy consumption. Biased analysts make the calculations without experience in decision-making. However, as is typical in policymaking, conservation sounds so good that its economic vacuity can be shouted down.

More generally, politicians crave methods that avoid facing the consequences of rising costs. The "no-free-lunch" mantra is warning that no sensible alternative exists. Subsidies of both conservation and more consumption that possible at unregulated prices will necessarily cost more than free-market pricing. Since the subsidies must be financed, the net distribution effect is unclear. However, again the direct-impact delusion motivates the imposition of these polices.

This complaint applies to the other components of 1970s energy policy and the 1992 Energy Policy Act. The alternative energy provisions of PURPA make no sense. Only the defects of regulation would produce

---

[6] For example, the New Jersey portion of General Public Utilities is not contiguous to its Pennsylvania potion. Moreover, the predecessor company was forced to divest New York holding that were contiguous to those in Pennsylvania. New England Electric's service territory was not contiguous and its merger with Eastern Electric did not cure that.

[7] The most blatant cases were in Texas. Texas Utilities, until it abolished its subsidiaries, limited their integration. Central and Southwest has subsidiaries inside ERCOT (the Electricity Reliability Council of Texas) and interstate subsidiaries in another pool. ERCOT is not interconnected to the rest of the United States because Texas Utilities and what is now Reliance Energy want to avoid federal regulation. This forced Central and Southwest to keep its ERCOT divisions separate from the others until pressures arose to establish connections.

inefficient choice among options. PURPA then asked the regulators who caused the bad choices to correct the effects of their regulatory errors.

In a similar spirit, the regulators took it upon themselves to add consideration of social costs. These included both environmental impacts and the alleged cost of dependence on oil imports. These seem matters that are handled already by other agencies. While these agencies are highly flawed, a public utility commission is unlikely to do better.

What matters here are that the tightening of regulations produced serious problems for electric power and the implementation of PURPA and NECPA made matters worse. Electric utilities were saddled with investments that were in danger of becoming unprofitable. These investments included expensive capacity, particularly nuclear, that the utilities had built. In addition, commitments were made to NUGs under long-term contracts based on utility-commission imposed expectations of continued rises in energy prices, and the recovery of investments in conservation was questionable.

Both the utilities and consumers (or at least those claiming to represent consumers) were dissatisfied with the situation. Enough of the excess cost was passed on to produce consumer discontent. Enough was absorbed by utilities that at least the ability to expand and possibly even the ability to survive were endangered.

4.   THE DEBATE OVER RESTRUCTURING

n response to these concerns, public utility commissions and legislatures began searching for new means to respond. As the debate developed, it became apparent that impossible demands were being made. The changes were supposed to give lower prices to consumers, higher profits to producers, and not upset existing interventions. This was a recipe for disaster. Economic commentary (e.g., Brennan, Palmer, Koop, Krupnick, Stagliano, and Burtraw, 1996) recognized that that eliminating inefficient requirements was critical to success.

The biggest debate among economists was over "stranded cost" recovery, how much of past investment could be recouped. The issue stressed was whether or not failure to recoup was an unanticipated confiscation of capital. Defenders (e.g., Baumol and Sidak 1996). of compensation saw a breaking of the (tacit) accord between utilities and regulators that profits would be guaranteed. Critics (e.g. Michaels 1996) replied that regulatory experience long ago warned that the compact had ended.[8]

The latter is both more plausible and only part of the story. The ability to measure stranded cost is questionable. Measurement depends upon more accurate accounting measures than are used and an ability to forecast that the regulators failed badly to display. Past errors are to be offset by the processes that caused them. It is no surprise that wildly different figures arise from estimates based on conventional accounting data, those based on security-market data, and those rising from powerplant sales.

Given that much of the stranded cost arose from bad arrangements with NUGs and conservation efforts, these last are another source of adjustment. Changes in energy contracts since the 1970s make clear that economic reality outweighs sanctity of contract. It is self-defeating to insist on performance of a contract that will ruin the other party. This applies to NUG and conservation arrangements.

Finally, it is unclear how much, if any, recovery is possible. The amount of monopoly profit possible is finite. The amount available may not equal that industry stranded costs or be available to those who incurred these costs.

Another aspect of the problem, the probable price, was universally handled incorrectly both in the early years of PURPA and during restructuring. In restructuring, the price of generation using natural gas in combined-cycle plants was treated as the marginal cost of electricity. Coal and nuclear costs were considered mere artifacts of bad past decisions. This overlooked the persistent need for coal and nuclear capacity. The current situation allowed low-cost additions to capacity with gas-fired combined cycles but precluded total replacement of coal and nuclear generation.

This emerges immediately from viewing the data. The energy used for generating electricity from coal and nuclear power (28 quadrillion Btu in 2000) exceeds *total* gas use (23.3 quadrillion). If the needed doubling of gas supply were so economic, it would have been effected long ago.

What was involved was that in the 1970s capacity expansion raced ahead of demand growth. That capacity was expensively stockpiled to serve growth in the eighties and nineties. This allowed limiting further capacity additions to gas-fired combined cycles and combustion turbines.

---

[8] Michaels is a negative review of a book by Baumol and Sidak advocating stranded-cost recovery. The cited Baumol and Sidak article is a rejoinder. The same volume of *Regulation* contains other articles and numerous letters on this debate.

5. AN OVERVIEW OF RESTRUCTURING

As a reader of Peltzman's version of the capture theory of regulation would expect, the regulators tried to please everyone and produced a failure. More precisely, the extent of restructuring differed radically. Some states saw no need to change. Those that did took different approaches. Among the areas of decision-making were the setting of wholesale and retail rates, policy towards vertical integration, and the status of conservation, "renewable" energy, and social costs.

The tendency was to free wholesale rates and rely upon formal organizations to bring buyers and sellers together. The three long-extant regional coordinating centers in the United States cover the Northeast.[9] California made creation of a new center a requirement of restructuring. When proposed, fears were raised that the states two largest utilities, Pacific Gas and Electric and Southern California Edison, would possess excessive monopsony power. Once operations began, charges were rife that the suppliers were deliberately withholding supplies to force up prices. While that seems implausible, observers purport to see such behavior.

Another characteristic is imposing price cuts on retail sales with guarantees to sellers and buyers that subsequent changes would be limited. This guarantee is *the* reason for present problems.

Much variation prevailed in the attitude towards vertical integration. Some states such as New York and Massachusetts pushed total divestiture of generation. California took an intermediate course. Sales of fossil plants in the state were required and effected. However, nuclear and hydroelectric capacity and interests in coal-fired plants in the Mountain states were at least initially retained. Pennsylvania encouraged, but did not require, sales. Some utilities complied; others did not. Other states such as Texas, Ohio, and Virginia were content with placing generation in a separate subsidiary of the utility.

The buyers of the divested plants were mainly the unregulated subsidiaries of established utilities.[10] For example, while AES, a leading

---

[9] One pool covers New England; another, New York; the third (PJM), New Jersey, Delaware, the District of Columbia, much of Pennsylvania and Maryland, and a bit of Virginia. The Allegheny Power system has indicated an intention to enter PJM. This would add another large part of Pennsylvania and parts of other states.

[10] Name changes were produced by actual and cosmetic changes in the electric utility business. The more predictable aspect is the provision of a name for the many newly-created entities. In addition, many companies have instituted a new parent for themselves and possibly for new businesses established or acquired. Finally, the utility or its new parent may have made a name change. The key characteristic of these changes is that they are, probably deliberately, far less descriptive than the original name.

NUG, bought the fossil plants of Southern California Edison, Pacific Gas and Electric's fossil plants were bought by subsidiaries of Duke and Reliant.[11] Then, the NUG subsidiary of Southern California Edison's parent bought the fossil generating capacity of Commonwealth Edison.

The separation was supposed to remove alleged incentives inefficiently to favor integrated facilities over NUGs. This is a dubious justification. The spot price would in any case be set no lower than the marginal cost of the highest-cost unit that served the spot-market demand. An integrated company would then operate only those facilities with running costs less than the spot price, This would produce inefficient results only if the spot price could be made higher than in a competitive market. This is infeasible. With integration, the efficient response is for the integrated firm to reduce open market purchases by exactly the amount it expects to produce efficiently. The same spot prices prevail as would occur without integration. The only way to get a higher spot price is to bid for more than is wanted. Then the bidder is stuck with extra supplies and cannot produce as much as is efficient, let alone too much.

In any case, predictably the transfer of ownership only produced a transfer of scapegoat. The new independent operators, as noted, are accused of price gouging. A policy spawned in response to the policy errors of the 1970s is inspiring repeating the same price-control errors.

Some states tried also preserve some of the commitments to conservation, covering social costs, and fostering alternative energy, Typically, surcharges are imposed or maintained on electricity rates to finance favored programs, For example, California stresses conservation. Texas emphasizes renewables. New York includes conservation, additional environmental monitoring, and subsidy of low-income consumers.

Much publicity was given the California ban on long-term contracts for electricity. As Jerry Taylor and Peter Van Doren (2001) of the Cato Institute have argued, concerns over this ban are misplaced. Taylor and Van Doren share the viewpoint expressed here. Contract prices cannot and should not be maintained. California utilities sold precisely the facilities, gas and oil plants, with which contracts would have been unsustainable because of the rise in natural gas and oil prices.

The critical problem in California is the radical disparity between free-market wholesale prices and the capped retail prices. The low price stimulates consumption when the underlying economics require powerful

---

[11] The two buyers were prime movers on different sides of the 1992 debate over changing PUHCA to ease utility entry in the NUG. Duke wanted liberalization and its participation as a NUG. Reliant was the spearhead of opposition seeking restoration of the health of the integrated-company model.

price incentives to reduce consumption. Efforts to maintain these low prices are part of California's continuing policy of trying to shelter users from the consequences of rising natural-gas prices. Thus, the excess demand situation is aggravated and expansion is chilled in the interest of questionable claims that the price rises are too painful.

This has involved requirements that out-of-state capacity ownership be retained and efforts by the state government to purchase hydroelectric and transmission facilities and, of course, subsidy of power purchases. It is, in short, a repetition of the errors of the 1970s.

Other states have smugly argued that they adopted changes that lessen the danger of a California style situation, a view too glibly endorsed in the Vice President's May 2001 energy plan. The problems are independent of whether or not the firms reorganize. Any state with a deregulated wholesale market and prices caps on retail sales can experience California-type profit squeezes. At least one eastern company, General Public Utilities operating mainly in supposedly soundly restructured Pennsylvania, complained about such a squeeze in its 10K for 2000.

A major reason what California is leading the country in power problems is that it led the country in restricting expansion. The state created an Energy Commission as large as and independent from its quite large state public utility commission. For three decades, the Energy Commission steadfastly prevented expansion. Lesser restraint elsewhere means lesser need to catch up. In addition, California has above average dependence on natural gas that increased its vulnerability to the rise in natural-gas prices.

6.   MERGERS AS A RESPONSE TO REGULATORY UNCERTAINTY

Another process that has accelerated with restructuring of electric power is many attempts, not all successful, at many types of mergers. Large electric companies have purchased smaller ones. Companies have merged with companies comparable to themselves. Independent power producers have purchased established utilities. Acquisitions were made of companies in the transmission as well as the distribution of natural gas. Almost all the main electric-power companies were involved in at least abortive efforts to merge.[12] Among the consequences is radical change in PUHCA

---

[12] Counting and classifying electric companies are problematic, and the statements in the text are expressed to minimize the impacts of adopting different concepts. The widely-cited counts made by EIA include subsidiaries of holding companies and of operating companies and joint ventures but exclude the holding companies. Thus, the latest database,

enforcement. New holding companies totally ignore both the single industry and the contiguity requirements.

The largest mergers are particularly blatant violations of the contiguous territory principle. In 2000, American Electric Power with a multi-state territory centered in Ohio acquired Central and Southwest, operating in the West South Central states. Unicom, the parent of Chicago's Commonwealth Edison, was allowed to acquire Peco Energy, the parent of Philadelphia Gas and Electric producing Exelon. Peco also is one of many combination companies allowed to become part of a regulated holding company. A merger between Florida Power and Light and Entergy (operating in Louisiana, Arkansas, Mississippi, and Texas) was abandoned in 2001 by the companies. Entergy is one of the original regulated holding companies with combination company subsidiaries and not contiguous to Florida Power and Light.[13]

Other examples of separated companies, some of which have gas activities, merging are Xcel and Progress Energy. Xcel combines Northern States Power (in Minnesota and Wisconsin) with Public Service Co of Colorado and Southwestern Public Service; the first two are combination companies. Progress Energy combines Carolina Power and Light with St, Petersburg–based Florida Power, both electric only.

At a smaller level, several mergers in the Northeast involve separated companies. After the merger of New England Electric with Eastern Utilities, the resulting company was acquired by Britain's National Grid. It then arranged a pending merger with Niagara Mohawk (another combination company). The creation of Energy East violates both the continuity and combination company aspects of PUHCA. It is composed of two electric companies–New York State Gas and Electric (another combination company) and Central Maine Power and two gas companies. A merger with another combination company, Rochester Gas and Electric, is pending.

---

for 1999, lists 238 private companies but only 146 independent entities existed; mergers through early 2001 reduced the number to 139; pending mergers would lower this to 133. Comparisons of EIA reports with Edison Electric Institute material suggest differences in inclusion criteria; however, these tend to involve the smallest companies. With mergers, the count in any case keeps changing. Different measures of size and different cut-off criteria for big companies are available. Another concern would be how far back the coverage should go. One major entity, the Southern Company, has not attempted a merger since its acquisition of Savannah Gas and Electric in the 1980s. Otherwise, the largest companies on which no merger attempts have come to my attention are Tampa Electric and Dayton Power and Light.

[13] Florida Power and Light then is the largest company not to complete a merger. Entergy acquired Gulf State Utilities.

The biggest merger of contiguous purely electric companies is First Energy. In 1997, it combined Ohio Edison, Cleveland Electric Illuminating, and Toledo Edison. A merger with General Public Utilities is pending. These mergers and those in the Northeast illustrate another aspect, consolidation in states with large numbers of companies.

The participants in these mergers differ widely in how restructuring altered their composition. For example, the expanded AEP retains and plans to retain its generating capacity; Exelon combines Commonwealth Edison that sold its fossil-fuel but not its nuclear plants and Peco that has sold nothing. National Grid and Energy East, in contrast, consist of companies largely divested.

However, the biggest ventures into gas were not by regulated holding companies. TXU, the former Texas Utilities, and Reliant Energy, the parent of the former Houston Lighting and Power were both allowed to acquire natural gas companies. CMS Energy, a combination company in Michigan acquired Panhandle Eastern, a major pipeline company. Dominion Resources acquired Consolidate National Gas. NiSource, the parent of Northern Indiana Public Service, acquired the Columbia Gas system.

One major independent power producer, AES, has acquired Central Illinois Light and Indianapolis Power and Light. Dynergy has acquired Illinois Power. Enron acquired Portland General Electric but a sale to Sierra Pacific is pending.

At least two forces may be at work. The pressures to change the industry are an impetus to seek new organizations better adapted to the alterations. However, the literature makes clear that as long ago as the 1920s many in the industry believed creation of giant companies was needed for efficiency. The Great Depression created PUHCA and other barriers to creating these larger entities. Regulation has clearly become less restrictive. This is a welcome retreat from the tendency to over control the industry.

7.   CONCLUSIONS

Restructuring was every bit the failure that should have been expected. Once again Posner's fears were proven too limited. Thus, it can be urged to undertake the steps Posner was too timid to propose of totally removing government from the regulation, ownership, and subsidy of electricity.[14] This means no more FERC or PUHCA and also no more Nuclear Regulatory Commission, the privatization of government-owner electricity operations,

---

[14] Since the early 1980s I have contended these changes merit examination. Subsequent experience has shown that the changes are clearly the preferable option.

and the end of rural electrification subsidies, demand side-management, and forced NUG access.

REFERENCES

Baumol, W. J., and J. G. Sidak. 1996. 'Recovering Stranded Costs Benefits Consumers." *Regulation* 19 (2): 12-15.

Brennan, T. J.; K.L. Palmer, R.J. Koop, A.J. Krupnick; V. Stagliano, and D. Burtraw. 1996. *A Shock to the System: Restructuring America's Electricity Industry*. Washington, D.C.: Resources for the Future.

Edwards, E. O., and P.W. Bell. 1961. *The Theory and Measurement of Business Income*. Berkeley and Los Angeles: University of California Press.

Fisher, F. M.; J.J. McGowan, and J.E. Greenwood. 1983. *Folded, Spindled, and Mutilated Economic Analysis and U.S. v. IBM*. Cambridge, MA: The MIT Press.

Gordon, R. L. 1982. *Reforming the Regulation of Electric Utilities*. Lexington, Mass.: Lexington Books, D.C. Heath.

Gordon, R. L. 1986. "Perspectives on Reforming Electric Utility Regulation." In *Electric Power: Deregulation and the Public Interest* edited by John C. Moorhouse. San Francisco: Pacific Research Institute for Public Policy.

Gordon, R. L. 1992. "The Public Utility Holding Company Act: The Easy Step in Electric Utility Regulatory Reform." *Regulation* 15 (1): 58-65.

Gordon, R. L. 2001. "Don't Restructure Electricity; Deregulate." *The Cato Journal* 20:3, p. 327-58.

Michaels, R. J. 1996. "Stranded Investments, Stranded Intellectuals." *Regulation* 19 (1): 16-17.

Peltzman, S. 1976. "Towards a More General Theory of Regulation." *The Journal of Law and Economics* 19 (2): 211-40. Reprinted in Peltzman (1998): 155-87.

Peltzman, S. 1998. *Political Participation and Government Regulation*. Chicago: University of Chicago Press.

Posner, R. A. 1969 "Natural Monopoly and Its Regulation." *Stanford Law Review* 21 (February): 548-643. (Reprinted 1999 with new foreword Washington: the Cato Institute.)

U.S. Energy Information Administration. 2000. The Changing Structure of the Electric Power Industry 2000: An Update.

U.S. Energy Information Administration. 2001. Status of State Electric Industry Restructuring Activity. (Regularly updated tabulation posted on www. eia.doe.gov.)

U.S. Federal Trade Commission. 1935. *Utility Corporations: Summary Report of the Federal Trade Commission to the Senate of the United States*, Nos. 72A and 73-A. Washington: U.S. Government Printing Office.

Section II

Competitive Pricing

Chapter #5

Challenges in Designing Default Retail Electric Service
What Regulated Retail Services Should Be Available Following Restructuring

John L. Jurewitz
Southern California Edison

1. INTRODUCTION

A key question in any state's electricity restructuring plan is whether some form of retail power service should continue to be offered on a regulated basis under certain circumstances to some customers. This question is both a short-term (transitional) and a long-term (post-transitional) issue. So far, in every state adopting retail competition, utility distribution companies (UDCs) have been required to offer retail service on a regulated basis at least during a transition period. Generally, these arrangements have been part and parcel of complex comprehensive political settlements involving many issues including UDC recovery of stranded costs. But the post-transition responsibilities of UDCs to provide regulated retail service have usually been deferred as issues "to be determined" in later regulatory proceedings.

In most states, UDCs have been required to offer retail service during a transition period at prescribed fixed or escalating prices (often called "standard offer service"). The clear political motivation for these regulated offerings is to provide a "safety net" of consumer protection during a transition period of typically 5 to 7 years. But therein also lies the source of increasingly apparent problems. There are clear tensions between the

A. Faruqui and B.K. Eakin (eds.), Electricity Pricing in Transition, 67-86.
@ 2002 *Kluwer Academic Publishers.*

objectives of protecting consumers and encouraging the development of competitive retail markets. In short, as some observers[1] have noted, we may be "killing the market with kindness." By shielding consumers from the volatility of competitive spot markets, we may leave them largely uneducated and incompetent to cope with these forces at the end of the transition period. Moreover, by doing so, we may also inhibit the development and marketing of essential risk management products by third-party retailers. This may make it inevitable that consumers and politicians will insist that some type of regulated retail service continue to be offered beyond the transition period. In other words, there may be no full transition. Moreover, UDCs may be placed in an untenable long-term position of offering retail services at fixed prices and exposing themselves to huge risks of undercollected wholesale procurement costs. Indeed, Wall Street rating agencies recently have even coined a term for this risk: "default provider risk."

But even if a "successful" transition can be achieved, will some entity still be required to offer a regulated retail service under certain circumstances even after this transition? Will it be the UDC or one or more other entities? What will be the nature of this required service and who will be eligible to receive it? In order to simplify logistics and minimize transactions costs, it may be desirable to require the UDC or some other entity to provide at least a "plain vanilla" service that simply passes through wholesale power costs to retail consumers. Perhaps following an appropriate transition in which consumers are adequately educated and electricity supply is sufficiently depoliticized, such a "plain vanilla" service may be all that is required from the default provider. But the recent experience in San Diego in the summer of 2000 is not encouraging. San Diego Gas and Electric was the first UDC in the U.S. to reach the end of its designated transition period. In the absence of explicit regulatory direction, it then continued to offer its customers plain-vanilla pass-through of wholesale spot-market prices – the barest minimum retail offering. But the subsequent extraordinary events in California's wholesale market proved too much for politicians to forebear intervention. By the end of summer 2000, a retail rate cap was re-imposed, raising serious doubts about the future of retail competition in California. On the one hand, one can argue that San Diego customers were not adequately prepared for retail access during the preceding transition period and that retail markets were not sufficiently developed. On the other hand,

---

[1] "POLR and Progress Towards Retail Competition – Can Kindness Kill the Market?," presentation made by Frank Graves and Joseph Wharton of The Brattle Group, at the Winter Committee Meetings of the National Association of Regulatory Utility Commissioners, Washington, D.C., February 27, 2001.

one can view this episode as an abject failure of restructuring and wonder skeptically whether a successful transition to full retail competition can ever be made and, if so, what groundwork will be required.

This paper focuses on the issues surrounding the continuation of regulated retail power service in restructured electricity markets. Such services go by many names: default service, standard offer service, provider of last resort, etc. While these words are used in different and not always consistent ways, they all involve closely related concerns. We will use the term "default service" as an umbrella term covering the full range of these services. Despite specific differences among various state restructurings, some common issues emerge. This chapter explores these issues.

2. IS ELECTRICITY DIFFERENT?

2.1 Political Considerations

Markets are seldom totally "free." Instead, they are both facilitated and constrained by political institutions. Nonetheless, the normal assumption is that consumer protection is best achieved by active competition among suppliers and that no single supplier or group of suppliers is needed to provide "safety net" service at regulated prices. Is electricity different and, if so, why?

One simple response is that there is no basic difference between electricity and other "necessary" commodities such as food for which there is no regulated default provider. Instead, what makes electricity different is mainly its past political history. Therefore, converting electricity to the status of a "normal" commodity is largely a matter of consumers and their political representatives having the political will to make this transition.

But, there is plenty of evidence that this political will is likely to be in short supply when the system is stressed by extraordinary circumstances. Previous regulatory practices have created a sense of consumer entitlements. Consumers are bound to complain if they feel that these entitlements are being eroded by restructuring. If these complaints are sufficiently strong, politicians will feel compelled to take them into account in restructuring. This may lead to restructuring rules that are internally inconsistent with the retail competition they are attempting to foster.

In general, regulators[2] can be expected to have several concerns about consumer protection that will color their thinking about having a continuing role for regulated retail service even following the introduction of retail competition:

1. **Universal Service**: Regulators have traditionally been interested in assuring that all consumers who meet certain minimum qualifications receive adequate electric service even though they may be low-income, costly to serve, or poor credit risks. Whether by designating a "provider of last resort" or by some other mechanism, regulators can be expected to continue to pursue this objective of providing universal service.

2. **Affordable Basic Level of Service**: Regulators have also traditionally come under effective political pressure to keep prices low for some basic quantity of electricity. Ideally, this entitlement should be income-conditioned. But in some jurisdictions this objective has simply been achieved through low initial price tiers in residential rate schedules. To the extent that regulators wish to perpetuate these income policies in the new market, they will have to do so through T&D rate structures or though special income-conditioned regulated commodity rates offered by a designated default provider.

3. **Price Stability**: Retail consumers traditionally have been shielded from the volatility of wholesale spot-markets. To the extent that regulators designate a default provider, this entity will likely be required to hedge some portion of its purchases through longer term contracts rather than simply pass through wholesale spot-market prices as in the recent San Diego experience.

4. **Reluctance to Involuntarily Assign Customers to New Retailers**: In general, electricity consumers are not extremely dissatisfied with their traditional utility retail suppliers. Moreover, where consumers have recently experienced involuntary reassignment of their telephone service, many have complained. Therefore, most regulators will be reluctant to pursue a retail transition plan that involves an initial involuntary reassignment of consumers to new retailers for the purpose of jump-starting retail competition.

---

[2] The term "regulators" is used very broadly throughout this chapter. It is intended to include not only public utility commissioners themselves, but also legislators and governors, and even the courts review of agency actions and laws.

2.2 The "Orphaned" Customer Problem

In addition to this long history of political involvement and regulation-based consumer entitlements, there is another very significant way in which electricity delivery differs from other commodities and makes the designation of a default provider desirable, if not practically essential. This involves the "orphaned customer" problem. Whenever a customer is interconnected to the grid, power flows to the customer and some entity must be made financially responsible for scheduling its load, purchasing power on its behalf, and collecting payment. Under normal commercial circumstances, the financially responsible retailer and customer engage in a mutually voluntary transaction. But in electricity, the special problem of the "orphaned customer" arises if the retailer (1) goes out of business, or (2) simply elects to discontinue service to the customer for whatever reason. Unless the customer is immediately disconnected, power continues to flow to the customer and someone must be immediately financially responsible for procuring this power at wholesale and collecting payment at retail.

One can imagine a structure in which each retailer is responsible for interconnecting and disconnecting its own customers. A retailer interconnecting a customer would remain financially responsible for the customer until the retailer disconnected it. This regime seems workable and would closely conform to most familiar commercial arrangements. But there are two problems that argue against adopting this regime: one economic and one political. First, physical disconnection and reconnection involve non-trivial transactions costs (probably $50-100 for each transaction). Policymakers must consider the potential impacts of these transactions costs as barriers to customers switching retailers as well as whether it is best simply to avoid these costs altogether by adopting an alternative institutional framework – such as one employing a default provider. Second, this disconnect-reconnect model also seems politically infeasible. Regulators will be very reluctant to grant disconnection authority to the universe of unregulated competitive retailers. Moreover, the "orphaned customer" is not necessarily a bad actor (e.g., a payment deadbeat). Instead, the customer may be momentarily orphaned simply by the changed business strategy of its retailer. Thus, the "orphaned customer" problem seems a compelling reason for regulators to designate a default provider to avoid politically troublesome service interruptions as well as large transactions costs and barriers to customers switching retailers.

2.3   Default Service: Providers Of First And Last Resort

If we choose not to adopt a retail structure that disconnects orphaned customers, then we immediately encounter the issue of "default service." The very term suggests a service that is provided "by default" – i.e., as a result of some failure or inaction. In this case, the inadvertence is a customer's failure to secure power from a willing competitive retailer. It is useful to distinguish two distinct circumstances in which this failure may occur.

The first category of default service is encountered at the initiation of retail choice. At this point, if the customer simply declines to take any action to select a retailer or cannot locate a willing retailer, then we need to decide the precise terms under which the customer will continue to receive service and who will be financially responsible for procuring its power and collecting retail payment. One option is simply to disconnect the customer for failure to make a choice. But, as just explained, this option is costly and politically unpalatable. The alternative is to designate some entity as the "provider of first resort" (POFR) and make it responsible for procuring power and collecting payments.

A second and potentially distinct category of default service is encountered downstream of the initiation of retail choice. If, after taking power from a competitive retailer, a customer is subsequently dropped by that retailer (e.g., due to a change in the retailer's marketing plan, or failure of the customer to pay its bills), then either the customer must be disconnected or a default supplier must be immediately assigned. We will call this default supplier the "provider of last resort" (POLR).

A few additional points are worth making before proceeding. First, the POFR and the POLR may or may not be the same entity. Second, the POFR may be assigned financially undesirable customers, which it may choose to transfer to the POLR almost immediately. Third, not all the customers assigned to the POLR will necessarily be financially undesirable (as the term POLR is so frequently used to imply). Instead, some POLR customers may simply be customers "in transition" in the sense that they have been dropped by their retailers, but are perfectly desirable customers capable of securing service from other retailers assuming a reasonable interval for search. Therefore, it may be useful to distinguish a time-limited "transitional POLR service" from a longer-term POLR service for chronically financially undesirable customers.

2.4 Why Not Simply Deregulate Retail Markets Outright?

Why not simply say that tomorrow customers will have retail choice under completely unregulated prices, terms, and conditions? One reaction is that we can and should proceed this way. But this is a minority view. Instead, the prevailing view is that there are serious POFR and POLR issues that require special attention:

1. **Avoidance of Windfall Retail Profits**: The first objection to simply deregulating retail prices involves perceived incumbent advantages and customer inertia. The argument is that if we simply deregulate retail prices and assign the POFR responsibility to the incumbent UDC, the UDC will exert its previously unexercised market power at retail, increase prices, and earn monopoly profits. The argument is that this will result despite a fully contestable market and free entry by retailers. The alleged sources of this latent market power are customer inertia, scale economies, sunk investments in existing retail systems, brand recognition and loyalty, etc. Thus, many constituents would object to simply designating the incumbent UDC as the POFR and immediately deregulating its retail pricing. This then leads to several policy alternatives for designating one or more POFRs and regulating or absorbing their potential windfall profits.
2. **Transitional Price Stability**: This objective is the most nakedly political and potentially troublesome in its inadvertent consequences. Specifically, politicians are simply afraid of the potential for embarrassing near-term price increases and, therefore, insist on fixed, capped, or gradually escalating service rates for a 5-7 year transition period following the introduction of retail competition.
3. **Temporarily Orphaned Customers**: Financially desirable customers who are suddenly abandoned by their retail suppliers need some sort of limited-term (e.g., one month) POLR service to maintain the continuity of their electric service while they are provided a reasonable opportunity to secure service from another retailer.
4. **Chronically Undesirable Customers**: Customers who are chronically undesirable (e.g., due to poor credit ratings) may need to be provided POLR service on a continuing basis.

3.   ALTERNATIVE APPROACHES TO DEFAULT SERVICE

3.1   Provider Of First Resort Issues

POFR polices involve the designation of one or more default suppliers at the outset of retail competition. The primary focus is on existing customers, but a secondary issue is the treatment of new customers. In each case, the questions are: (1) What entity should be assigned initial retail responsibility for a customer who has not actively chosen a competitive retailer? (2) What should be the pricing and wholesale procurement basis of this default service? (3) How long should this POFR service be made available? (4) Who should be eligible for this POFR service?

3.1.1   What Are We Trying To Accomplish Through POFR Service?

Although some restructuring plans do not clearly distinguish POFR and POLR services, it is important to recognize that they are potentially different services with different objectives. Failure to recognize these differences can lead to overworking a single default service in an attempt to achieve conflicting objectives that can best be achieved through multiple default services. In general, POFR policies are intended to serve one or more of the following objectives:

1. Provide customers with initial regulated safety-net service during a transition period (usually 5-7 years);
2. Allow time for competitive retail markets to develop before forcing certain customers into them;
3. Remove the incumbent advantage of the UDC while "jump starting" retail competition by initially reassigning customers away from the UDC;
4. Remove the windfall profits associated with customer inertia;
5. Gradually "push" customers into the competitive retail market by making POFR increasingly less desirable.

Clearly, these possible objectives are neither mutually exclusive nor necessarily mutually consistent. Policymakers should carefully consider exactly which of these objectives they are designing POFR service to accomplish.

3.1.2   Alternative models for POFR Service

Alternative frameworks for providing POFR service differ based on the answers to several key questions:

1. Who should be designated as a POFR?
2. Should the pricing, terms and conditions of POFR service be regulated?
3. Should the wholesale procurement practices of the POFR be regulated?
4. Should the profit margin on POFR service be regulated?

If regulators are interested in minimizing the degree of commercial disruption for customers, then designating the UDC as the POFR makes the most sense. Various models are possible:

1. **Simple deregulation**: Under this model, the incumbent UDC is simply allowed to secure wholesale power however it wants and to charge whatever prices it chooses. As previously discussed, many parties will object to this model due to their expectation of UDC windfall profits based on incumbent advantages and customer inertia.

2. **Profit And/Or Price Regulation**: This model has several variations. For instance, the UDC might be free to purchase wholesale power as it wishes but required to pass through the cost of this power plus a regulated retail profit margin. This maximum profit margin might be increased annually to gradually phase out regulation. Alternatively, POFR service prices themselves might be capped at fixed or graduating levels. Directly controlling retail prices provides customers with the greatest protection against price instability, but with very risky side-effects. It threatens to send consumers retail price signals that are grossly out of touch with wholesale market realities. It also risks large UDC cost undercollections and financial distress.

3. **Upfront UDC Payment For POFR "Franchise"**: An alternative approach is to absorb any UDC windfall profits by having it pay an upfront fee for the POFR "franchise." Of course, the value of this franchise will depend upon the restrictions placed on the pricing, terms and conditions of POFR service. If POFR pricing is simply deregulated, the UDC's upfront fee might be large. If, on the other hand, onerous pricing terms are imposed (e.g., retail price caps), the appropriate fee could be small or even negative. If the UDC is designated as the POFR, then quantification of an appropriate upfront fee will be contentious since it will be determined administratively or through negotiations rather than through open competitive bidding among alternative suppliers.

If regulators are willing to disrupt traditional commercial relationships and assign POFR responsibilities to an entity other than the incumbent UDC, then a number of additional possibilities become available:

1. **Simple outright assignment**: One approach is to initially assign customers to one or more POFRs. These POFRs might be permitted to charge whatever price they pleased or might be constrained to provide service under specified pricing, terms and conditions. If pricing is

regulated, these terms would be increasingly relaxed over time to achieve an eventual free market. Designating several POFRs would certainly "jump start" retail competition. However, the initial problem would be deciding how to select these POFRs since many entrepreneurs of varying competencies would likely indicate their interest unless the specified pricing and terms of service were onerous. Simple assignment of POFR responsibilities to one or more third-party POFRs seems fraught with numerous political risks including customer dissatisfaction with being involuntarily assigned to an unfamiliar retailer, immediate increases in retail prices, and potential incompetence of some POFRs.

2. **Assignment through an auction**: Under this approach, the right to be the designated POFR for specific customers would be auctioned.[3] There could be multiple winners so long as each customer has only one designated POFR so that POFR responsibility for each customer is unique and unambiguous. This approach solves several problems encountered in the prior model. Identifying the potential POFRs is straightforward and potentially incompetent retailers will be less likely to be designated as POFRs. Auction revenues can be rebated to consumers to offset the tendency toward any initial price increases.

3. **Assignment through competitive success**: Under this model,[4] consumers are told that they must choose a retailer other than their UDC by a specific date (e.g., within six months) or they will be assigned to a retailer. There then ensues an "open session" in which competitive retailers vie for customers. At the end of this "open season," all remaining customers are assigned to retailers in proportion to the number of customers they have previously attracted voluntarily. This provides retailers the incentive to offer customers attractive retail terms during the "open season." One possible drawback to this approach is the potential for a fly-up in retail prices when initial retail "loss leader" contracts expire.

---

[3] Bidding approaches for default providers have been mandated in Maine and Texas. In Pennsylvania, PECO voluntarily conducted an auction and awarded approximately 300,000 residential customers to The New Power Company to supply Competitive Default Service. In Pennsylvania, GPU also attempted to conduct an auction for Competitive Default Service but received no bids because its designated rate cap was below the prevailing wholesale market price and negative bids were not allowed.

[4] This model was legislatively imposed on Atlanta Gas Light as part of natural gas industry restructuring in Georgia.

3.2 Provider of Last Resort Services

By definition, POLR service is regulated retail service available to a customer who has been dropped by its competitive retailer, has not or cannot locate another voluntary retailer, and is not eligible for POFR service because the customer has previously left POFR service to take service from a competitive retailer or because POFR service has been phased out and is no longer available. It is useful to identify four different types of POLR service because they each serve distinctly different purposes:

1. **Transitional POLR Service**: This refers to service from which customers can freely come and go during the initial transition period. By definition, this service is not available following the end of the transition period. At the end of the transition period, any customer remaining on this service is transferred to a totally deregulated service.

2. **Emergency POLR Service**: This service is automatically provided to any customer who has been dropped by its retail provider but has not yet located another retailer. It is available for only a very brief period (e.g., two weeks). Alternatively, it may be available for a longer or indefinite period, but price regulation protection would disappear after a brief initial period.

3. **High-Risk Customer POLR Service**: This category of POLR service is designed for customers who are too commercially undesirable to find a competitive retailer willing to sell them power under "reasonable" terms. These consumers may have bad individual credit profiles or may suffer from "neighborhood effects" which make them appear undesirable to competitive retailers. This is the type of POLR service that most people have in mind when they use the term "provider of last resort."

4. **Continuing Routine POLR Service**: This category of POLR service, if it is created, would be generally available to at least certain broad categories of customers (e.g., all residential consumers) on a continuing basis even after the end of the transition period. The offering of this type of POLR service is based on the judgement that some customer groups are best protected by the continuing availability of a regulated safety-net service option.

Of these four POLR categories, Emergency POLR Service is the most necessary and benign. Transitional POLR Service is the next most necessary and can have reasonably benign or even salutary impacts on the development of retail competition so long as it is designed to become significantly less desirable over the course of the transition period. Otherwise, it risks never being terminated and instead becoming Continuing Routine POLR Service -- the most competitively troubling category. High-Risk Customer POLR Service can also have adverse consequences if its terms are too attractive and entry is not restricted to include only qualified customers. Ideally, for customer entry to be appropriately self-policing, most ordinary customers must find this service to be less attractive than available competitive offerings. If instead, High-Risk Customer Service is heavily subsidized, then customer entry will have to be effectively policed to allow entry of only qualified customers. Otherwise, this service will effectively become Continuing Routine POLR Service and will be especially adverse to the development of competition due to its subsidized terms.

All of these functional categories of POLR service will not be present simultaneously in any single restructuring plan. For instance, if regulators choose (unwisely) to have a Continuing Routine POLR Service, no additional categories are functionally necessary. Alternatively, regulators might choose to have only Transitional POLR Service (or POFR Service) offered during a transition period, and then only Emergency and High-Risk Customer POLR Service offered thereafter. Such a differentiated and targeted set of default services seems the best way to achieve multiple goals while minimizing inadvertent side effects.

3.2.1   Alternative For Assigning POLR Responsibilities

The alternatives for designating one or more POLRs are similar to those discussed for POFR service. There are three broad candidates for designation as POLRs:

1. The incumbent UDC only;
2. All or some competitive retailers;
3. All retailers including the incumbent UDC.

If the UDC is designated as the POLR, it faces several hazards. First, under High-Risk Customer Service, it faces the problem of recovering costs in the face of adverse selection by customers. Second, it would apparently need to maintain sufficient slack retailing capacity to accommodate orphaned and voluntarily returning customers. It must establish an understanding with regulators as to just how much slack capacity is reasonable to maintain.

An alternative is to designate some or all retailers (possibly including the UDC) as POLRs. This POLR "pool" might include: (1) all retailers as a condition of licensing; (2) all retailers who volunteer to share the costs; or (3) only competitively selected low-bidding retailers. Under the first two options, it would seem that a regulated profit margin would have to be imposed.

4.   POTENTIAL PRICING AND PROCUREMENT REGULATION FOR DEFAULT SERVICES

To the extent that regulators mandate default service, it is likely that they will be concerned not only about its *physical availability* but also its *pricing* Their concern about price regulation is also likely to make them interested in both *profit regulation* and *wholesale procurement regulation*. Indeed, they are likely to view appropriate price regulation as being composed mainly of a combination of profit regulation and procurement regulation.

4.1   Some Possible Wholesale Procurement and Pricing Bases For Default Service

Because price regulation can be thought of as a combination of procurement regulation and profit regulation, regulatory attention will be focused on the procurement practices of the default service provider. The following are several ways in which regulators might structure the wholesale procurement bases of default service:

1. **"Plain Vanilla" Pass-Through of Spot-Market Power**: This would be the most competitively benign of the alternative procurement requirements. Service to small customers could be passed through on a load-profiled basis. Customers larger than a specific size could be required to install hourly meters. The main potential disadvantage of this model is that the results may become politically unacceptable if wholesale spot-market prices are overly volatile or rise sharply as they did recently in California.

2. **Price-Stabilized Pass-Through of Wholesale Spot-Market Power**: To address the potential price volatility of the prior option, various mechanisms might be used to dampen price instability. For instance, limits could be placed on month-to-month retail price changes and temporary overcollections and undercollections could be tracked in balancing accounts for later recovery. To protect the default provider, price-change limits might be over-ridden by limits on maximum balancing account undercollections. Unfortunately, no such framework is likely to avoid later political adjustments if wholesale spot-market prices become sufficiently extreme.

3. **Short-Term Fixed-Price Power Contracts**: An alternative means to address retail price stability is to require the default provider to offer default customers a fixed-price tariff for a designated time period (e.g., 6

months or one year).[5] This default service obligation could be periodically auctioned to willing default providers, or the UDC could periodically seek wholesale suppliers willing to provide the necessary power on a "requirements" basis at a fixed price.[6] Compared to the previous two options, this framework has the advantage of shielding the default provider from uncompensated risks.

4. **Pricing Based on Actual Purchases into One or More Regulated Portfolios**: Under this option, default services are based on passing through the costs, plus a regulated profit margin, of wholesale power purchased into a regulated procurement portfolio. The reasonableness of wholesale purchases would be subject to regulatory oversight, preferably emphasizing *ex ante* approval of procurement plans rather than *ex post* review of specific purchases. There could also be two or three portfolios and customers could be free to commit to one (e.g., spot-market portfolio, two-year portfolio, five-year portfolio).

4.2   Profit Regulation Of Default Service

While regulators will likely want to regulate the profit margin earned on default services, the default provider must be allowed to make a rate of return commensurate with its risks. There are several possible profit/risk structures that might be adopted depending on whether the default provider is the incumbent UDC or a designated third-party retailer.

If the UDC is designated, then profit regulation might be conducted in one of three ways:

1. **Simple Pass-Through Without Profit Margin**: Under this model, the UDC purchases power at wholesale and simply passes these costs through without a retail profit mark-up. There are two problems with this approach. First, to the extent the UDC incurs any market or regulatory procurement risks (as it generally will), the UDC will be undercompensated for providing default service. This, in turn, is the cause of the second problem: the price of default service will essentially be subsidized and, therefore, discourage the development of competitive retailers.

2. **Performance-Based Ratemaking for Default Procurement**: Under this approach, the UDC would earn profits (or incur losses) based on its wholesale purchasing performance relative to some external benchmark of

---

[5]   This is a fairly typical structure for "Standard Offer Service" in several New England states (e.g., Massachusetts, Rhode Island).

[6]   Such as in the auctions in Maine and Pennsylvania (PECO and GPU).

wholesale price indices. Profits or losses relative to the benchmark would likely be divided between shareholders and consumers according to some sharing formula. The main shortcoming of this approach is the current lack of robust and liquid wholesale price indices. Moreover, any aggregation of indices would create a performance benchmark that might be inferior to a regulated portfolio approach. Nonetheless, this approach does provide a basis for a retail profit margin and a basis for third-party retail competition.

If instead of the incumbent UDC, a third-party retailer is designated as a default provider, then profit regulation might be conducted in one of three ways:

1. **Directly Regulated Profit Margins**: If default service customers are simply assigned to certain third-party retailers, then one approach would be to directly regulate their profits through something like conventional cost-of-service regulation. This model also seems likely to drag the regulator into scrutinizing the reasonableness of procurement practices. It seems very unlikely that third-party retailers will be eager to volunteer to become default providers under this approach.

2. **Indirect Profit Regulation Through Pricing Benchmarks**: An alternative to direct profit regulation would be to pursue profit regulation indirectly through price regulation by establishing some benchmark procurement index plus some allowed profit margin. Default providers would then be free to procure power in whatever way they wished and make whatever profit margins they could. Getting third-party retailers to volunteer `for this arrangement without creating windfall profit opportunities would be a matter of fortuitously setting the benchmarks "just right."

3. **Auction-Determined Retail Profit Margins**: The best means of regulating the profit margins when third-parties are designated as default providers is to let these margins be established through an auction process. All the obligations to be imposed on the default provider must be publicized ahead of time. Under a slightly different model, the UDC remains the default service retailer but solicits competitive bids from wholesale suppliers willing to supply full-requirements power for a fixed period (e.g., one year) to groups of customers. The wholesale supplier takes all the market risk. The UDC then simply passes through these procurement costs, leaving the wholesaler's auction-determined profit margin as the only "retail" profit embedded in the default service price.

5.   PROBLEMS EMERGING WITH DEFAULT SERVICE

In virtually every state that has restructured its electricity industry, the UDC has been required to provide service under specific conditions at least for awhile. While this arrangement has some immediate appeal, it is rapidly becoming clear that it also has problems that can threaten the very existence of retail competition. For restructuring to succeed, the desire to protect customers must be deftly balanced with the goal of encouraging retail competition. In doing so, we must also be careful to provide for appropriate compensation for the default provider, and to reconcile customers' freedom to switch suppliers with the default suppliers' responsibility to hedge forward.

5.1   Conflicting Objectives Of Consumer Protection And Retail Market Development

Two of the main objectives of default service are to provide a degree of consumer protection and to facilitate a transition to full retail competition. Most of the problems currently being encountered with default service arise because these goals are conflicting. Achieving a satisfactory level of both objectives requires careful design of default service structures and is proving far more difficult than originally anticipated. Moreover, extreme wholesale price performance may render simultaneous satisfaction of both objectives simply impossible.

In many cases, default service has been made available at fixed prices, either at levels currently exceeding wholesale prices or scheduled to escalate to levels expected to exceed wholesale prices.[7] Fixing default prices above wholesale prices protects customers from price spikes while also allowing competitive retailers to attract customers with lower average prices. But fixed-price default service – even at high prices – is financially risky for the default provider and provides a safe haven for customers that discourages the development of competitive retailers. Moreover, by shielding customers from wholesale price signals, fixed-price default service discourages their investment in demand management technologies as well as behavioral demand responses. In turn, this tends to reduce price elasticity of wholesale demand and worsen price volatility in wholesale markets, thereby

---

[7]   For instance, for most utilities in Pennsylvania, the retail "price to beat" generally exceeds the wholesale price. Also, prices for Standard Offer Service in Massachusetts and Rhode Island were originally set at levels below prevailing wholesale prices but scheduled to gradually escalate to levels expected to exceed wholesale prices.

reinforcing the political demand for preserving their protective default service.

By discouraging the development of competitive retailers, overly protective default service also tends to inhibit the development of forward markets and innovative retail risk-management products. Moreover, development of competitive retailers will be arrested or reversed if rising wholesale prices put fixed-price default service "in the money" causing customers to voluntarily switch back, or be put back by their retailers, to the default provider.[8] In any event, customers may reach the scheduled end of the transition period largely uninformed about market dynamics and how to protect themselves in the retail market. Failure of effective retail competition to develop is likely to require the continuation of protective default service as a sort of self-perpetuating institution.

5.2 Inconsistency Between Customer Switching Freedoms And Default Procurement Obligations

Providing consumers with a degree of transitional protection may be a laudable and politically necessary objective of default service. However, some protection mechanisms may be rife with unintended consequences and best avoided. One especially troublesome arrangement is to give consumers too much flexibility to come and go from default service. In some states, this has become so institutionalized that retailers have actually offered "donut" contracts to customers, actively returning customers to fixed-price default service during summer periods when wholesale prices are expected to be high or volatile and then afterward returning them to the competitive retailer.[9]

It goes without saying that every consumer wants a "free option." But the option to come and go freely from default service, although a "free option" from the consumer's viewpoint, is hardly free from a broader social perspective. Default providers will find it very expensive, if not impossible, to provide consumer protection through purchases of forward price hedges if consumers are allowed to come and go freely, and do so depending on whether the default suppliers' hedges are "in the money" or "out of the money." For default service to be provided on any reasonable terms, limits must be placed on customers coming and going. This can be accomplished

---

[8] For instance, in California, almost two-thirds of competitive retailer loads have migrated back to UDC default service as a result of high wholesale prices and capped UDC default service prices. Also in Pennsylvania, GPU expects that three-quarters of competitive retailer loads will revert to its UDC default service on June 1, 2001

[9] Such "donut" contracts have been used by retailers in Pennsylvania.

by having minimum notice periods (e.g., two years) for customers leaving default service once they return to it. Alternatively, for POFR service it may be appropriate to preclude customers from returning once they take service from a competitive retailer. This structure also solves the problem that the UDC otherwise must maintain enough slack capabilities to accommodate all orphaned and voluntarily returning customers on demand.

5.3   Imbalanced And Costly Default Service Responsibilities

Although provision of default service, especially the provision of transitional default service by the incumbent UDC, can look deceptively simple, it is complex and expensive. UDCs and their regulators are not accustomed to risk management of a large procurement portfolio. Cost-of-service regulation was designed to deal with large physical investments – not financial hedging. Many UDCs will view the default provider role as a lose-only proposition that they accept temporarily as part of a comprehensive settlement. But they may be dismayed to find later that it is a non-compensatory "tar baby" from which they cannot shake free.

A significant problem with simply assigning the default provider role to the UDC is that this burden thereby never gets "market valued." The alternative is to auction the responsibility to a voluntary third party. The problem with this is that customers thereby get involuntarily reassigned and they generally find this objectionable. Moreover, some default provider solicitations have attracted no bidders (apparently because the responsibilities have been too onerous or risky) while other solicitations have revealed such high costs that bids have been rejected.[10] Indeed, these failed auctions illustrate the likely size of the default burden that is frequently just routinely placed on the UDC by regulators without much evident concern.

---

[10]   In Pennsylvania, GPU's recent solicitation produced no bidders because the designated retail rate cap was below the wholesale price of power and negative bids were not allowed. In Maine, in 1999, the regulators rejected all the bids it received for providing Standard Offer Service to Central Maine Power and Bangor-Hydro Electric saying they were "unreasonably high." The Maine regulator again rejected all default supplier bids in December 2000.

6. TEN RECOMMENDED DESIGN PRINCIPLES

Based on the preceding discussion and experiences to date, it is recommended that the following principles be followed in designing default service institutions:

1. Be careful not to overwork a single default service in pursuit of multiple objectives. Consider having several types of default service and targeting each of these different services to different objectives (e.g., Transitional Service, Emergency Service, High-Risk Customer Service).
2. Be extremely skeptical about desires to have a broadly available Continuing Routine POLR Service. This service is fundamentally inconsistent with moving toward retail competition. If stakeholders believe that such a service is necessary for some customers (e.g., residential), then it is probably best not to adopt retail access for that group.
3. Make sure that transition default service becomes progressively less attractive over time. This service may be motivated as an initial safety net, but it must become continually less attractive if a transition is actually to be achieved. Phasing out this service solely by means of a one-way customer switching rule is probably not sufficiently aggressive to accomplish a sufficiently expeditious transition and avoid problems.
4. Err on the side of adopting restrictions against customers switching back and forth from default service. Viewing this right as a customer protection is an illusion. Serious consideration should be given to adopting a POFR service that becomes gradually less attractive and has an exit-only rule.
5. Give strong deference to excluding large customers (i.e., larger than 500 MW or 1,000 MW) from all but a very short transition (2 year maximum). Large customers should have no problem locating willing retail suppliers.
6. Give careful consideration to having High-Risk Customer POLR Service served through a voluntary "risk pool" of third-party retailers. Make sure that this service is at least moderately unattractive (as it will be if competitively priced through a voluntary risk pool).
7. Do not use default service to pursue the objective of assisting low-income customers. Low-income customers are best assisted through direct government subsidies such as the Low Income Home Energy Assistance Program or income-conditioned energy vouchers. If a special default service is used to assist low-income customers, entry into this service must be well policed. It may be best to bid this service to voluntary retailers in order to assure that subsidies are fully monetized

and paid for within the system rather than being unfairly imposed on UDC shareholders.

8. Make sure that all default services contain demand management options including the ability of customers to purchase hourly meters and receive hourly pricing. These service options should be voluntary for small customers but mandatory for large customers.

9. Make sure that default service is adequately price-stabilized based on wholesale purchases that are hedged forward. Over-reliance on spot market purchases has already proved catastrophic for default service in California.

10. Assure fair compensation for providers of default service. To the extent that the UDC is designated as a default service provider, it should be allowed to transfer most risks to wholesale providers. Moreover, this procurement process should be subject to before-the-fact – not after-the-fact – reasonableness standards.

REFERENCES

Faruqui, Ahmad. 2000. "Electric Retailing: When Will I See Profits?" *Public Utilities Fortnightly*, June 1: 30-39.

Faruqui, Ahmad and Ken Seiden. 2001. "Tomorrow's Electric Distribution Companies." *Business Economics*. 36 (No. 1, January): 54-62.

Flaim, Theresa. 2000. "The Big Retail 'Bust': What Will It Take To Get True Competition?" *The Electricity Journal* 13 (March): 41-54.

Goulding, A.J., Carlos Ruffin, and Gregory Swinand. 1999. "The Role of Vibrant Retail Electricity Markets in Assuring that Wholesale Power Markets Operate Effectively." *The Electricity Journal* 12 (December): 61-73.

Graves, Frank C. and James A. Read. 2000. *Residual Obligations Following Electric Utility Restructuring*. Edison Electric Institute.

Graves, Frank C. and Joseph B. Wharton. 2001. "POLR and Progress Towards Retail Competition – Can Kindness Kill the Market?," presentation made at Winter Committee Meetings of the National Association of Regulatory Utility Commissioners, Washington, DC, February 27.

Graves, Frank C. and Paul Liu. 1998. "Price Caps for Standard Offer Service: A Hidden Stranded Cost." *The Electricity Journal* (December): 67-76.

Herbert, John H. 1999. "The Gas Merchant Business: Still a Place for LDCs?." *Public Utilities Fortnightly*. 137 (No. 13, July).

Kahn , Alfred E. 1999. "Bribing Customers to Leave and Calling It Competition." *The Electricity Journal* 12 (May): 88-90.

O'Rourke, Patrick. 2000. "Redundant Restructuring: How the Dual-Retailer Model Makes Electric Markets Too Complex." *Public Utilities Fortnightly* 138 (no. 13, July): 54-60.

Stavros, Richard. 2000. "Risk Management: Where Utilities Still Fear to Tread." *Public Utilities Fortnightly* 138 (No. 19, October).

Tschamler, Taff. 2000. "Designing Competitive Electric Markets: The Importance of Default Service and Its Pricing." *The Electricity Journal* 13 (March): 75-82.

Chapter #6

Energy Markets and Capacity Values
How Complex Should Pricing Be?

Robert J. Michaels
California State University at Fullerton

Abstract: Efficient tariffs for reliable ("firm") electricity service in theory must charge
              separately for capital costs ("capacity") and operating costs ("energy"), while
              interruptible service should carry only an energy charge.   Competition,
              however, often occurs in short-term exchanges with one-part pricing and no
              explicit accounting for capacity.   The movement to one-part energy prices is
              consistent with increased competition and the lower transaction costs that
              characterize competitive markets. Installed capacity requirements and energy
              surcharges will be inefficient and quite possibly anticompetitive.   Option
              pricing models facilitate the extraction of capacity values from data on energy
              prices.

Key words: electricity; two-part tariff; capacity; energy

1.   INTRODUCTION

Until recently, efficient electricity rate design was one of the least controversial areas of public utility economics.   Electricity was (and is) a non-storable commodity produced by durable, industry-specific capital.   By the 1960s a long chain of theoretical development had produced numerous variations on the theme suggested by these characteristics: efficient pricing of reliable ("firm") power required a consumption ("energy") component and a capacity ("demand") charge.[1]   If demand fluctuates quickly and

---

[1]   For a summary of the earlier literature on two-part tariffs based on short-run and long-run
      marginal costs, see Kahn (1989), 87-109.

A. Faruqui and B.K. Eakin (eds.), Electricity Pricing in Transition, 87-98.
@ 2002 *Kluwer Academic Publishers.*

randomly, users for whom reliable service necessitates investment in new capacity should be charged for both the capacity and energy. Users willing to risk interruption should pay only the operating cost of that capacity.

With the restructuring of the industry has come an emphasis on short-term energy transactions and the eclipse of capacity as a relevant market.[2] Energy transactions often take place in bid-based markets such as the (now defunct) California Power Exchange (PX) or the pool arrangements in the United Kingdom and parts of Australia. Numerous generators supply these markets and the identities of the units operating at any moment are of interest primarily to their owners and the system operator. Marketers who do not own generation can supply energy on the same terms as owners in some trading environments. Energy is also traded under a range of bilateral arrangements, which sometimes specify production by particular generators or rights to their capacity. Transactions that do not specify capacity sources are increasing relative to those that do.

Energy contracts and markets price their product on the basis of output flows with no capital cost components. Competitive markets are theoretically efficient, but can this actually be the case if the energy prices arising in them are inconsistent with established results on the efficiency of two-part rate designs? Energy prices alone may suffice for efficiency in interruptible service, but if energy markets are competitive are claims on capacity still needed to produce reliable service? Competitive energy markets may allocate existing generation efficiently, but do energy prices alone give proper signals for new investment in light of the industry's unique reliability problems?[3]

If the costs of negotiating and enforcing arrangements on reliability are low enough, economic theory informs us that self-interested agents will reach an efficient solution.[4] The assumptions that underlie earlier economic work on two-part tariffs are no longer accurate descriptions of reality. In particular, reliability and energy are now separable services, both of which can be competitively procured. "Energy-only" markets are the analogues of regulated interruptible service and warrant one-part pricing. In ancillary services, the availability of capacity to ensure reliability is valuable, and capacity must carry a price. Requirements that participants in some energy markets hold claims on identifiable capacity are applications of a logically

---

[2]   The Federal Energy Regulatory Commission's (FERC) standards for the approval of electric utility mergers concentrate on short-term energy markets to the exclusion of nearly all others. See FERC (1997). Short-term energy is generally defined as power scheduled to flow no more than a single day ahead. A transaction may or may not specify that a particular generator is to produce the output.

[3]   This chapter supersedes and modifies certain of the views expressed in Michaels (1997).

[4]   See Coase (1960) and Demsetz (1988).

correct economic theory to institutions that no longer satisfy its underlying assumptions. The separation of energy and reliability is an important move toward greater competition. Lowered costs of metering and organizing markets point the way toward differentiation of service quality, new financial instruments, and increased consumer choice.

2. THE VALUE OF CAPACITY

2.1 How Different is Electricity?

Producers in any capital-intensive industry must commit themselves to investments in durable capacity that can only be reversed at substantial cost. Whether the producer is a competitor or a monopolist, its income is affected by uncertainty about future competition and the costs of other inputs. However important these investments, non-utilities seldom if ever quote two-part prices consisting of a capacity component and a component that reflects the costs of variable inputs. If electricity is produced competitively, does any case for two-part pricing remain?

An affirmative answer usually begins with an appeal to differences between electricity and other goods. In particular, [1] power cannot be held in inventory to equate supply and demand when capacity is short, and [2] a seller that fails to own or contract for capacity adequate to instantaneously meet commitments to its customers can imperil reliability for the entire interconnected grid it operates on. Regarding inventories and storage, electricity differs only in degree from industries (e.g. perishable goods, emergency services) that do not charge explicitly for production capacity. Some storage technologies (pumped storage hydro, compressed air, flywheels) that alleviate the need for instantaneous generation either exist or are on the horizon. Improved metering and control technologies are increasing the ability of users and suppliers to make contracts that adjust demand quickly and in response to market factors. Other markets equilibrate without explicitly accounting for capacity, while technology and competition are increasing the potential for power markets to do likewise.

In the past, reliability could only be a responsibility of the monopoly utility, inherent in its obligation to serve by producing in plants that it owned or had under contract. In today's markets the entity that controls the grid, whether a utility or an independent system operator (ISO), can maintain reliability without owning generation. ISOs operating under rules that accept bilateral transactions can require that sellers schedule sufficient power to meet the loads they have contracted to serve. Evolving technology makes

it possible to directly bill standby service (or cut off power) to those whose
suppliers have not adequately provisioned themselves.

2.2   Regulation

The operations of regulated utilities are changing with the emergence of
competition, but important elements of their traditional service obligations
remain.   A utility must sell to all customers in its franchised territory at rates
set by the costs of serving them, in return for which it recovers its costs plus
a reasonable profit.   To fulfill its obligation the utility must build or contract
for production capacity in anticipation of demand growth.   Assuming that
it's allowed rate of return on investment is adequate, the utility will generally
build for future demand without much regulatory compulsion.   At least in
theory, overcapitalization as a rational response to rate-of-return regulation
is as much a risk as underinvestment.[5]
The models underlying the theory of an efficient two-part tariff are
increasingly poor descriptions of today's electricity industry. The tariff was
imposed on a firm that was a natural monopoly in all of its relevant
activities. Whatever pricing arrangements would evolve under competition
with capacity and reliability constraints were not relevant alternatives. The
possibility of a competitive market for bulk energy only became relevant
after the 1970s, when economies of scale in generation ran out and new
smaller-scale technologies became economic. The design of tariffs for
default service by utilities whose customers have competitive alternatives
remains a largely open theoretical question. (Brennan (1990)) The metering
technologies assumed (and actually available when the theory originated)
were costly and crude. Meters that tracked demand in real-time and allowed
information to flow bidirectionally did not exist. They could record a user's
overall peak consumption, but not necessarily its consumption coincident
with system peaks as the theory required for computation of capacity
charges. Third, the two-part tariff assumed highly inelastic demands,
consistent with a regulatory view of a fixed revenue requirement.
Importantly, users were assumed unable to be able to lower their capacity
charges by shifting their times of use.

---

[5]   In practice, actual interruption of users on interruptible rate schedules has been extremely
      infrequent. Most regulatory commissions have required that utilities load some capacity
      costs into interruptible rates. When interruptions became frequent in California during
      2000-2001, customers attempted to return to firm service against the wishes of utilities and
      regulators. For a summary of the economic literature on overcapitalization under rate-of-
      return regulation see Bailey (1973).

2.3   Markets

Competition first emerged in the 1970s with bilateral contracting among utilities for surplus energy. Joint ownership of capacity and contracting for shares of capacity followed. Utilities gradually moved transactions and investments beyond the franchised territories in which they were responsible for reliability. With few exceptions, utilities controlled their individual transmission grids and owned enough local generation to ensure self-sufficient operation in all but extreme emergencies. Utilities chose reserve levels and operating practices in accordance with engineering standards that accounted for the greater risks of imported energy supplies. If external power markets somehow vanished, the effects would be almost the equivalent of those caused by a large but bearable generation or transmission outage. Utilities continued to benefit from strong legal barriers to the entry of non-utility competitors, and a culture that encouraged coordination to maintain reliability whenever it was threatened. This situation would become less sustainable as the independent power and power-marketing industries grew over the 1980s and 1990s.

Even if a bilateral contract identifies the capacity that will produce its energy, third parties may bear reliability risks. A seller that cannot replace a failed energy source in real time may endanger the entire regional system. If left to themselves, the buyer and seller may take inadequate precaution, because the cost of the outage to them will be less than the cost to all parties on the grid who lose their power. This externality takes place under conditions of extremely high transaction costs. To provide the right incentives for the contracting parties a bilateral contract must specify contingent payments based on damages suffered by all who are affected by the outage. The second-best solution of a liability rule entails its own transaction costs in using the courts. (Frech1979)) Here the initial assignment of responsibility matters because the costs to all affected parties of negotiating an alternative solution are high relative to its benefits.

In a competitive energy market, the seller in a bilateral contract will in principle be able to obtain replacement energy at the market-clearing price if its own generator suffers an outage. The need to ensure reliability in real time, however, means that the energy must be found instantaneously, and that its producer has no incentive to exercise short-term market power. Available sources of energy may in reality be few and their owners may attempt to act opportunistically even if contracts for contingent operation already exist. Except in the most extreme textbook cases, the benefits of finding deliverable energy and deterring opportunistic behavior will be lower for the bilateral contracting parties than for the system as a whole.

Competition by itself may not provide an adequate backstop for governance by contracts.

3.   CAPACITY REQUIREMENTS

3.1   Responsibility for Reliability

The economic theory of external costs and benefits provides prescriptive rules for situations like this one. To illustrate, take product liability as an analogy.   The law may either specify that persons harmed by a given product are responsible for their own injuries, or it may require that the manufacturer pay compensation. Assume that transaction costs are high: nearly all injuries result from discretionary misuse of the product in ways that the manufacturer did not intend and cannot prevent, e.g. by redesigning it. The economically efficient rule minimizes the total cost of injuries plus adjudication. In this case it would most likely make users responsible for their injuries, since they are "cheaper cost avoiders" than manufacturers.[6]

In electricity the identity of the cheapest cost avoider is changing with the advent of competitive markets. As this happens, one-part pricing of energy and an abandonment of capacity-related rules cease to be efficient. Assume at the outset that energy is traded on a PX whose price equates supply and demand over each interval. Both owners and non-owners of generation can submit bids, and need not show that they hold claims on capacity prior to bidding. Absent some provisions for reserve management, energy traded in this PX is best viewed as interruptible, despite its market-clearing price. Absent unrealistically low transaction costs, outages will be excessive. A rule requiring that a supplier whose unit goes down compensate other market participants will be too costly to enforce. Alternatively, those who lose as the result of a blackout can make their own provisions for emergency supplies. Here market participants must on balance hold redundant capacity whose aggregate amount could be lessened by either direct intervention or by the formation of a market for reliability.

The broadening of markets for energy has lowered the cost of market provision of reliability, relative to direct intervention by a generation-owning entity. Constraining ancillary services to generation owned by a single utility or an ISO ensures that the cost of maintaining reliability will be higher than under market provision, but only if markets are economic alternatives to the centralized system. The efficient mix of resources to produce ancillary

---

[6]   The term is from Calabresi (1970), 135-173.

services will vary with system conditions (including geographically) and the preferences of market participants. To handle a full range of contingencies will require that the ISO or utility operator own some redundant resources. Only if there is a market will it be able to dispose of the output of some unused resources, but with a market there is no obvious need for it to own any resources at all.

3.2   Market pricing for reliability

Unlike the energy markets, those for ancillary services require that the ISO have command over identifiable capacity resources. Energy may be available, but from plants whose location does not allow them to be of help in maintaining reliability under current system conditions. In effect, giving the system operator the right to take bids from specific capacity resources and to use them in defined ways makes ancillary services firm, as opposed to interruptible power produced in the energy markets.

At least prior to the demise of California's PX, its system for providing ancillary services through the ISO priced them efficiently. The ISO accepts day-ahead and hour-ahead bids for four categories of ancillary services, regulation (load-following) and reserves at three levels of readiness. A bid consists of three elements: [1] the amount of capacity being offered to potentially supply any service; [2] the minimum price the supplier will accept for dedicating capacity to each of the services; and [3] a minimum payment the owner will accept per kwh for the energy produced if the capacity runs.[7] The ISO selects capacity bids in ascending order, but the generator only operates if its energy bid is at or below the price that clears the real-time energy market.

Thus ancillary services in California are priced (approximately) by the efficient two-part rule: the ISO effectively pays the market price for energy plus a surcharge for capacity availability. The bid for that capacity will depend on its alternative value in the energy market. Where capacity is generally in excess, its equilibrium value will be low. Its price will rise with scarcity, as the value of the reliability services it produces become higher. California has experienced difficulties in providing contractual incentives for generators that have location-specific market power to bid into energy markets when doing so would foreclose them from providing potentially more lucrative ancillary services under their contracts.

---

[7]   A generator whose day-ahead bid is not accepted can bid into the hour-ahead PX energy market (now gone with the demise of the PX), the hour-ahead ancillary services markets, or directly into the real-time market. It can also make bilateral transactions. See Quan and Michaels (2001).

4.   VALUING CAPACITY IN MARKETS

4.1   Capacity requirements for energy trades

Whether energy-only pricing is efficient depends on the state of market development, favoring those markets with more numerous participants whose trading institutions convey richer information.   Markets in the Pennsylvania – New Jersey – Maryland (PJM) Interconnection are among the most developed in the country, containing many buyers and sellers, pricing transmission nodally, and having established delivery points and protocols.   The PJM region has significant import-export capabilities and supports a mix of bilateral and pool transactions.   Ancillary services are currently centrally provided, but the organization plans to put markets for them in place.   Despite the state of market development, sellers in PJM must adhere to an "installed capacity" (ICAP) standard that requires offers of energy to be associated with capacity.   The seller may either own its capacity or purchase it on PJM's ICAP market.[8]

The reasoning behind an ICAP requirement is questionable.   Before markets became extensive a prudent utility held capacity in excess of load because there was no other source of reliability.   Holding the requisite capacity, however, does not imply that a seller's energy will be reliable. ICAP does not require any margin of capacity above that needed to produce the amount of energy promised.   Moreover, the PJM operators (the pool is centrally dispatched) will have already acquired commitments for reserves and other ancillary services that will be charged to the trading parties.   Even if there were no centralization of reserve purchases, holding of excess capacity by all sellers would be an inefficient way of achieving reliability. With a market as rich in alternatives as PJM, a prudent supplier whose plant unexpectedly goes down can easily make arrangements to obtain energy elsewhere.

In PJM the combination of ICAP and energy price is in effect a two-part scheme where efficiency warrants no more than energy-only pricing.   The capacity component will vary with market conditions and probably be positively correlated with the energy component.   In practice the ICAP market taxes sellers who do not own generation.   Utilities that continue to own a large share of the area's generation can dedicate it to exports and then recall it at strategic times to be sold as ICAP.   Ancillary services charges are

---

[8]   See PJM Capacity Credit Market (2001) and Flaws with the ICAP Process (2000).   For a deense of the ICAP market based on externality reasoning, see Comments on a PJM Energy Only Market (2000).

collected on an uplift basis and are priced independently of any capacity holdings. ICAP both stifles competition and fails to improve reliability.

4.2   Administered capacity value premia

In a competitive market that is short of capacity, equilibrium energy prices can be substantially above booked costs, particularly if demand is highly inelastic. Unless the premium results from restrictive conduct due to market power, it provides evidence on the value of investing in new capacity. If there is a reliability externality, however, even a high one-part energy price will underestimate the value of new capacity. Prior to the introduction of new trading arrangements in March 2001, the U.K. pool added an amount to its market-clearing energy price (financed by uplift charges) in hours when demand is close to capacity. The premium was set using an estimate of the value of load at risk of outage, a very difficult figure to estimate.

The added payment is the equivalent of a Pigouvian tax to account for the difference between the private and social values of electricity. In practice, however, it took effect for an unpredictable but relatively small number of hours and its value was highly sensitive to small changes in generator availability. (Green (1998)) Generators with market power were alleged to have manipulated their availability in order to earn higher prices for power from their units that remain in operation.

This type of administered payment is an attempt to salvage some of the principles of the efficient two-part tariff in the context of a one-price market. In the U.K. the payment has not led to discernible increases in generation capacity, unsurprising in light of its randomness. Other factors, most importantly the growing availability of inexpensive natural gas, have apparently driven investment in that area.[9] Even if the adder is an accurate measure of the average value of reliable service, it cannot reflect the marginal value of electricity to different users, or differences in the value of additional output from different types (e.g. fuel, location) of generation.

4.3   Value-Differentiated Call Options

Imposing a capacity requirement on energy bidders is inefficient in theory and manipulable in practice. Strong but arbitrary assumptions are necessary to calculate an energy price premium based on value of lost load.

---

[9]   In many cases the newer plants had bilateral contracts that committed significant parts of their output to forward sales.

Where market power is absent it increases the volatility of energy prices, and where market power is present it may be manipulable by generation owners. As an alternative to capacity rules and questionable premia, it may be possible to estimate capacity values from data contained in energy prices. If so, new possibilities open up for individualized management of reliability risks with little direct regulatory intervention.

The value of new generation capacity cannot be estimated as the discounted value over a time horizon of the difference between the expected future spot price of energy and the generator's operating costs. Because future prices and system conditions are volatile, investing in a generator is the equivalent of purchasing a call option on energy. That option will be in the money at high energy prices and will go unexercised at low prices. Holder of such an option (a right to generation capacity) gains the right to purchase energy at a strike price equal to the variable cost of producing it in the plant.

The theory of option pricing allows us to conclude that the value of capacity is higher [1] the lower is its variable cost; [2] the longer one has the right to use it; [3] the higher the market price of energy; and [4] the greater the variability in the market price of energy.[10] The value of capacity depends on market conditions looking forward. That value is only coincidentally associated with the cost of building a generator or the value of lost load. It will however, vary with the price of fuel, the major component of the option's strike price.

A generator represents a real option, but the energy it produces is a commodity on which financial options can be written as well. The values that these instruments fetch in option markets can be informative about both expected future energy prices and the value of new capacity. Users who place different values on reliability and price certainty will hold differing portfolios. Those who are easily interruptible will commit themselves to sell their rights to energy at relatively low costs, while those who value reliability and price stability will buy calls that give them the right to energy at the strike price. As long as consumption levels and plant operations can be monitored, energy-only markets will develop instruments such as these without explicit regulatory intervention.[11] As they do, market-based capacity choices will become better informed and more efficient.

---

[10]   The basics of option pricing theory appear in Cox and Rubinstein (1985). For more on the equivalence of rights to capacity and options on energy, see Graves and Read (1997).

[11]   For information on the relatively slow growth of such instruments in New Zealand's small but evolving market, see Graves *et al* (1998).

5.   SUMMARY AND CONCLUSIONS

The technology of electricity creates both operational difficulties in maintaining reliability and the need for large investments in industry-specific and site-specific capital. These industry attributes served to rationalize the treatment of electricity suppliers as public utilities and rate designs that separated capacity and energy charges. Two-part rates are giving way to market prices for energy that do not contain explicit values for capacity. If competitive energy markets do not acknowledge capacity and the reliability externality that stems from inadequate investment, their efficiency is in question.

The theoretical two-part pricing rule specified that those taking firm service pay rates that include a capacity component, while those who take interruptible service pay only energy charges. Some critical assumptions underlying the derivation of these rates are no longer relevant, including natural monopoly in generation and highly inelastic demands. Reliability problems remain whether electricity is centralized in utilities or decentralized into markets. High transaction costs of negotiating an efficient solution also persist, suggesting that a centralized entity that holds rights to sufficient generation remains the preferable source of reliability services.

The coming of energy markets, however, lowers the transaction costs of using market mechanisms to provide for reliability. These latter markets will price and trade claims on capacity. Transactions in the energy market are best viewed as interruptible services, for which a one-part price is efficient. To maintain reliability, the entity that operates the ancillary services market must purchase rights to identifiable sources of capacity as well as the energy they produce. The ancillary services markets are the remnant of traditional utility service that must remain, and two-part prices are efficient in them.

Regulatory interventions to deal with the reliability externality in energy market regimes are unlikely to perform efficiently. Capacity requirements on energy bidders turn a one-part rate into a two-part rate, but do not by themselves remedy any reliability problem. In practice they appear to provide generation owners with opportunities to exercise market power. Energy price premiums based on lost load are difficult to compute, operate sporadically in practice, and can also be manipulated by sellers with market power. It has recently been shown that forward-looking energy prices contain information that allows estimation of the market value of new capacity by option pricing methods. These methods offer the possibility of market-based instruments that will offer users new choices in physical reliability and financial predictability. They are best thought of as privately produced components of two-part tariffs, as valuable as ever for efficiency but produced by the market this time.

REFERENCES

Bailey, Elizabeth E. 1973. *Economic Theory of Regulatory Constraint.* Heath Lexington.

Brennan, Timothy J. 1990. "Cross Subsidization and Cost Misallocation by Regulated Monopolists." *Journal of Regulatory Economics* 2: 37-51.

Calabrese, Guido. 1970. *The Costs of Accidents: a Legal and Economic Analysis.* Yale University Press.

Coase, R.H. 1960. "The Problem of Social Cost." *Journal of Law and Economics* 3: 1-44.

Cox, John C., and Mark Rubinstein. 1985. *Options Markets.* Prentice-Hall.

Demsetz, Harold. 1988. "When Does the Rule of Liability Matter?" In *Ownership, Control, and the Firm,* edited by H. Demsetz. Basil Blackwell.

Frech III, H.E. 1979. "The Extended Coase Theorem and Long-Run Equilibrium: The Nonequivalence of Liability Rules and Property Rights." *Economic Inquiry* 17: 254-268.

Graves, Frank C. and James A. Read Jr. 1997. "Capacity Prices in a Competitive Power Market." In *The Virtual Utility,* edited by Shimon Awerbuch and Alistair Preston. Kluwer Academic Publishers: Boston, MA.

Graves, Frank C. *et al.* 1998. "One-Part Markets for Electric Power: Measuring the Benefits of Competition." In *Power Systems Restructuring: Engineering and Economics,* edited by Marija Ilic *et al.* Kluwer Academic Publishers: Boston, MA.

Green, Richard. 1998. "The Political Economy of the Pool." In *Power Systems Restructuring: Engineering and Economics,* edited by Marija Ilic *et al.* Kluwer Academic Publishers: Boston, MA.

Kahn, Alfred E. 1989. *The Economics of Regulation,* Reprint of 1971 Ed. (MIT Press, 1989).

Michaels, Robert J. 1997. "MW Gamble: The Missing Market for Capacity." *Electricity Journal* 10 (December): 56-64.

Pennsylvania - New Jersey - Maryland (PJM) Interconnection. 2000. "Capacity Credit Market Policies," at www.pjm.com.

PJM Interconnection. 2000. "Comments on a PJM Energy Only Market / Restructuring the Existing ICAP Market," at www.pjm.com.

PJM Interconnection. 2000. "Flaws with the ICAP Process 10/13/99," update 02/08/00. at www.pjm.com.

Quan, Nguyen T., and Robert J. Michaels. 2001. "Games or Opportunities: Bidding in the California Power Exchange." *Electricity Journal* 14 (January): 63-73.

U.S. Federal Energy Regulatory Commission. 1997. Inquiry Concerning the Commission's Merger Policy Under the Federal Power Act, Order No. 592-A, June 12, 1997.

Chapter #7

How to Make Power Markets Competitive

Samuel A. Van Vactor[1] and Seth A. Blumsack[2]
Economic Insight Inc. [1] and the University of Cambridge, [1] and Carnegie Mellon University[2]

Abstract: In the face of high transactions cost and externalities, firms often seek to
internalize resource allocation through *vertical integration* – the integration of
multiple stages of the production process under one roof. This paper discusses
the applicability of economic theories of industrial organization to the electric
power industry, and suggests that an integrated grid operator with ownership
and control of balancing resources, combined with active bilateral trading in
prescheduled markets, may be the most efficient means to achieve a
competitive power market. The alternative, markets for balancing energy,
must operate twenty-four hours per day and involve complex accounting and
scheduling procedures. Furthermore, market-based congestion pricing, too,
has proven costly. Large numbers of bidding rounds and adjustment periods
would be needed to attain optimal scheduling and pricing, since decisions
regarding market transactions and physical scheduling of load are
interdependent and interwoven, with each influencing the other.

Key words: Competition, vertical integration, transactions costs, restructuring, markets,
electricity.

1.   INTRODUCTION

For a century, most private utilities in the United States have exhibited a
straightforward industrial structure. The companies were "vertically
integrated," in the sense that they produced, transmitted and distributed
electricity to customers from the same organization. In almost all cases
private utilities faced cost-plus (return-on-investment) regulation, skewing
managerial choices towards capital-intensive projects. Combined with the

A. Faruqui and B.K. Eakin (eds.), Electricity Pricing in Transition, 99-111.
@ 2002 *Kluwer Academic Publishers.*

"obligation to serve," these incentives led to gold plating of the electric infrastructure, in which reliability, rather than least cost, was the main objective.

In the last three decades a number of public policy changes have worked to loosen the tight structure. The Public Utility Regulatory Policy Act (PURPA) of 1978 introduced the notion of independent power producers (IPPs), interstate transmission lines were built to connect local grids, and the Federal Energy Regulatory Commission (FERC) relaxed regulations in interstate commerce beginning with the Western Systems Power Pool (WSPP), which allowed market-based transmission rates and market-based wholesale pricing. Despite the overtures towards liberalized electricity markets, imperfections remained as incumbent utilities were thought to be restricting access to their transmission lines and favoring the use of their own power generators rather than buying (possibly cheaper) power on the open market.[1] In the view of many regulators and analysts in California, the integrated structure was inhibiting the formation of a competitive market and resulting in unnecessarily high retail rates.

When California began the process of restructuring its electricity industry in 1996, it took as its model the experiment in the United Kingdom. The U.K. had sought to privatize a nationalized monopoly and create a competitive market for electricity. However, the situation in California was quite different. The utilities were already private companies, owning generation resources and various components of the distribution and transmission grid in regulated service areas. While the U.K. broke apart a government-owned utility into generation, transmission, and distribution companies, California forced the divestiture of generating assets from its utilities in order to reduce market concentration. The utilities also relinquished control of their separate pieces of the transmission grid to a centralized grid operator. California's private utilities agreed to the restructuring because the new system allowed for the recovery of "stranded costs," i.e., the recovery of capital from investments and contractual obligations no longer thought to be competitive in the new structure.

The experiment in the U.K. had its problems – there were too few generating companies and the mandatory power pool proved to be unnecessarily rigid and constraining. In April 2001, the U.K. shifted to a more liberal system of bilateral trading and private exchanges. Ironically, the new U.K. power market has evolved toward a structure that in many respects is similar to the system of trading in the WSPP that was abandoned by California in 1998.

---

[1]   Such barriers to trade have been observed by the FERC (1998), particularly in the Midwest.

In the meantime, California's market restructuring has failed, blown apart by unlucky timing, mismanagement, poor market design, defaults, bankruptcies, and claims of market power abuse. In response, California's political leaders are moving towards consolidating the state's electricity infrastructure into a centralized State Power Authority, modeled after the New York Power Authority created in the 1920s by then-governor Franklin Delano Roosevelt. If some or all of the assets of California's private utilities are consolidated within the new Power Authority (which seems a distinct possibility), the state will end up with an electric industry structure similar to the one dismantled by the U.K.

With hindsight, the California energy market structure is easy to criticize. Retail power rates were frozen, while wholesale rates were allowed to vary; utilities were discouraged from hedging spot market purchases; load balancing generators were sold by the incumbent utilities, while base load resources were kept; natural gas buyers were too dependent on the spot market; gas storage was inadequate, emission controls were excessive and often irrational, and California depended on unreliable hydropower from the Pacific Northwest to balance its load. Each of these issues is worthy of further investigation and analysis, but this paper concentrates on an issue that has largely been overlooked. The consolidation of the California's privately owned transmission systems into the California Independent System Operator (CAISO) has been extremely costly, has led to bungling and mismanagement, and may have been unnecessary. This is an important issue, as federal policy has encouraged consolidation of individual utility operating areas into centralized grids. These initiatives, however, could prove destabilizing, leading to unexpected problems and slower liberalization of the power market.

Breaking apart vertically integrated electric companies has revealed hidden costs. For years engineers have managed the grid in real time. Through experience they frequently know how to optimize the system—to dispatch the least cost set of generators and avoid congestion. Dividing distribution, generation, and transmission activities into separate companies requires that some sort of market replace the minute-to-minute decisions of a utility's experienced dispatchers. But, the grid balancing market is one like no other, it must run twenty-four hours a day and requires continuous costly information if decentralized decision makers are to make rational bids and the market is to be competitive. Moreover, the balancing market was meant to be tiny – never accounting for more than 5 or 10% of the total energy delivered. Operation of the small and highly complex real-time market has involved extremely high transactions costs, and it is not at all clear that the existence of such a market has yielded net benefits, either in the form of lower prices or enhancements in reliability. More to the point, are there lower cost alternatives that would allow competition between generators and an efficient wholesale market?

This paper will first explore the classic reasons why companies choose to vertically integrate. It will examine the specific characteristics of the power industry and conclude that in many circumstances it would be more cost-effective for the grid operator to own and control sufficient resources to balance the grid, rather than attempt to operate a real-time market. Furthermore, such a system is not inconsistent with policy objectives aimed at promoting competitive wholesale and retail markets.

2.   VERTICAL INTEGRATION

In 1937, Ronald Coase, who would later go on to win a Nobel Prize, questioned the very existence of a firm. Why, Coase asked, if markets work so well, do firms organize in the first place? Why are there multiple production processes (such as the manufacture of component parts and assembly of the final product) taking place under the same roof? The answer, he reasoned, was that firms existed because use of the price mechanism (i.e., the market itself) imposes costs on its users. Final products in the retail market could be the result of hundreds of individual transactions (and prices), with separate contracting for each tier of the chain of production, or the activity could be housed in a single firm in which the resources are allocated on the basis of internal decisions.

Economists have long focused on two broad approaches to explain why firms vertically integrate. One school of thought analyses the problem with the conventional tools of neoclassical economics. Firms, for example, may integrate in order to take advantage of economies of scale or scope, or to diversify in the face of uncertainty.

The second school of thought has emphasized the role of "transactions" costs. These are, according to Arrow (1974), the "costs of running the economic system." The transactions cost approach focuses on the implicit or explicit costs of participating in a market. Transactions cost may include the cost of information, brokerage fees, enforcement costs, and certain types of insurance. Economic theory often ignores transactions costs. In many economic models, for example, auctions are held without having to pay the auctioneer or coordinate a group of auction participants. Most economists were ready to admit that such simplifying assumptions were false but maintained that such costs were small, in any case.

Perry (1989) discusses types of transactions costs that, when large, are likely to lead to vertical integration. Such costs include gathering market information, the risks associated with incomplete or incorrect information, and investments in firm- or industry-specific assets.

Information costs are a powerful motive for vertical integration. Commodities that enjoy a broad market, are homogenous, and cheaply stored are easier to market than those with narrow markets and conspicuous heterogeneity. The agriculture industry is typically not integrated because farming is constrained by geography and climate. It requires a great deal of specialized knowledge to optimize output from a particular farm. On the other hand, the products a farmer typically producers – grains, meats, and so forth – are most economically processed and marketed in larger-scale operations. Information on the value of various agricultural commodities is cheap and easy to obtain, due to mature futures markets and information systems.[2]

Asset specificity, as defined by Williamson (1975), and discussed by Perry (1989) and Dietrich (1994) is the investment by firms in assets that have little use or value other than to the task to which they arc applied. Markets for specialized assets are inherently narrow, and often times result in "bilateral monopoly," in which the "market" consists of a single seller and a single buyer. Such a narrow market creates an incentive for opportunistic behavior, where, for example, the seller may threaten to withhold supply or the buyer may threaten to close down. Firms whose products depend on specialized assets might integrate backwards or forwards (say, by extracting raw materials themselves rather than purchasing from someone else or by constructing their own retail marketing outlets) to shield themselves from such opportunistic behavior and ensure themselves a steady supply of key production inputs.[3]

Asset specificity need not be limited to physical assets, such as machines or raw materials. Human capital specialized to a particular industry or firm can involve sufficiently high costs as to incite vertical integration. Technologically intense industries with an array of sophisticated products are naturally run by managers who have extensive knowledge of the industry and the intricacies of each step in the supply process. For example, in the oil industry, production companies know quite a bit about the crude oil they

---

[2] Various satellite systems designed specifically for farmers give minute-to-minute details on spot and futures prices of all major agricultural products. These systems are available for as little as $45 per month.

[3] Adam Smith captured this idea in *The Wealth of Nations* with the doctrine, "The division of labor is limited by the extent of the market." George Stigler's 1951 essay in *The Journal of Political Economy*, which shares the same name as Smith's doctrine, noted that Adam Smith was only partly correct. The level of maturity of the industry, as well as its cost structure, also influence the extent of firms' vertical integration. Perhaps the maxim of "If you want something done right, do it yourself" (sometimes attributed to Benjamin Franklin) is even more succinct.

extract from the ground. The composition of the oil, its density and type of contaminants, affects how it will be refined. A refining company will not have the same level of knowledge as the producer and would have to constantly check the quality of crude oils being purchased. Therefore, refiners may want to own (or share ownership and operation of) oil fields and pipelines to ensure the quality and reliability of raw materials.

Firms may also integrate in order to avoid the costs or capture the benefits associated with externalities (benefits or costs of a good or service not reflected directly in its price). Computers, operating systems and applications, for instance, can be viewed as three tiers of a production process. They are also "complementary" products. That is, the usefulness of computing is derived from all three products taken together. Apple has always included its own operating system with its computers, but relies on third-party vendors to supply applications. Its principal motive to produce an operating system is to sell computers. In contrast, PCs are produced independently and standardized by Microsoft's Windows operating system. Microsoft, however, is also a major producer of application software for its own operating system. Microsoft is highly motivated to develop and market applications software, because it enhances the value of Windows. That is, consumers derive a benefit from the seamless integration of applications and operating system, because it saves time and minimizes confusion.

A similar motive for vertical integration can be observed in other industries. For example, brand-name restaurant companies (such as McDonalds, Burger King, etc.) are vertically integrated in order to standardize the quality of their products. Unhappy experiences by consumers in one outlet could reduce demand across the whole retail chain.

Diversification is often a motive for vertical integration. All industries are prone to cyclical changes, but the impact may not be consistent across all segments. For example, increasing crude-oil prices often causes refinery margins to decline. Higher crude-oil prices are passed through to the product market and the drop in demand for gasoline, diesel, jet fuel, and other products results in spare refining capacity, which leads to declining margins. If a firm is balanced between crude oil production and refinery capacity, increasing profits in one segment offset declining profits in another. Even if profitability in any given segment of an industry is independent of profitability in other segments, vertical integration allows a firm to diversify while retaining its core identity and expertise.

Not all of the classic reasons for vertical integration are applicable to the electric power industry, whose structure is as much the result of regulation as it is natural economics. Nonetheless the industry has certain unique characteristics, which may make operating a real-time or balancing market for the industry prohibitively costly and inefficient. In an unregulated environment, firms would likely seek to internalize these costs, through vertical integration. However, most economists also believe that transmission grids are natural monopolies. If utility companies integrate upstream through investments in generation assets, and integrate downstream through monopoly control of transmission assets, what then becomes of a competitive market for electricity? As will be argued, the economic impetus for utility companies to vertically integrate, combined with the natural monopoly of the grid, need not sound the death knell for competition in electricity markets. Before exploring this further, however, it will be useful to review the industry's unique characteristics.

3. SPECIAL CHARACTERISTICS OF THE POWER INDUSTRY

Five aspects of the power industry suggest that there may be substantial economies in a partially integrated structure — one that encompasses grid operation and sufficient generating resources necessary for load balancing. First, the information and contracting costs of operating real-time or balancing markets are extremely high. Second, the economies of scale in trading twenty-four hours per day suggest a natural monopoly market structure in the provision of real-time power. Third, any efficiency gains that might be obtained from auctions and other bidding mechanisms are likely nullified by externalities in the management and pricing of transmission congestion. Fourth, efforts to value congestion differentials in real time will inevitably fragment the market, enhancing the market power of some generators. Finally, in a real time environment load is far less flexible than generation. This makes the grid operator continuously vulnerable to the exercise of market power by suppliers.

The widespread penetration of computers has reduced information costs and made real time trading possible, but it is still a costly undertaking. In a conventional setting, dispatchers have at their command minute-to-minute information on load, output of generating units, unit availability, weather, and other factors likely to influence the demand for base-load and balancing energy. Ideally, traders should have the same information set in order to make rational decisions about scheduling generation or load. However, a grid manager responsible for procuring balancing energy will likely not want to share this information. The information asymmetry gives grid operators a bargaining advantage, and ultimately reduces market efficiency as real-time traders are forced to make costly investments to obtain or estimate the system conditions known by the grid operator.

Invoicing and settling real time markets is an accounting nightmare. A good share of the costs incurred by the CAISO and the CalPX arose from the accounting adjustments involved in running the real time market. The greater the number of small generators, or a move to nodal pricing (as some have suggested) would make the accounting activity more complex and costly, whittling away any net benefits.

From the point of view of generating companies or load managers there are substantial economies of scale in real-time trading. This is because the real-time trading desk has to be manned twenty-four hours a day with a continuous flow of complex data. This has concentrated the number of active firms and makes it all but impossible for speculators to participate. That is, only a few utilities and large marketing companies have the resources to maintain active real-time desks. Instead, the responsibility is frequently contracted to another company. With so few participants in the real-time market, the incentives for opportunistic behavior by sellers are high, particularly in emergency conditions such as unplanned generation outages.

As discussed in the previous section, externalities may explain why some industries tend to vertical integration. Electricity markets have a particularly strong supply externality that arises from transmission congestion, in the sense that final congestion prices are uncertain and difficult to forecast. Consider how a decentralized market actually works. Buyers and sellers must decide how much to buy or sell and at which delivery point, based on bid and offer prices. But, optimal allocation is unknown until load and resources are actually scheduled. Whatever the system used for determining congestion differentials, locational marginal pricing (LMP) or California's system of adjustment bids followed by zonal increments and decrements, market participants should be given enough time to change their bids and offers and readjust planned deliveries. This may require a large number of bidding rounds and substantial re-contracting before an efficient solution is found. Even a brief review of location differentials in either the CAISO or PJM, as shown in Figure 1, suggests that congestion pricing is inefficient; outcomes are frequently unpredictable and seem to bear little relationship to the natural flow of energy.

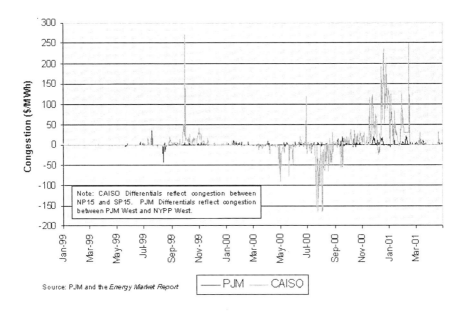

Figure 1. Congestion Differentials in California and PJM

Attempts to calculate congestion prices based on nodes or zones naturally increase asset specificity in each node or zone. In the extreme, the several thousand nodes in the PJM system and even the broader zones of the CAISO grid all suffer from the problem of "bilateral monopoly" discussed in the previous section. For example, assume that congestion exists between two nodes or zones, unless *all* the load-balancing generators in the first area are running. Further, assume that one company owns all the balancing capacity in the area. A small reduction in output by that company may earn it a huge premium. Both theory and practice suggest that the grid operator would be much better off to integrate upstream through ownership of balancing resources.

As a second-best alternative, grid operators whose balancing resources have already been divested (as in California) should consider signing contracts for real-time energy. However, as Klein, Crawford, and Alchian (1978) have noted, asset specificity and bilateral monopoly give rise to a high-cost contracting process, as each side attempts to extract pricing concessions from the other.[4] Klein, Crawford, and Alchian also pointed out an obvious problem inherent in contracts: they are not always honored. Incentives for opportunistic behavior are present in any system of rules,

---

[4] This game of "chicken" may have particularly high stakes when the assets in question have little value in alternative uses. This is certainly the case with balancing resources. Since the real-time market represents the shortest-term market, there is no practical alternative use for a generator once it enters the balancing market.

regulations, and contracts, so the provision of proper punitive measures becomes vital. Indeed, one of the biggest problems with the California system was its implicit encouragement of opportunistic behavior in its real-time markets by failing to take steps to encourage use of the day-ahead or other forward markets. The bilateral market in the WSPP uses a system of "liquidated damages" in its firm, prescheduled contracts. If a seller fails to deliver firm energy to a buyer, for example, the seller must then bear the specific replacement cost of that energy, no matter how high. California would have done well to implement a similar system in its own markets instead of aggregating transmission costs and real-time prices. Overuse of the real-time market or large deviations from forward schedules imposes costs on grid operators and system planners (as well as consumers and other market participants), and should be severely punished.

The bilateral monopoly problem of the real-time market is compounded by the fact that last-minute electricity demand is almost perfectly price inelastic. This makes it very easy for shrewd suppliers that have more flexibility to take advantage of the grid operator. The extreme prices seen in California, PJM, and the Midwest have, for the most part, been for electricity supplied to the real time market. Buyers with adequate time may choose to cut back demand or can do a more complete survey of alternative suppliers.

4.   HOW TO SIMULTANEOUSLY INTERNALIZE COSTS AND PROMOTE COMPETITION

The market structure transformations in the U.K. and in California were radical and neither experiment has proven to be a success. In contrast, the much slower evolutionary process at PJM appears to be producing superior results.[5] Although analysts have asserted that PJM's success is due to its method of managing congestion pricing, there may be a much simpler explanation. The utilities that participate in the pool have retained their integrated structure and the expertise garnered in managing day-to-day operations has been developed and enhanced as its market has evolved. Far and away the most abrupt and revolutionary change occurred in California, where a whole new grid operation and management process was designed and implemented in fifteen months.

---

[5]   It is important to note, however, that the PJM system has not been without its own disadvantages. The PJM system is largely self-contained, and opportunities for export to other areas of the Eastern Interconnect have been limited by restrictions on when energy can be sold outside of the pool. Fewer export restrictions, for example, may have been able to soften the impact of the price spikes in the Midwest in the summer of 1998.

All experimental power markets have suffered from an undue emphasis on the real time market and congestion pricing. Ninety-five percent of the effort has been focused on what ought to be no more than five percent of the market. Base load and intermediate resource costs have historically been the main determinant of regulated retail rates, since these costs are far more stable and predictable. Base load resources are usually not suited for load balancing since they face large start-up costs and ramping times. Load balancing resources have volatile costs, resulting in extremely volatile spot prices. If balancing resources are allowed to set prices for all electricity delivered, then electricity prices are unacceptably and impractically volatile to consumers. Such a pricing scheme, based solely on short-run marginal costs, would be unique among fixed-cost network industries, and with good reason. Airlines, for example, generally follow longer-term marginal cost pricing and adjust load factors of specific flights using a variety of techniques – stand by passengers to cover "no shows" and cash and other benefits to persuade flexible passengers to give up their seats.

Operating load balancing with an integrated structure – the grid manager owning and/or controlling key resources – does not eliminate the possibility of efficient and competitive power markets. The present regulatory emphasis on consolidating operating areas into ISOs and ISAs, congestion pricing, and other radical restructuring may be inhibiting rather than encouraging a practical market structure. While the standardization of transmission protocols may have its advantages, the transactions costs involved in implementing any such scheme will certainly be large, and will not solve the problems of congestion pricing uncertainty or localized market power at specific nodes or in specific zones.

It is instructive to observe which market structures have performed well. The WSPP was a model of an efficient power market before the California restructuring tore it in half. In this system competition was achieved through the interconnection of multiple operating areas. Integrated companies, controlling both transmission assets and balancing resources, managed each operating area. Originally, the system had its disadvantages – retail prices did not reflect the movement of wholesale energy costs, and the vertically-integrated structure of the utilities (along with regulation) was an effective barrier to entry into retail markets. Since then, however, the FERC has opened transmission access and individual states are allowing more and more retail buyers to purchase power directly from wholesale sellers. Similar market structures have been implemented in several European countries (see Table 1).

Table 1. *European Wholesale* Electricity Market Structures

Key: D Day W Week E Weekend M Month Q Quarter S Season Y Year B Baseload P Peak	Exchange	Bilateral market							Futures
	Spot market	Day ahead	Week block	Weekend block	Month block	Quarter block	Season block **	Year block	Futures market
Austria	✘	1DBP	1W BP	1EB	3M BP	-	-	-	✘
Germany	Hourly Day A/H	1DBP	1WBP	1EB	6M BP	4QBP	-	3YBP	✔
Netherlands	Hourly Day A/H	1DBP	1WBP	-	2M BP	4QBP	-	2Y* BP	✘
Nordic Area	Hourly Day A/H	-	-	-	-	-	5SB	3YB	✔
Spain	Hourly Day A/H	-	1WBP	-	2M* BP	2QBP	-	Rem of yr	✘
Switzerland	✘	1DBP	1WBP	1EB	3M BP	-	-	-	✘
England & Wales	30 Mins Within day	1DBP	1WBP	-	3M BP	2QBP	2S BP	1Y	✔

* 2 full years plus balance of current year . ** There are three seasons in the year in the Nordic area (winter 1, summer and winter 2) but only two in England and Wales (winter and summer) ⁎ 2 full months plus balance of current month
Source: Caminus

5. CONCLUSION

Markets are generally very good at allocating resources and directing investment, but when the costs of running a market become too onerous, internal decisions by a vertically integrated company or a set of enlightened rules can often be preferable. On paper, operating a market for real-time energy is very appealing. A well-designed auction, with multiple firms bidding should result in efficient pricing. However, this idealized scenario ignores the transactions costs involved in running such a market, and these costs have proven extensive both in California and in the U.K. Moreover, the geographically segmented structure of the grid (zones in California and nodes in PJM) has reduced the number of sellers in any particular area. Instead of facing a multitude of sellers, congestion problems may result in the grid operator dealing with several monopoly or oligopoly suppliers. Facing such high transactions costs and uncertainty, upstream vertical integration by the grid operator is surely more efficient than trying to operate an inherently flawed market.

If anything has been learned from the California experiment it should be that there are often unexpected costs involved in radical restructuring. To a large extent, the California market was over-designed. It was the product of a federal and state regulatory process that attempted to balance a host of conflicting special interests. In contrast, other successful markets have evolved and matured over long periods of time with modest of experimentation and adaptation. It took the WSPP over a decade to develop an efficient market with reliable and transparent electricity pricing. It took California regulators only a few days to dismantle it.

REFERENCES

Armour, H.O., and D.J. Teece. 1980. "Vertical Integration and Technological Innovations," *Review of Economics and Statistics* 62:490-494.

Arrow, K. J. 1974. *The Limits of Organization*, New York: Norton

Coase, Ronald. 1937. "The Nature of the Firm." *Economica* NS-4: 386-405.

Dietrich, Michael. 1994. *Transactions Cost Economics and Beyond.* Cambridge University Press.

Federal Energy Regulatory Commission. 1998. *Causes of Wholesale Electric Pricing Abnormalities in the Midwest in June 1998.* Available at www.ferc.gov.

Klein, B., R.G. Crawford, and A. A. Alchian. 1978. "Vertical Integration, Appropriable Rents, and the Competitive Contracting Process." *Journal of Law and Economics* 21: 297-326.

Klein, Benjamin, and Kevin Murphy. 1988. "Vertical Restrictions as Contract Enforcement Mechanisms." *Journal of Law and Economics* 31.

Perry, M.K. 1989. "Vertical Integration." In *Handbook of Industrial Organization,* edited by Richard Schmalensee and Robert Willig. New York: Elsevier Science.

Stigler, George J. 1951. "The Division of Labor is Limited by the Extent of the Market," *Journal of Political Economy* 59: 185-193.

Williamson, Oliver. 1975. *Markets and Hierarchies: Analysis and Antitrust Implications.* New York: Free Press.

Williamson, Oliver. 1989. "Transaction Cost Economics." In *Handbook of Industrial Organization,* edited by Richard Schmalensee and Robert Willig. New York: Elsevier Science.

Chapter #8

Is Market-Based Pricing a Form of Price Discrimination?

Kelly Eakin and Ahmad Faruqui
Laurits R. Christensen Associates and Charles River Associates

Abstract: This paper finds that market-based pricing is a form of price discrimination.
But market-based pricing embodies the transistory discrimination that is a
fundamental part of the competitive process. Is market-based pricing
undesirable? Absolutely not. On the contrary, besides being a fundamental
part of the competitive process, market-based pricing also involves the
introduction of new and better products and giving customers a wider choice
among those products.

Key words: Price discrimination; anti-trust laws; Lerner's index; Product Mix model

1.   INTRODUCTION

Market-based pricing is being introduced to the retail electricity industry
and other formerly regulated industries. Market-based pricing involves three
fundamental changes from the cost-of-service pricing that characterized
regulated industries. First, the foundation for market-based prices is
marginal cost, rather than average embedded cost. Second, the mark up of
market-based prices is driven by competition and customers' willingness to
pay, rather than by a revenue recovery requirement. Third, consumers enjoy
a choice among a variety of differentiated pricing products, each of which is
priced according to the market. That is, market-based pricing brings
together traditional supply forces (reflected in marginal cost) and demand

A. Faruqui and B.K. Eakin (eds.), Electricity Pricing in Transition, 113-122.

forces (reflected in the customers' willingness to pay) to determine a market price.

One of the initial consequences of market-based pricing is that different customer groups may pay different prices for similar products. Is this tantamount to price discrimination?

Price discrimination has typically been viewed as unfair and potentially anti-competitive. The general framework of U.S. antitrust laws has prohibited price discrimination in any case where it *may* lessen competition. Likewise, major pieces of legislation and regulations affecting the electricity industry expressly prohibit discriminatory practices in the buying and selling of wholesale electricity.

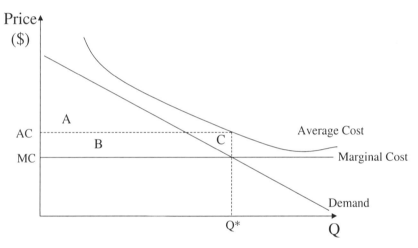

Figure 1. Natural Monopoly and Price Discrimination

Price discrimination, however, has been a useful tool to improve efficiency in pricing in a natural monopoly situation. Figure 1 provides a compelling case on the usefulness of price discrimination in a particular natural monopoly situation. In this case, the demand curve falls everywhere below the monopolist's average cost curve, yet over some range of output the total benefit from consumption exceeds the total cost of production. No uniform price allows the recovery of cost. With marginal cost pricing, the optimal output Q* is consumed, but the monopolist suffers losses equal to area B + C. If no production and consumption were to occur, foregone surplus would be area A + B while the producer avoids losses B + C. The net social loss of not producing is area A − C, which is greater than zero. Thus, prohibition of price discrimination in this case would mandate that a socially beneficial activity be foregone. Only by allowing some degree of price discrimination, would production be feasible.

Likewise, Ramsey pricing uses price discrimination as the regulator's tool to increase efficiency (and presumably social welfare) in a natural monopoly situation. Ramsey pricing is the solution to the following constrained optimization problem: maximize consumer surplus subject to the monopolist breaking even. The Ramsey pricing solution has the markup in each market such that

$$(P_i - MC_i) / P_i = -[\lambda / (1 + \lambda)](1 + \varepsilon_1)$$

Where the subscript indicates a customer group, P is price, MC is marginal cost, ε is the price elasticity of demand and λ is the LaGrange multiplier on the breakeven constraint. λ increases in value the greater the revenue shortfall from marginal cost pricing. Interestingly, the Ramsey pricing solution gives precisely the same pattern of pricing that a profit-maximizing price discriminating monopolist would set. The only difference is that the Ramsey pricing regulator would mark up prices only a fraction of the monopolist's mark-up.

Thus, as electricity markets are restructuring and market-based pricing is replacing cost-of-service pricing, concerns about price discrimination are being raised. However, the concept and practice of price discrimination is not new to the industry. In this chapter, we take yet another look at price discrimination and address two key questions: (1) is market-based pricing a form of price discrimination? And (2), If so, is market-based pricing therefore undesirable?

2. PRICE DISCRIMINATION

What constitutes price discrimination? A seemingly straightforward definition is selling the same goods or services to different buyers at different prices. Indeed, this is essentially a paraphrase of the definition found in Black's Law Dictionary (1979, p. 1070):

> **Price discrimination.** Exists when a buyer pays a price that is different from the price paid by another buyer for an identical product or service.

This seemingly simple definition, however, leads to ambiguities, confusions and inconsistencies. Two key shortcomings to the simple definition are the lack of any cost basis in the definition and the inability to compare the pricing of similar but slightly different products.

Most students of price theory would find a more meaningful definition of price discrimination to be that different customer groups are charged different percentage markups for similar products. That is, $(P_i\text{-}MC_i)/P_i \neq (P_i\text{-}MC_i)/P_i$ implies price disrimination.

What is illegal price discrimination? The key prohibitions against price discrimination can be found in the Section 2 of the Clayton Act (1914) and in Section 2(a) of the Robinson-Patman Act (1936). Both of these acts prohibit price discrimination if it may lessen competition or injure competitors. The language of these two acts are linked to Section 2 of the Sherman Antitrust Act (1890) which makes it illegal for a firm with market power to display intent and purpose to exclude other competitors. Price discrimination on the basis of race, sex or religion is prohibited by the Civil Rights Act (1965). Most states have price discrimination laws similar to the Robinson-Patman Act. Likewise, federal energy legislation and regulations prohibit discriminatory behavior in buying and selling wholesale electricity and require non-discriminatory access to transmission networks.

Since its passage sixty-five years ago, the Robinson-Patman Act, and particularly its prohibition on price discrimination, has been a controversial topic in law and economics. The strongest criticism of the law comes from the "Chicago School" of economics. The essence of the attack is that price discrimination does not lessen competition or worsen monopoly power and that the laws prohibiting price discrimination, while ineffective in preventing price discrimination, in fact themselves greatly harm competition. As Bork wrote (1978, p. 382):

> One often hears of the baseball player who, although a weak hitter, was also a poor fielder. Robinson-Patman is a little like that. Although it does not prevent much price discrimination, at least it has stifled a great deal of competition.

Examples of legal price discrimination abound in every day life. The price you paid for your last airline ticket was likely different from the price paid by the person nearest you on the plane. It was likely different from a person with the identical itinerary. Public universities typically charge a large differential for out-of-state tuition. Often doctors have charged patients on an ability to pay basis. Movies and restaurants have lower prices for seniors and for early birds. And, as mentioned above, price discrimination has been practiced and mandated by regulators of natural monopolies. Declining block pricing is a leading example of second-degree price discrimination in the electricity industry.

Legal or not, there are three necessary conditions for the practice of price discrimination. First, the seller must have some market power. That is, the seller cannot be entirely a price taker. The classical measure of market power is the Lerner Index of Monopoly Power (Lerner 1934):

$LIMP = (P - MC)/P.$

The seller must be able to raise price above marginal cost.

Second, there must be some meaningful diversity between customer groups. This diversity can be in their price responsiveness and in marginal cost differences.

Third, it should be possible for the seller to segment into meaningful groups. Furthermore, for price discrimination to be profitable, the cost of identifying, separating and keeping separate different customer segments needs to be less than the extra revenue generated from the exercise.

3.   MARKET-BASED PRICING

What is market-based pricing? Market-based pricing brings both supply and demand considerations together in the determination of the price. Often menus of differentiated products are developed that reflect the diversity in customers' needs and preferences. Marginal costs, especially wholesale market prices, provide the foundation for pricing. The products, especially in retail electricity, may be de differentiated by who bears the financial risks. The risk-differentiated products are derivative products. The breakeven prices for these products are derivatives of the underlying asset price (wholesale price) and its uncertainty. It is the breakeven price that is a market version of marginal cost. The breakeven price provides the foundation for market-based pricing. It is the price, which if charged, would result in zero expected profits. Thus it is not the price to charge customers. Instead, it is the backstop below which price should not be set.

Market-based pricing using the breakeven price as a foundation from which to markup price according to market conditions. The ability to markup price above breakeven price depends upon the customers' willingness to pay for the product in general, the customers' level of risk aversion and the availability and pricing of competitive alternatives.

The necessary conditions for market-based pricing are: knowledge of variations of price responsiveness across customer segments; knowledge of variation in risk aversion across customer segments; and limited possibilities for resale and arbitrage. With choice and self-selection, identification and separation of specific customers is not required.

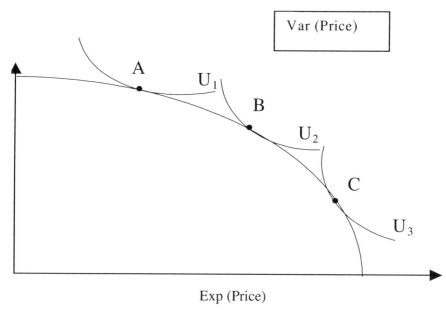

Exp (Price)

Figure 2. The Efficiency of Market-Based Pricing

Market-based pricing and customer choice leads greater efficiency as customers choose a product with characteristics that best meet their needs and preferences. Figure 2 illustrates that choice lets customers pick the expected price/risk combinations that are best for their situations. In this figure, a risk/price locus shows the product possibilities and the trade-off between expected price and price risk., as measured by the variance of price. The figure also shows indifference curves for three distinct customers are presented. Customer A is risk neutral and has considerable flexibility on the timing and intensity of electricity consumption. Customer B is equally flexible, but risk averse. Customer C is both inflexible and risk averse. Each customer picks the price-risk combination that best matches their preference, and in doing so, attains the highest possible indifference curve, shown as U_1, U_2, and U_3 for customers A, B and C respectively.

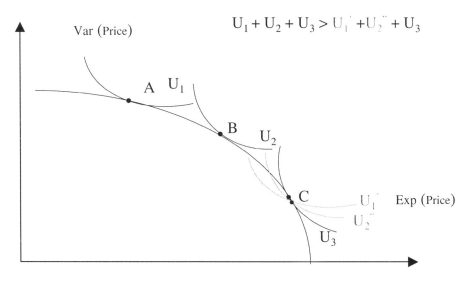

$$U_1 + U_2 + U_3 > U_1' + U_2'' + U_3$$

Figure 3. The Inefficiency of Uniform Product Pricing

Figure 3 shows the inefficiency when a single product is offered to all customer segments as a uniform price. In this case, the product being offered is the guaranteed price product. This is best for Customer C but is inferior for Customers A and B. thus, customer C is able to attain U3, but customers A and B are forced onto indifference curves U1, 2.

3.1 A Numerical Example

We now present a stylized example to illustrate the efficiency gain from market-based pricing. This example, developed using EPRI's Product Mixing Model, assumes uncertain wholesale prices, uncertain customer loads and three distinct customer segments. The assumptions and descriptions are reported in Table 1.

Table 1. Example Assumptions

Wholesale Price	Expected Price = $20/MWh
	Wholesale Price Uncertain
Customer Segments	Three Distinct Segments
	inflexible and risk-averse
	flexible but risk-averse
	flexible and risk-neutral
Load Profiles	Expected Flat Load Shapes
	Customers Actual Loads Uncertain
	Load Uncertainties Correlated with Wholesale Price
	Uncertainty
Status Quo	Single fixed price product 4 ¢/KWh

The example posits three distinct customer segments. Table 2 reports the profit from and price markup to each of these groups under the status quo of a 4 ¢/KWh guaranteed price product. Table 2 provides the baseline for comparison to the results involving product choice.

Table 2. Baseline: No Choice; Retail Price = 4 ¢/KWh

CUSTOMER SEGMENT	PROFIT ($ Million)	Price Markup
C: Inflexible and Risk-Averse	$3.33	44%
B: Flexible but Risk-Averse	$3.33	44%
A: Flexible and Risk-Neutral	$3.33	44%
TOTAL	**$10.0**	**44%**

The example analyzes the impact of offering three risk-differentiated products, rather than the traditional guaranteed price product, to the different customer segments. Another product now being offered is a spot product that is marked up 100% above wholesale price. Finally, the spot price product is offered with price cap protection at 5¢/KWh. The additional cost of the price cap protection is $2,500/mo. Table 3 summarizes the risk-differentiated products.

Table 3. Product Variety

PRODUCT	PRODUCT PRICE
Guaranteed Price	4 ¢/KWh
Spot Price	2 x wholesale price
Spot + Price Cap	2 x wholesale price but not more than 4 5¢/KWh; $2,500/mo. Cap fee

Table 4 shows how market-based pricing and customer choice lead to an increase in both profit and customer benefits. That is, value has been created. Examination of Table 4, particularly the "Price Markup" column suggests that market-based pricing may lead to a pricing pattern appearing to be price discrimination. However, customer choice distinguishes the situation from one of classic price discrimination. It is also interesting to note that the customer segment facing the largest percentage markup (Customer Segment A) is also the segment, which receives the greatest benefit from choice. Also, the spot + cap product has a lower markup than does the pure spot product, but generates more profit.

Table 4. Gains from Market-Based Pricing

CUSTOMER SEGMENT	Product	Change in Customer Benefit ($ million)	Change in Profit ($ million)	Price Markup (Lerner Index)
C: Inflexible and Risk-Averse	Guaranteed Price	$0	$0	44%
B: Flexible but Risk-Averse	Spot + Cap	$0.50*	$0.62	47%
A: Flexible and Risk-Neutral	Spot Price	$0.55	$0.35	50%
TOTAL		$1.05	$0.97	47%

Market-based pricing makes all customers better off, compared to uniform product pricing. It als make the supplier better off.

4.   SUMMARY

Market-based pricing has similarities to price discrimination. Specifically, price is set above marginal cost in inverse proportion to the customer group price elasticity of demand. However, the practice of market-based pricing is reflective of the competitive process rather than indicative of captive monopoly pricing.

Market-based pricing has two fundamental differences from classical price discrimination. First, it involves introducing new products. Second, it involves giving customers choices and allowing self-selection of products.

Market-based pricing does result in apparent price discrimination, as defined by equation (2). However, these price markups have been called transitory discriminations. It has been argued that, far from being anti-competitive, these markups are essential to the evolution of the competitive process. Adelman (1959, pp. 164-165; also found in Bork 1978, p.388) writes:

> With ever-changing supply-demand conditions, new profit discrepancies are constantly being created and destroyed. These transitory discriminations are incentives for better resource use, because they occur as part of the competitive process which eventually liquidates them.

Likewise, Bork (1978, p. 388) argues that:

The evanescent discriminations of competitive markets are the sellers' antennae. This adjustment to shifting costs and demand is socially desirable, and it is best that appropriate responses be made as quickly and sure-footedly as possible.

We conclude by answering the questions posed at the beginning of this paper:

Is market-based pricing a form of price discrimination? We say yes. But market-based pricing embodies the transitory discrimination that is a fundamental part of the competitive process.

Is market-based pricing undesirable? Absolutely not. On the contrary, besides being a fundamental part of the competitive process, market-based pricing also involves the introduction of new and better products and giving customers a wider choice among those products.

REFERENCES

Adelman, Morris. 1959. *A &P: A Study in Price-Cost Behavior and Public Policy.* Cambridge: Harvard University Press.

Black, Henry C. 1979. *Blacks Law Dictionary*, Fifth Edition. St. Paul: West Publishing Company.

Bork, Robert H. 1978. *The Antitrust Paradox: A Policy at War with Itself.* New York: Basic Books.

Lerner, Abba P. 1934. "The Concept of Monopoly and the Measurement of Monopoly Power." *Review of Economic Studies* 1: 157-175.

Stigler, George J. 1987. *The Theory of Price*, Fourth edition. New York: MacMillan: Publishing Company.

Chapter #9

Use of Market Research in a Competitive Environment

Michael V. Williams and Ziyad Awad
Wales Behavioral Assessment and Awad & Singer

Abstract: Market research is a critical component of marketing and pricing strategy. Applying the methods and mechanics of attitude research, competitive analysis and conjoint modeling is the means of fulfilling that role. Two examples are provided describing the markets, necessary data, and the pricing approach applied. Finally, a description of the strategic process implied by the approach is provided.

Key words: Market Research, Strategy, Mass Markets, Market Data, Competitive Advantage, Segmentation, Attitude, Product Attributes, Value.

1.   OVERVIEW

The primary objective of pricing strategy is to support two marketing objectives. The first is properly positioning a product within a desired market segment. The second is to ensure that the profit objectives for the product are met.

Pricing for power in the energy business is generally based on one of two theoretical foundations: rate studies and commodity pricing theory. Neither of these theories perfectly fits the needs of a firm in a newly competitive market seeking to develop areas of competitive advantage. This chapter presents an empirical approach to pricing using market and customer data. The approach focuses on the underlying needs of buyers that drive purchase decisions among available products to set prices.

A. Faruqui and B.K. Eakin (eds.), Electricity Pricing in Transition, 123-141.
@ 2002 *Kluwer Academic Publishers.*

In our opinion, pricing is a strategic variable and this chapter reflects that view. Little is said about the direct mechanics of pricing, a review of trade-off or revealed preference work will provide the mechanics, the focus of this chapter is on the use of market data to price, or stated another way, how your extract competitive advantage through market data and pricing

The chapter outlines two distinct methods within the broad empirical approach, each aimed at a particular market segment: mass market pricing and custom market pricing. For both methods, the relevant market is described, the data necessary is defined, and the pricing approach is demonstrated. The conclusion is a short description of the strategic process implied by the approach and a demonstration of the general approach using actual data.

2.   GENERAL DISCUSSION OF PRODUCTS

Products are typically described from the seller's point of view in the energy industry. Because this approach is based on understanding the customer needs satisfied by products, it is important to start by describing products from the customers perspective. The products available for purchase in a market have attributes from which buyers derive value:

$$Value = \sum_{j=1}^{n} V_j \qquad (1)$$

where, V_j is the value of an attribute (the jth attribute), n is the number of distinct attributes. The total value of the good for a buyer would then be the sum of the attribute values.

The attribute value can be thought of as having two components: the amount or degree of the attribute the buyer believes to be present, and the importance of the attribute to the buyer:

$$V_j = I_j x_j \qquad (2)$$

where, V_j is the value of attribute j, I_j is the importance of the jth attribute and x_j is the amount of the jth attribute present in the product. The total value of the good for a buyer would then be the sum of the attribute values, each attribute having distinct utility to the buyer.

This description of goods provides a foundation for thinking about customer choice in a competitive market. Customers view products as an aggregation of attributes that provide value. All other things being equal, customers should purchase the good that provides the greatest value, by delivering the highest aggregation of valued attributes. This high valuation might be due to providing a great amount of a single highly valued attribute, it might be due to having all or nearly all attributes represented, or some combination of both.

A market can be thought of as the interface between customers who are seeking goods that have similar attribute sets and suppliers who are providing those goods. However, if the goods in a market are arrayed by their attributes, it is possible that the goods at one end of the market have no attributes in common with goods at the extreme other end. The implication is that all products will not satisfy every buyers needs. Only the products that have the right combination and amount of the key attributes will satisfy the customer.

A simple example using personal cleaning products illustrates this point. At one end of the spectrum is Betadine scrubs, used by physicians, at the other Pond's cold cream, used to remove makeup. There is a continuum of products between these two, but a customer buying Betadine would have no interest in cold cream and visa versa. Products then may differ from each other to the point where they are not seen as directly fungable, and yet they lie within a general market category.

Table 1. Attributes and Product Distribution

Attribute	Product
Disinfectant	
Antibacterial	Betadine
Water Soluable	Soft Soap
Attractive Scent	Dove
Moisturizing	Pond's Cold Cream
Removes Makeup	

3. GENERAL DISCUSSION OF CUSTOMER CHOICE

Unfortunately, customer evaluations of product attributes cannot be directly observed. Instead, we ask customers to express their attitudes toward specifically defined and identified product attributes. To value the product, the customer must have a need that the product is viewed as being instrumental in meeting. The nature of these needs is that they define the importance that the customer gives to particular attributes.

The need, then, is related to something that the customer will use the good to fulfill, to an intent that the customer has formed that the good is useful in meeting. The need is, therefore, congruent with an attitude, and attitude toward a specific attribute of the product.

Customer's generate purchase intents based on their attitudes toward particular product attributes. Attitudes are defined as intents and can be modeled as a combination of what the customer believes about an attribute and the customer's regard (either positive or negative) toward the attribute. It can be expressed as:

$$A_j = \sum_{t=1}^{m} r_t * b_t \tag{3}$$

where, A_j is the customer's attitude toward the jth attribute, r_t is the customer's regard for the tth trait of the attribute, and b_t is the customer's belief about the tth trait of the attribute.

By using attitude, it is possible to develop an importance weight for each attribute of a product, for example, the attribute "attractive scent," discussed above. The customer might form an attitude toward the scent by several beliefs, the use of scent is femine, the use of scent adds extra chemicals to the mix, the use of scent masks other odors. Each of these beliefs might have positive or negative regard from the individual customer. For example, a person with many allergies might view the addtion of unnecessary chemicals negatively, where another might find that fact indifferent or even positive. Together these attitudes combine to form the overall importance of the attribute to the customer.

4.   MASS MARKET PRICING

Few power firms focus on the mass market although in some respects the mass market is closest to the traditional utility marketing problem. The key to the mass market for retail pricing is segmentation. In this section, two methods of segmentation are described, the data necessary to drive such segmentations are defined, and we demonstrate how the information is applied to extract competitive advantage.

4.1   Description of the Mass-Market

A mass-market can be defined in a number of ways, but for the purposes of this discussion it will be defined as a market within which no individual buying unit provides enough margin to the producer to make product customization profitable. Using this description the products must be essentially "off the shelf" from the view of the seller. In addition, within mass markets, purchase decisions are modeled for groups of customers (segments) rather than for individuals.

4.2   Mass-market segmentation 1

Segmentation divides customers into groups; our approach is to base those groups on similar choices. One approach to segmentation is the direct application of differences in value (Figures 1 and 2), this leads to customers that find similar values in product attributes. Understanding where and how value is extracted from the product, it becomes a direct matter to price the product within the market. It also becomes possible to avoid either competing against, or ignoring, entrants in the market that are attracting your customer base.

4.2.1   Understanding the market

The first step is to understand the range of product attributes available in the market. This is a straightforward process using qualitative research. Simply stated, you talk to a group of customers and ask them why they buy at all and why they buy some products and not others. This process yields an overall picture of the underlying dynamics of a market based on the product attributes used by customers to make purchase decisions.

In addition, the customers are asked to indicate the importance of each attribute. This allows the analyst to develop estimates of the value of each attribute for each person in the study. A value estimate is developed for each specific attribute for all products sold into the market. All products available in the market and all their attributes must be described.

4.2.2   Generating the segmentation data

Once the market is described qualitatively, it should be measured quantitatively. This means that the taxonomy of attributes identified in the qualitative phase is used to develop research instruments to measure the importance of various product attributes in making purchase decisions.

There are a variety of ways in which this might be done. An experiment can be designed where individuals gauge their preference for products containing one set of attributes as compared to another. A similar experiment might be designed where customers gauge their preferences for alternative attributes. These approaches are variations on a general class of experiments called "trade-off" experiments.

Alternatively, data can be gathered on purchase patterns within the market. Prices commanded by alternative products, with various attributes can be compared statistically to extract estimates of the individual values those attributes provide. In general this is referred to as revealed preference analysis. Revealed preference analysis is built on the assumption that individuals make choices designed to extract the greatest value from their purchases.

4.2.3   Market segments

Customers differ in their views of which attributes deliver value. By understanding how customers differ, and the amount of value that various customers put on the product attributes, it is possible to identify specific groups of customers that value attributes similarly and to develop products designed for each group. The estimates of value correspond to the importance coefficients in Figure 2. Thus, customers are grouped with others for whom similar importance is derived from similar attributes.

In energy, a well-documented segment is the Green Power segment. These are customers who derive value from connecting their power purchase to environmentally friendly production methods. Other segments that have been found in various markets are Local Power segments, who value local power companies; Low Cost-Low Reliability customers, who will trade reliability for price reductions, Conservation customers, who value conservation advice and service, etc.

Each of these segments has a core set of attributes from which the members derive significant value, and this value proposition is generally different for each segment. In a statistical sense, these customers are significantly more like other members of their segment than they are like the members of any other segment in the manner in which they derive value from the product.

4.2.4   Pricing

Pricing becomes a simple matter of determining the overall value of the product (the sum of the importance of each attribute). Figure 4 illustrates an increase in value to Green Buyers created by offering a green product.

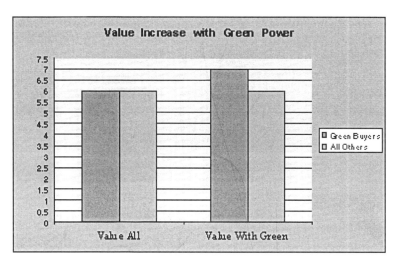

Figure 1. Green Power Segment

4.3   Mass-market Segmentation 2

A second approach to segmentation is by analysis of differences in attitudes toward the various attributes. As described at the beginning of this chapter, attitude can be conceptualized as a decomposition of importance. The reason for such decomposition is that it gives insight as to why customers give various attributes the importance weights they do.

By understanding the underlying motivations of customers, it is much easier to communicate with them. In the personal cleaning products example described above, some customers value antibacterial properties. They might believe that antibacterial properties protect their children from colds, or they might believe that antibacterial properties protect their hands from infection, or, they might believe antibacterial properties increase the cleansers value as a deodorant. Clearly, the nature of the communications effort would be different in each case.

Attitude segmentation, allows the market strategist to identify customers with similar motivations in selecting particular attributes.[1] This refinement on importance segmentation allows both value extraction and communication to particular customer groups.

---

[1] In the first segmentation, customers are grouped according to the importance given to attributes, in this case they are grouped by their motivations (different motivations can lead to the same attribute importance rating.

4.3.1   Understanding the customer

The first step is to understand what motivates customers to buy a particular product. As with understanding the importance of attributes, this process starts with qualitative research. Qualitative research is based on talking to customers about their beliefs and regard for particular attributes.

Utilizing a structured interview process with customers, the researcher develops a description of the beliefs customers hold about attributes. Each attribute in the market is assessed, at the end of the process all beliefs held in relationship to all attributes of the product should be defined.

4.3.2   Generating the segmentation data

The data for segmentation are developed from the statistical association between beliefs and regard for attributes and the importance rating of the attribute (see Figure 3). These data are then used to provide attitude scores for each respondent in a survey of the market.

The statistical analysis develops estimates of the salience and direction, or regard, for each belief as it relates to the importance of a particular attribute in choice.[2] From the attitude scores, customer segments are developed that have similar importance ratings due to similar underlying attitudes.

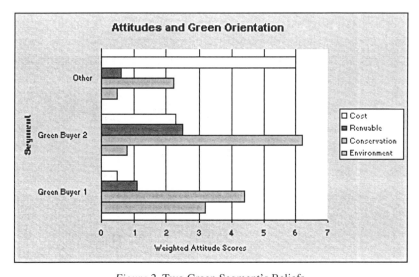

Figure 2. Two Green Segment's Beliefs

---

[2]   Salience is defined as the relative importance of the belief in developing an importance. Direction is defined as whether the belief is viewed as positive or negative in choice.

4.3.3   Market segments

Because attitude is made up of the beliefs and regard for particular elements of attributes, it is possible that individuals may rate the overall value of a product similarly, but for different reasons. Consequently, it is possible that very different groups want or desire the same attribute for very different   reasons.   Figure 5 presents data on Green Power where two different groups have an interest in the same attribute - greenness - but for different reasons.

The importance of particular attributes is primary in communicating with customers.   Here are two groups both of whom are interested in Green Power, neither of whom is basing this choice primarily on a belief in renuables.   Therefore, it is very important to understand the underlying attributes and the importance of each for different groups of customers.

In general, attitude information informs the pricing strategist as to why a customer is making a choice.   Attribute information informs the market strategist as to what the customer is making a choice for within a complex product. Strategy, especially strategy focused on share, benefits from both pieces of information.

4.3.4   Pricing

Pricing becomes more complex when attitude data are entered into the strategic mix.   Attribute values are split among potential target groups. Figure 6 presents data on the different value given to Green Power by the two Green Buyer segments and then all others in the market. Notice that the price is different for the two segments, with one having a value over $70/MWh and the second segment under $70/MWh.

This finding is expected and reflects the to price elasticity of green power.   Assuming the intent is to maximize return, determining an appropriate price point is a function of the potential·revenue lost from one group and the potential revenue made from the second group when, for example, a price drop is used to attract a second group.

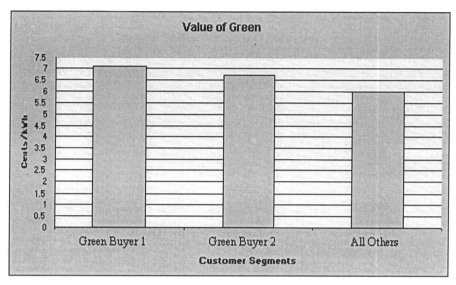

Figure 3. Role of Targeted Marketing

4.4 Competitive Advantage in mass-markets

Strategic issues in mass-market pricing revolve around development of competitive advantage. Competitive advantage is extracted where some type of barrier to competition can be developed. In marketing, competitive advantage springs from the ability to provide an attribute to customers at a consistent cost advantage.

4.4.1 The basis of the data analysis

The strategic analysis that underlies competition involves discovering attributes that critically influence customers' choices that your firm can provide at a consistent cost advantage. It can be further informed by information on attitudes towards those attributes. It is important to understand that this information underlies competition; therefore, to the extent possible, the strategist wants both costs for their own firm as well as costs for the competitor firm.

In this case, we extend the analysis for Green Power. Figure 7 presents the value ratings the two customer segments give to Green Power, plus the costs of Green Power (all-in costs are presented for this example.) Notice that the competitor's cost meets the first segment's cost requirement, nearly meets the second segment's cost requirement, and beats our cost. Unfortunately, this is not an atypical situation in the beginning of an analysis.

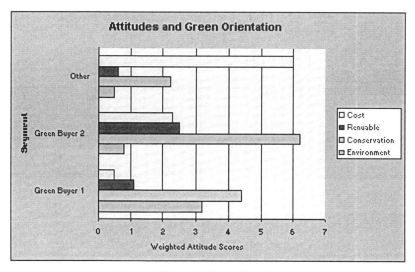

Figure 4. Value of "Green" by Segment

4.4.2 The role of targeting

Reviewing the segmentation data, in particular, the attitude data, reveals that both segments are less interested in the environment *per se* than in conservation. Knowing this, a review of the resource mix in the Green product might be in order. In the case of this firm, introducing hydroelectric power to the mix could produce a "Greenish" product.

Introduction of hydro power had two decided advantages: first, the competitor did not have access to hydro because their broader agenda required them to stay perfectly green; second, addition of hydro power had a decided benefit to the product's cost.

4.4.3 Competition

Developing competitive advantage does not require that a firm have overall cost superiority to its competitors. Rather, it requires that the firm have specific cost advantage defined by key attributes for key segments. The battle need not be joined where the competition would have it joined, selection of attributes and segments allows the firm to define the competition in its own terms, and thus define the keys to competitive advantage, also, in its own terms.

Maintenance of competitive advantage requires constant vigilance. Customers' beliefs are dynamic, changing with new information and new experience. To maintain competitive advantage, the firm must move with agility as the ground over which they contest, the customer's mind, changes and grows.

5.   CUSTOM MARKET PRICING

A custom market is a market where the margins are great enough that it is economic to develop a specific product for a specific client. Therefore, segmentation does not drive product design. Rather, product design is driven directly from individual customer value. Segmentation drives specialization, allowing the extraction of competitive advantage.

5.1   Custom Market Product Development

The key to customization is to develop a quantitative understanding of individual customers. Because these customers or "key accounts" are large, frequently international businesses, it is necessary to understand them in several varying domains: their use of energy, buying process, and, their business. It is also essential to understand their view of the energy product, its attributes and value, as well as their attitude towards energy.

While much of this is analogous to the mass market, it is considerably more complex. In addition, the choice made by a large firm is rarely made by a single individual, therefore, it becomes necessary to develop a quantitative understanding of decision processes involving several decision making individuals.

5.1.1   Understanding how customers extract value from the product.

The first step is to understand the customers' business; this is done using secondary research. Like qualitative research, the intent with this effort is to understand the broad base of businesses. However, it is supplemented by analysis of specific firms through generally available information. At the end of such analysis the strategic analyst should have a both a broad picture of the industry, and, as well, a reasonable understanding of the specific firm.

The second step is to understand the uses and values of product attributes. This step is accomplished by an analysis of the processes and applications of energy in the industry. This is also undertaken through analysis of the engineering practices in the industry.

The result of these efforts is a general picture of the customer. This picture is used to frame a data collection effort that will be used to develop a customized product. This effort is accomplished by the development of quantitative data for individual customers businesses.

5.1.2   Generating quantitative data

Data on the firm's use of energy is best gathered by an on-site review of equipment. This analysis allows a review of the processes employed as well as the flow of production. Together this information can provide insight into system level synergies that could be lost in a more superficial analysis.

Data on the business should cover three areas: firm level strategy, production strategy, and financial strategy. All are important, as each will affect the procurement process in a unique way.

Firm level data on strategy can be gathered at several levels of management, including executive management, financial management, as well as production management. The views of these different management perspectives will vary. This variation leads to two complications, one is in developing the structure of questions that will be asked, the second is in developing specific questions to elicit underlying attitudes for each manager.

Production strategy and financial strategy information is developed primarily from the executive in charge of each. Those elements of these strategies that fall under the guidance of executive management should course be gathered in the executive manager interview, however most of the data is best elicited directly from the responsible executive. While overall business strategy should influence production and financial strategies, the specifics of how that influence is made manifest may not be straightforward.

Developing information allowing the melding of these three decision domains is itself difficult. Information on group decision style and the specific numerical representation of that style require their own set of data.

While large datasets are available for mass markets, this is not the case for individual firms. Quantitative estimations of decision likelihood are possible through application of techniques such as adaptive conjoint analysis where attitudes or predispositions to particular choices can be explored. Conjoint analysis provides a "part-worth" or importance weighting for each product attribute.

5.1.3   Product design and pricing

As with the mass markets, product design and pricing for the custom market is based on the value of the attribute mix to the customer. Based on the information developed in the conjoint analysis it is possible to create a custom set of attributes that uniquely meet the needs of the customer. By combining the importance weights for each product attribute it is possible to price a product appropriately.

The critical issue is of course to pick the best set of attributes for the customer. This exercise cannot be done effectively with out a constant review of the competitors offers. To be effective, the product offering must provide more value per cost than any alternative offering.

This mix of attribute is absolutely critical; the attribute mix provides the basis of competitive advantage. As with the mass market competitive event pitch is developed and maintained by providing attributes for which you have a cost advantage. This must be done with reference to those attributes that are important to your customers.

5.2   Competitive Advantage in Custom Markets

Development of custom markets is an ongoing effort to identify attributes that provide extraordinary value for the customer and that a supplier can develop with a consistent cost advantage over a competitor. Therefore, competitive advantage can only be gleaned and maintained by an ongoing dialogue between the supplier and their key accounts.

5.2.1   The basis of the data analysis

Competitive advantage can show up in a number of interesting ways in a market where multiple decision makers are involved. This occurs because a critical attribute to one decision maker may be of very little interest and to another decision maker within the decision process of the same key account. Figure 8 provides information developed from a specialty steel manufacturer. This information illustrates both the power of quantitative analysis at this level as well as the importance of understanding all layers of decision-making.

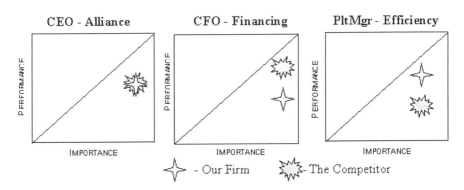

Figure 5. Comparisons of Competitive Offers

Notice that the plant manager has a focus on the efficiency that the equipment provides. The finance manager is focused on the ability to finance equipment improvements. The CEO is interested in developing a corporate alliance.

The reason for these discrepancies is that each manager has a different and unique standard of performance. The basis of these standards of performance can only be understood through qualitative analysis. But the value that each brings can only be fully understood through quantitative analysis.

The plant manager's focus is primarily on output. His budget, staffing, and operations are based on efficiency measures that have as their basis the factory utilization rate. Therefore they are interested in equipment that will allow them to maintain operations during periods of high product demand.

Figure 8 shows the three executives' most important attributes in determining their choice for fuel source on a new furnace. The horizontal axis graphs the attribute importance and the vertical axis the relative performance on these attributes of our product versus our competitor's product. Notice that in two cases, financing and efficiency, there is a significant spread in these attributes, we out perform the competitor in efficiency and they out perform us in financing. In the remaining attribute, there is a dead heat. In all cases, our products perform at a level less than the importance of the attribute to the customer; this means that neither our competitor nor we are meeting the expectation of the customer.

The finance manager's focus is on terms that minimize current period cash requirements. At the point these data were collected the economy was growing and the demand for steel was very high. The chief financial officer, being a good one, was skeptical that this condition would continue. He was interested in developing reasonable returns while he had an opportunity and did not want to pour all his cash out the door.

The CEO had a reasonably long-run view of his firm's performance. In the qualitative interview, he expressed an interest in doing business with firms that had an equally long view. He expressed this in terms of payment requirements during down turns and margins during the up times, expressing a willingness to trade the former for the latter.

Figure 9 presents our strategic opportunity. Here is a case that allows us to win without even adjusting the price. The CEO's version of alliance is one that requires their allies to be sensitive to their downturns with reciprocity on the up turn. By offering a price that hedges commodity steel market price, we ride the swings with them, do not have to change the substance of the finance package and provide the plant manager with a more efficient product. By adding this characteristic we move our delivery on alliance up, as is indicated in the figure, and beat our competition.

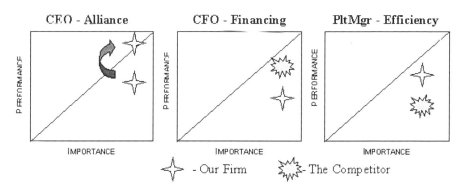

Figure 6. Strategic Opportunity Identified

5.2.2   The role of targeting

Targeting in the mass market allows the firm to develop competencies that are the basis of a competitive advantage. Each energy firm cannot be equally good at developing all product and service attributes. For example, green power may be an issue for some firms. Clearly, the resource base to provide a cost competitive certified green product will not be available to all firms. Emissions credits, cross commodity hedging, weather products, backup power, and capital financing are a part of a long list of products and services large customers may require. The ability of a particular firm to provide the attribute efficiently defines its competitive advantage within the large customer class. Once this competency is defined, those customers that value that set of attributes for which you have a competitive advantage become your targets.

5.2.3   Competition

As with the mass market, competitive advantage cannot be viewed as a static thing. The frontier of innovation is always proceeding, and your competitive advantage may quickly be eroded by newer and more attractive attributes. For this reason, it is important to constantly update your customers' desired attribute set, as well as their view of your competence and that of your competition.

6.   MARKET STRATEGY AND MARKET DATA

Strategy is only meaningful where there is a contention for resources. Marketing strategy is most often found where the resource under contention is market share (though other instances, such as contention for investment capital also occur.) The example given in this chapter will be the contention for share.

Energy firms, with a few obvious exceptions, too often focus simply on the delivery of value. Where market is contested, the field is similar to any competition in theory, if not in the ruthlessness of execution. A useful reference is the ancient Chinese philosopher Sun Tsu, whose work The Art of War is a classic[3] and regularly referenced by market strategists.

Sun Tsu listed a series of strategic options; this example focuses on the first two of his principles. The first is "To subdue the enemy without fighting is the acme of skill." The second is "what is of supreme importance in war is to attack the enemy's strategy" (*The Art of War* by Sun Tzu circa 300 B.C. translated by Gen. S. B. Griffith, U.S.M.C. (ret.), 1963.

Sun Tsu's most important advice is to win a competition by avoiding conflict. A marketing conflict, however completely won in the end, always degrades revenue and earnings.

As discussed in the introduction, it is our belief that price should be used as a tool of strategy, not simply taken passively from the market. In this section, we describe this approach.

6.1   Description of the market

Our hypothetical firm (Our Firm) was the original monopoly provider in a new opening market. Our strategy was to focus mostly on the peak and shoulder markets represented by large commercial, light industrial, and retail resellers. Nonetheless, starting from a position of providing for the whole market our firm had a complete array of resources.

A new firm (The Competitor) entered the market by purchasing a 10-year contract for the output of a large base load generating station. The intent of this new market entrant was to cover their lease by making below market price contracts for base load power. Once the lease cost was covered, they would sell the remaining power bilaterally to commercial customers and to resellers, and in the spot market as opportunity allowed.

This strategy suggested to us that the competitor intended to make its profits and take the peak and shoulder markets we were targeting. In doing so they would have covered their risk in the original base load sales and be

---

[3]   I make this reference with the full, and humble recognition of the aptness of the statement by Roger Staubach in reference to similar analogies made about football. Mr. Staubach said, "I have seen football and I have seen war. Believe me, football is not war." Similarly, I do not mean to imply that marketing is war; however, strategic principles developed in the ultimate competition are useful.

totally unencumbered by profit requirements as they competed for peak and shoulder sales. It is the market strategists' problem to respond to this threat.

6.2   The traditional response

The traditional response to this threat would be to develop estimates of the value of base load power sales and to price the product at that value. The method of doing this is well described elsewhere in this volume.

An alternative approach is to offer a more elaborate product with the characteristics described earlier in this chapter. In this way one can extract more value for bulk sales by adding attractive secondary attributes.

But what does Sun Tzu suggest? His suggestion is that both of these direct methods would cost the company far too much. He suggests an indirect method, a method of attacking the company's strategy.

6.3   What is the nature of the analysis

In analyzing how to price in this market it is important to begin by generating a complete understanding of our strategy and the competitor's strategy.

In our strategy we never anticipated making our living through base load power sales. Base load power was viewed as a basic cost of doing business to retain market share; strategically, we were more interested in peak and shoulder sales. We expected our greatest margins would come from the flexibility of our generation mix. The primary threat from the competitor was that he would take peak and shoulder sales.

The greatest concern of our competitor was to insure that he covered his lease. It was his intent to fatten his return in our market but his primary focus would be the original base load sales.

Therefore, the traditional response is of no value to us. By a traditional contest for his market entry we were bound to lose, he would consistently undersell us and undersell at a large enough profit to meet his basic objective – covering his lease.

6.4   What is the nature of the non-traditional response

The non-traditional response is to attack the competitor's strategy. The key to his strategy is the margin created over cost in his original bulk power sales. We knew that these sales would be under market value. But market value was not the issue; the issue was the cost of his least.

Every kilowatt consumed in the market entrant's effort to meet his lease cost was a kilowatt that could not be sold as shoulder or peak power.

Therefore every mil that we expended in contesting his bulk market was more than compensated in the higher profits available to us in shoulder in peak periods. We therefore decided to contest his original sales as vigorously as possible.

Expending our base resources to contest the market entrant cost us very little. This is true if for no other reason than because the new market entrant could not really afford to lose and because we had knowledge of his base contract, a matter of public record, thus we knew exactly how low he was willing to bid. Therefore, even in contesting these contracts we knew we'd lose very little.

So, by expanding marketing resources and cutting the competitor's potential profit by a factor of two to three we have of cut their ability to compete in our market by 50 to 65%. This was accomplished with very little cost to ourselves in either resources or true loss of our core market.

7. SUMMARY AND CONCLUSIONS

This last example is one of a true strategic use a price. In it we had little to no need for market price studies. In it we have little or no concern for the economic value of the good. Our whole concern was on the strategic intent of our competitor and the value to us of thwarting that intent. We used price as a tool to confound our enemy's strategy with little to no cost to ourselves. This is a very different way of looking at price.

Section III

Demand Response and Product Design

Chapter #10

Price-Responsive Electric Demand
*A National Necessity, Not An Option**

Romkaew Broehm and Peter Fox-Penner
The Brattle Group

1.   INTRODUCTION AND OVERVIEW

California's electricity crisis has ignited a national debate over electricity deregulation and national energy policy. Thousands of words have been written on the role of transmission limitations, load growth, generation ownership, abnormal weather, poor market design, insufficient supply, and shortage of gas, and all of these factors are important. But one of the most important lessons emerging from this calamity clearly is a long-overdue realization that policymakers can no longer ignore the need to make electricity demand responsive to price. Without this critical ingredient, electricity markets will never function properly, electricity costs will be billions of dollars more than necessary, and a new generation of energy-saving technologies will sit on the shelf. Moreover, price-responsive demand can provide relief from the worst electricity price spikes at least as

---

* Associate and Chairman, respectively, *The Brattle Group*, Cambridge, Massachusetts (www.brattlc.com) [email – Romkaew_Broehm@brattle.com]. The opinions expressed herein are the authors' and not necessarily those of *The Brattle Group* or its clients, which include utilities, generating companies, electricity retailers, and independent system operators worldwide. The authors thank Greg Basheda, Severin Borenstein, Darrell Chodorow, Joel Gilbert, Eric Hirst, Art Rosenfeld, Gary Swofford, Joe Wharton, and Jim Wolf. All errors and omissions are the authors.'

A. Faruqui and B.K. Eakin (eds.), Electricity Pricing in Transition, 145-164.
@ 2002 *Kluwer Academic Publishers.*

quickly as new plants can be built, with lower costs and environmental impacts.[1]

Virtually every regulatory agency and economic expert who surveys the power industry agrees that greater demand responsiveness is needed. The Federal Energy Regulatory Commission's November 1, 2000 report on Western power markets noted:

> Demand side is critical element of the market. When consumers can receive price signals and have the ability to respond to those price signals by reducing demand, it reduces the overall cost of electricity in the market and reduces the electric bills of all consumers, not just those that responded with a load reduction.[2]

Another respected expert, University of California Professor Severin Borenstein, recently noted:

> . . .[I]t is . . .remarkable how little flexibility has been accommodated on the demand side of the market. While the technology to meter consumption on an hourly, or even 10-minute, basis is widely available, and has even been installed at many industrial and commercial locations, no electricity market in operation today makes substantial use of real-time pricing, *i.e.*, charges a customer time-varying prices that reflect the time-varying cost of procuring electricity at the wholesale level.[3]

In this paper we explain the reasons why price-responsive demand is critical for electric markets and review the literature on its benefits. We then survey the state of adoption of price-responsive demand programs, including fast-breaking developments in the West, as well as barriers to greater adoption. We conclude with recommendations for federal and state policymakers.

---

[1] We do not intend that our focus on price-responsive demand be seen as a lack of support for other cost-effective efficiency measures. These, too, are needed to address the western power crisis as well as the long-run energy service needs of the global economy.

[2] Staff Report to the Federal Energy Regulatory Commission, *Western Markets and the Causes Of the Summer Price Abnormalities: Part 1 of Staff Report on U.S. Bulk Power Markets*, November 1, 2000, p.48.

[3] Borenstein, Severin; "The Trouble with Electricity Markets (and some solutions)," *Program on Workable Energy Regulation (POWER)*, University of California Energy Institute, January 2001, pp. 2-3.

2. PRICE-RESPONSIVE DEMAND AND REAL-TIME PRICING

The first law of economics is that demand goes down when price goes up. In normal markets, the adjustment to higher prices occurs in two stages. Immediately following a price increase, consumers reduce their use a little bit, depending on their ability to use less of the product or quickly find substitutes. Over a longer period, much greater reductions in demand occur because new suppliers can enter the market and consumers have time to discover new ways to reduce consumption or go without. Moreover, if a good is storable, and its price varies over time, consumers will stockpile when prices are low and buy less when prices seem high.

The demand for power is somewhat different. First, electricity isn't economically storable, so it can't be stockpiled when prices are low. Second, a great deal of our electricity use is considered a real-time necessity – we simply must have power for our refrigerator, our street lamps, and so on, and we have little choice but to pay the going rate. Third, few of us even know how much electricity we use at any given time, or what its price is at that time – we get a single monthly bill that tells us the cost of all of our consumption in all of our electric devices, from our all-but-invisible central air conditioner to the electric pencil sharpener on our desk.

Under these conditions, it is not surprising that when electricity prices go up for just a short period of time – say, for a few hours a day – most consumers have few ways to adjust their use. To change our daily use, we'd have to monitor hourly prices and then decide which appliances we'd turn off (and for how long) if prices got above a certain level.

During the last 70 years, few companies or families went to this trouble. Thus, electricity markets have grown to have almost zero *short-run* adjustment to changing prices. The electricity-using technology in our homes, offices, and factories – indeed virtually the entire electricity-using marketplace – simply is not built to monitor prices and shift use patterns in the short run.[4]

---

[4] In the long run, electricity demand does go down when prices increase for significant periods. This occurs because higher power rates prompt buyers to purchase more energy-efficient appliances when they replace broken ones or add to the installed base. This reduces consumption in response to the price increase, but the typical time required to adjust to a price increase is four or five years. See Dahl, Carol, "A Survey of Energy Demand Elasticities in Support of the Development of the NEMs," Department of Mineral Economics Colorado, prepared for the U.S. Department of Energy, Oct. 19, 1993.

3.   FLAT RATE PRICES MASK THE TRUE HOURLY COST OF POWER

Before we go much further we should discuss just what it means for customers to face a "flat rate" for their power purchases. Our power bills are the sum of the costs of generating power, transmitting it over high-voltage lines, distributing and metering it locally to households. Transmission and distribution costs continue to be regulated in all deregulated power markets and do not vary much over time. So it is only the cost of generation that has been deregulated.

Whether we are talking about a fully regulated electric industry or one that is deregulated, the true marginal costs of producing electricity vary substantially by hour and season, much like the price of a stock varies throughout a trading day. The demand for power increases by as much as 60% over the course of a day in most areas, but the supply of power plants is fixed over the course of a year. As demand increases, successively more expensive plants are called upon to produce power. This feature alone causes production costs to triple or more within the course of a single day. In other words, even for a regulated utility, its cost of producing one additional kilowatt-hour of power varies between a fraction of a cent and perhaps ten cents a kWh, sometimes within the space of a single day.[5] (See Figure 1.)

If this price variability has always been true under regulation, why is it that we pay flat rates for power? Simple: utilities just took the *average* level of their costs over a year and made this their rate. A typical utility might charge five cents a kilowatt-hour for all generation regardless of time used, five cents being the average of all the high-priced and low-priced hours that year. (See Figure 1)

---

[5] It is important to note that the degree of this variability when utilities were regulated was NOT an indication of market power. The fundamentals of electric markets mean that marginal costs will vary considerably and rapidly without any market power whatsoever. However, when generation is deregulated the price of power in each hour is not set by actual costs, but rather by the market. In this case it is possible that some of the variability of prices is due to market power. But if this is the case, the price volatility induced by market power layers atop the natural volatility that has nothing to do with market power.

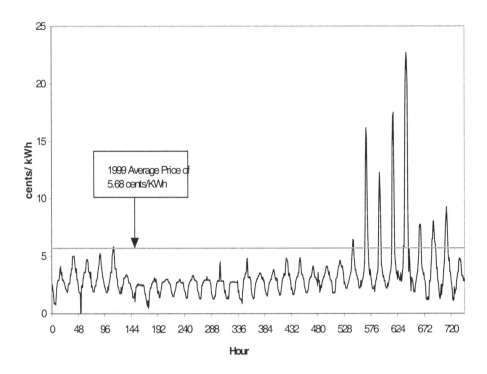

Figure 1. August 1999 Hourly Day-Ahead California PX Prices vs 1999
Average Price

In truth, regulated utilities have long been providing retail consumers a
price smoothing service as part of their retail tariffs. Regulators approved
rates in the form of average prices and generally were not enthusiastic about
utilities charging "real time" or time-varying prices instead of the annual
average.[6] And because consumers never saw varying prices over time, no
one had the desire or incentive to invest in technology that allowed easy
short-term demand adjustment. (For example, a dishwasher with a timer.)

When generation prices are deregulated, the laws of supply and demand
determine the price of power in the marketplace. Of course, a power buyer
may be buying under a long-term contract at a fixed price – much as
regulated utilities smoothed their price. However, all electric markets have
at least some power purchased in a short-term spot market. Prices in this
market are very sensitive to the level of demand, available supply (not

---

[6] It is also the case that many utilities have offered "interruptible" rates for many years.
They are voluntary and for large customers only. These rate schedules are generally flat
rates, but in exchange for the right to turn service off to the customer up to twenty or so
times a year, the utility charges a lower flat rate.

already purchased under contract), and the marginal costs of the available suppliers. It is the market that has exhibited "price spikes," which are the focus in California.[7]

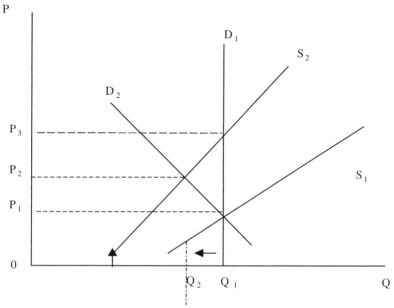

Figure 2. Unresponsive Demand vs. Responsive Demand

Figure 2 shows the enormous benefits that can be achieved if demand adjusts to price. In a particular hour when the demand is unresponsive to price as illustrated by the vertical line D_1, any changes in prices resulting from supply scarcity (shown as the supply curve shifting from S_1 to S_2) will lead to a price spike (from P_1 to P_3). However, if demand can and is allowed to response to higher prices, rational demand will reduce its consumption along the downward sloping demand curve, as illustrated by D_2. Thus, prices will not rise as much. In this case, the energy price will rise to P_2 instead of P_3.

The benefit of bringing demand into play in the electricity market can be enormous. Moreover, as Figure 2 illustrates, the benefits accrue to all users, not just those who reduce their use, since the subset of customers responding to price signals would result in a lower overall average market clearing price. Caves, Eakin, and Faruqui (2000) report that if 10 percent of retail customers can reduce their usage by 5 percent in response to a price rise, market prices will decline from their zero-response levels by 11% to 73%, depending upon

---

[7]   California had several spot markets. One was operated a day in advance by the California PX; the second was operated per hour by the California ISO. The PX markets will be abolished in April 2001, but the ISO's market will remain.

the observed prices. In their simulations, the observed price spikes of $7,500/MWh and $1,000/MWh could have been reduced to $2,171/MWh and $515/MWh, respectively. A study for the U.S. Department of Energy by Science Applications International Corporation (SAIC) estimates a 4.2 to 13.5 percent drop in average spot market prices as a result of load shifting during tight capacity conditions.

In California, a more recent analysis by Braithwait and Faruqui (2001) finds that California electricity customers could reduce summer peak prices by 6-19% via price-responsive demand, reducing statewide power costs by $0.3 to $1.2 billion for the coming summer. Joel Gilbert, a pioneer in promoting demand reduction markets, further estimates that a 10% reduction can reduce transmission system losses by 19% for all customers on the grid.[8]

To summarize, creating price-responsive demand is a win-win for customers who do reduce demand and for those who don't – everyone benefits from lower prices. The benefits, which come in the form of reduced volatility and lower overall power bills, are enormous.

4.   HOW TO CREATE PRICE-RESPONSIVE DEMAND

If price-responsive demand is such a win-win idea, why don't we see more of it?

Regulated utility markets have also developed interruptible and load management programs that mimic actual price-responsive behavior. However, these programs require a utility to monitor power markets continuously and to make a central decision as to whether to interrupt a customer. The customer never sees the varying prices, nor does the customer control the technology that turns their power off for a brief while. In other words, load management programs are a "central planning" alternative to price-responsive demand. They are certainly far better than nothing, but the whole idea of deregulation is to allow individual customers to choose how much they are willing to pay for a specific degree of price volatility or supply risk. By shoe-horning all customers into the same load management program, a utility offers these customers only a single level of price break and degree of interruptibility.

For a number of years, traditional utilities have made limited efforts to spur and take advantage of price-responsive demands. Their main way of doing this was to introduce prices that varied by hour along with their costs.

---

[8] "How Big is Your Business Opportunity?" Joel Gilbert, *The Demand Exchange*, 1999 (www.demx.com).

This pricing approach, which is the cornerstone of price-response demand, is known as "Real-Time Pricing" (RTP).[9] Part of the reason utilities target certain size of customers is that doing so increases utilities' operational efficiency. An RTP program, at the start, requires significant resources and access to data that traditional utilities have retained on only their larger customers. Additionally, large customers can see tangible benefits from their participation. As utilities gain experience with RTP, customers size limitation will dissipate. As an example, Georgia Power's RTP, the largest RTP program in the country, has also recruited customers with peak demand as low as 250 kW. If they could locate customers who were able to reduce their demand quickly – typically industrial customers who were able to shift their production a few hours backwards or forwards – these customers could reduce the utility's cost of producing power during expensive hours. These cost reductions could be shared with the customers making the reductions.

However, RTP is hardly widespread. There are two simple reasons: First, even in deregulated markets the majority of customers are still insulated from the true, highly volatile market prices because the "standard offer" or "default" service options offered in deregulated markets are mostly flat rate. Second, competitive energy service providers have found little interest in price-responsive programs among their customers.[10] Thus, we have another "chicken and egg" conundrum. Customers have no incentives to adopt any price-responsive programs if their power cost is one flat rate or an after-the-fact rate. Concurrently, energy service providers have little incentive to introduce price-responsive programs if they perceive that their

---

[9]   As of 1998, existing RTP programs include Wisconsin Electric, Virginia Power, Cinergy, AEP (Columbus and CSW), Georgia Power, Alabama Power, Oglethorpe Power, TVA, Duke Power, SCE&G, FP&L, Gulf Power, NIPS, Wabash, IP&L, BG&E, PP&L, Indiana Michigan Power, PG&E, SCE, SDG&E, SMUD, Empire District, OG&E, KCP&L, PSOK, Consolidated Edison, Commonwealth Edison, Illinois Power, HL&P, TU, West Texas Utility, Otter Tail Power, and Gulf States Utility. Most of these programs are either under experiment or pilot programs, most of RTP programs are a longer notice RTP, *i.e.,* 24-hour "forward" price. A utility/energy service provider can offer a shorter notice RTP such as an hour-ahead price if its Power Exchange has an hour-ahead market. Some utilities in New York, for instance, provide the energy service to residential and small business customers at an after-fact monthly weighted average price of New York Power Pool's Power Exchange spot prices. Since the late 1980s many traditional utilities have developed one form or another of these price responsive programs or RTP programs, to broaden their customers' service choices and to lower customer bills. However, many utilities RTP programs limit the number and the size of participants, in part to simply gaining regulatory approvals. Therefore, the impact of the aggregate RTP programs on reducing regional peak demand are minimal.

[10]   One of our favorite examples is a recent text, "Shopping Centers and Other Retail Properties" (J. White and K. Gray, Wiley & Sons, 1999). Large retail buildings are excellent candidates for price-responsive technologies. Nevertheless, in several hundred pages the text devotes a total of one sentence to energy-savings technologies.

customers are unable to respond to changes in power prices and therefore uninterested in getting price signals that vary by hour.

Fortunately, times are changing. It is now trivial to send electricity customers hourly price signals over the internet, and many power traders look at hourly prices all day long. The much more difficult element is getting hardware installed that will allow a decent segment of customers to adjust their demand easily, and to be credited their avoided costs on their power bill. Thus, the barriers to price-responsive demand no longer include the invisibility of the real-time price. Instead, the roadblocks lie in metering, billing, and our electric-consuming hardware. We call these the three missing ingredients of price-responsive demand.

5.   LARGE-SCALE PROGRAMS FARTHEST ALONG

Thus far the greatest progress in price-responsiveness has come in the form of "demand-side bidding" features of deregulated power exchanges. In these programs, competitive electricity providers look around for large customers who are willing to reduce large chunks of load when prices reach a prespecified level, in exchange for the exact hourly savings they create by not using the power during such expensive periods. For example, some large factories are willing to shut down their equipment for one shift rather than pay very high electricity prices.

When competitive providers have bundled up enough of these customers' offers to reduce load, they in turn make this offer to a power market that usually takes offers for supplies. Instead of offering supply, however, these retailers are making an offer to reduce demand, hence the name demand bidding.

This type of demand-side bidding is present in nearly every electric market. Table 1 shows that, in the U.S., every formal regional power market allows some form of demand bidding. (Appendix A describes the programs in more detail.) The Federal Energy Regulatory Commission has urged all Independent System Operators and Regional Transmission Organizations to expand these mechanisms, and in some cases it has directed participants to create demand bidding programs.

These large-scale programs are highly worthwhile if they are properly designed and executed. The most likely reason why these large-scale programs are limited is that the amount of this kind of demand response has thus far been limited to the few large customers who are able to adjust their demand in large enough blocks to make it worthwhile. In other words, of the three missing ingredients, it is the technical inability of electric consumers to adjust that is most constraining.

Table 1. Demand Bidding Programs

Market	Demand Bidding Program	Most Recent Response
California Power Markets	California ISO took bids for Summer 2000 demand reduction on a trial basis. Voluntary Load Curtailment Program for Summer 2001.	180 MW bid for Summer 2000
ISO-New England	November 2000 - pilot program using internet-based Load Management Dispatcher and will extend for Summer 2001.	250MW signed up for the summer 2000 program. 500MW for the November 2000 to March 2001 program.
New York ISO	Proposes Emergency Demand Response Program and Day-Ahead Economic Load Curtailment Program and to be effective May 1, 2001.	NA
PJM	Proposes PJM Economic Load Response Pilot Program that participants submit their expected load reduction via the internet.	NA

6.   SMART BUILDINGS AND PRICE-RESPONSIVE DEMAND

The vast majority of load is not at the fingertips of someone who constantly watches power prices. Most lighting, heating and cooling (HVAC) systems in buildings, office electronics, factory processes, and refrigeration do not have automated price response capability. Such systems require significant human effort, and a set of guidelines to turn equipment and fixtures on and off, or to reschedule production. As an example, the Dayton-Hudson Corporation (parent company of Target department stores), has decided not to control its load in response to price signals until automatic price-responsive technologies are more widely available.

Over the past two years a number of suppliers have developed communications and energy control technologies so that RTP customers can adjust their energy consumption automatically with short-term notice. The technologies have become easier to use, more sophisticated, and less expensive. The more sophisticated equipment receives real-time prices and transmits them electronically to computers in customers' buildings, which modify building operation in response to projected prices. These "smart building" products provide continuous load forecasting and equipment scheduling based on predefined pricing thresholds and RTP pricing information. "Smart building" equipment vendors include Honeywell's Building Automation System (BAS), Johnson Controls's RTP control solutions and Metasys EMS, and SAIC's ADEPT. Equipment is even available for homes–Carrier Corporation and EMW sell internet-addressable thermostats for residential use.

Actual Daily Reduction in Energy Use and Cost in Hotel RTP Pilot

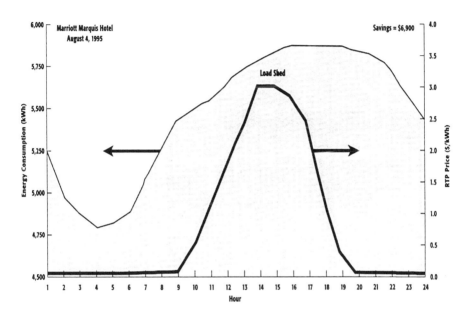

As real-time prices rose from under 10 cents/kWh to a high of 300 cents/kWh, the RTP Control System reduced loads below what they could have been without RTP. The result was less use during high-cost periods and a savings of $6,900 in electricity costs for the hotel this day.

Figure 3.

The impact of this "smart building" technology on customer costs can be stunning. Figure 3 shows the actual daily electricity usage pattern on August 4, 1995 of a large hotel in Manhattan during a 1995 RTP pilot project. When real-time prices rose from under 10 cents/kWh to a high of 300 cents/kWh, the RTP Control System reduced loads below what they could have been without RTP. The hotel was able to save about $6,900 in electricity costs on that day. Another example, West Bend Mutual Insurance, one of Wisconsin Electric Power Company's RTP customers, saved about $3,500 last year in its energy costs, and expects higher savings of nearly $10,000 this year (about 5 to 8 percent of the company's total energy bill).[11]

7.   THE MARKETS AND GOVERNMENT RESPOND

With savings like this, why aren't smart building systems in greater use?

We've already noted that the primary barrier is the chicken-and-egg problem of customers not wanting to invest in smart control technologies without time-varying savings choices and power sellers not willing to offer this product until customers will buy it. There are recent signs that both market and government are attacking these barriers, but much more can be done.

First, electricity prices have gotten sufficiently high and volatile as to induce electricity retailers to install real-time technology and act as market-makers between retail customers they physically control and the spot market. AES NewEnergy (AES), an energy service provider, has incorporated this automated RTP equipment into its wholesale trading practice.[12] AES signed an exclusive agreement with Powerweb Technologies for Omni-Link Internet energy platform. This platform links AES' s industrial and commercial customers to its real-time energy trading floor. When a power spot price is high, AES can automatically start up customers' distributed generation equipment and implement pre-planned demand-side management strategies. The excess energy will immediately be resold into the market at the spot price. Another sophisticated retailer, Enron Energy Services, envisions using its real-time control of commercial buildings to turn them into "virtual peaking power plants," adding power to the grid at times when it is most needed.

Regrettable as it may be, the western U.S. electric crisis has helped accelerate awareness of the importance of price-responsive demand. The California Energy Commission has accelerated a pilot program for "real-time rebates."[13] California utilities are looking at large real-time metering installations, and Assembly Bill 1388, now pending, provides $85 million for implementing price-responsive programs.

Other energy service providers and utilities are following suit by promoting RTP or at least installing advanced metering that enable RTP. For example, Puget Sound Energy has wired over 400,000 customers with

---

[11] ESource, *From RTP to Dynamic Buying Communication, Analysis, and Control Tools for Managing Risk,* EIC-7, October 1999

[12] *Power Markets Week*, September 18, 2000, p. 5.

[13] "Demand Responsive Load Management," Art Rosenfeld, Commissioner, California Energy Commission, November 9, 2000. (www.energy.ca.gov).

meters that allow a home to view its own electric use on its PC in near real-time. The program recently won the Washington Utilities and Transportation Commission approval to charge higher prices during peak periods in Summer 2001.

8.   POLICY RECOMMENDATIONS

Although equipment vendors can make smart building technology available, all new products take time to diffuse into the market. Building designers are not familiar with price-responsive systems and are reluctant to change to a substantially more complex system that costs more to install. By some estimates, it will take another century to turn over our existing commercial building stock. Under these conditions carefully targeted federal and state programs could do much to accelerate the greater adoption of smart building technologies in the marketplace.

There are several areas where government can assist the market without over-subsidizing price-responsive technology. Foremost among them are policies that require electricity sellers- -public utility or private- -to offer real-time pricing as an option. There are even good reasons to mandate that all large commercial and industrial customers face time-varying prices,[14] but there can be no excuse for failing to make the option available.

Utility distribution companies, which remain regulated or customer-owned, should also contribute by installing and recovering rates metering that allows for RTP.[15] The Penn Future Group, a group of pro-market ex-regulators from Pennsylvania, argues that:

> . . . [C]ustomers must be given the ability to change their demand in response to price in real times. As a result, customers must have time-of-use meters and control devices installed on appliances like air conditioners and water heaters. Utilities should be required to install this equipment over the next five to 10 years. Only with this equipment can customers gain financial incentives for changing demand.[16]

---

[14] See Borenstein (2000) for a discussion of the advantages of mandatory RTP over voluntary programs. In brief, Borenstein notes that voluntary programs will incentivize companies that might install RTP even without incentives, while failing to induce those who most need the encouragement to act.

[15] Whether such metering should be installed and operated as a competitive service is a controversial topic beyond our scope. Regardless of who does the work, it is critical that this get done for large customers.

[16] "Survivor: Electric Competition," John Hanger, John Rohrbach and Peter Adels, Penn Future[3], January 18, 2001, Vol.3, No. 2, p. 3.

Another appropriate role for government is demonstration and information diffusion. The U.S. Department of Energy (DOE) and its state counterparts should immediately find or create buildings that demonstrate RTP (and other) technologies and educate property developers, building designers, and building owners as to the benefits of smart buildings.

While the market-wide benefits of RTP are understood, there is a shocking dearth of research on the specific costs and benefits of demand-adjustment technologies for those who adopt them. Efforts like these can accelerate market acceptance of new technologies without favoring one vendor or platform or slowing innovation. DOE should consider hosting a summit on smart building technologies, bringing together leading building technology vendors, developer designers, energy retailers, and utilities, and following this up with sustained support.

Finally, state and federal legislators should consider targeted tax credits for RTP equipment retrofits and new installations. As with all tax incentives, the recipient and level of the incentive must be determined carefully. Nonetheless, with tax legislation joining energy prices at the forefront of our national political dialog, measures at the intersection of these two issues deserve serious attention.

It has never quite been acknowledged, but competitive electric markets have been living on borrowed time. Markets for nonstorable necessities without price-responsive demand meet neither the test of economic efficiency nor political acceptability. In California, the clock ran out, but this need not be the case in the rest of the nation if we adopt wise, aggressive policies to promote intelligent use of electric power- -our scarcest and most valuable form of energy.

REFERENCES

Borenstein, Severin. 2001. "The Trouble with Electricity Markets (and some solutions)," *Program on Workable Energy Regulation (POWER)*, University of California Energy Institute, January 2001, pp. 2-3.

Branscomb, Lewis M. 1993. *Empowering Technology,* Cambridge, MA.: The MIT Press

Braithwait, Steve and Ahmad Faruqui. 2001. "The Choice Not to Buy: Energy Savings and Policy Alternatives for Demand Response." *Public Utilities Fortnightly* 239 (No. 6, March): 48- 60.

Brown, Marilyn A., and C. Robert Wilson. 1992. "Promoting the commercialization of energy innovations: An evaluation of the energy-related inventions program." *Policy Studies Journal* 20-1: 87-101.

Cohen, Linda R., Roger G. Noll, Jeffrey S. Banks, Susan A. Edelman, and William M. Pegram. 1991. *The Technology Pork Barrel,* The Brookings Institution, Washington, D.C.

Caves, M., K. Eakin and A. Faruqui. 2000. "Mitigating Price Spikes in Wholesale Markets through Market-Based Pricing in Retail Markets." *The Electricity Journal,* April.

Dahl, Carol. 1993. *A Survey of Energy Demand Elasticities in Support of the Development of the NEMS.* U.S. Department of Energy, October 19, 1993.

Geller, Howard, and Scott McGaraghan 1998. "Successful government-industry partnership: The U.S. Department of Energy's role in advancing energy-efficient technologies,' *Energy Policy* 26-3: 166-177.

Hirst, Eric and Brendan Kirby. 2000. *Retail-Load Participation in Competitive Wholesale Electricity Markets,* Edison Electric Institute and Federal Energy Regulatory Policy, December.

Kocagil, Ahmet E. 1997. "Portfolio choice of government incentives: The case of commercialization of a new coal-based technology." *Energy Policy.* 25-10: 887-896.

Neij, Lena. 1997. "Use of experience curves to analyze the prospects for diffusion and adoption of renewable energy technology." *Energy Policy* 12-13: 1099-1107.

ESource, *From RTP to Dynamic Buying Communication, Analysis, and Control Tools for Managing Risk,* EIC-7, October 1999.

Newall, Richard G., Adam B. Jaffe and Robert N. Stavins. 1998. "The induced innovation hypothesis and energy-saving technological change." *Quarterly Journal of Economics,* October: 941-975.

Powell, Jeanne W., *Development, Commercialization, and Diffusion of Enabling Technologies, Progress Report for Projects Funded 1993-1995,* National Institute of Standards and Technology, NISTIR 6098, December 1997.

Schmalensee, Richard. 1980. "Appropriate government policy toward commercialization of new energy supply technologies," *The Energy Journal* 1 (No. 2, April): 1-40.

APPENDIX A: LARGE-SCALE DEMAND-RESPONSE PROGRAMS IN THE U.S.

ISO New England Load Response Program[17]

Because of its experience during the summer of 1999, ISO New England, Inc. has viewed a load response program as necessary during emergency conditions.   With   the   Federal   Energy   Regulatory   Commission's authorization, ISO New England implemented a load response program during the summer of 2000 (June 1 through September 30, 2000). Under the summer 2000 program, load-serving and distribution entities in New England Power Pool (NEPOOL) entered into agreements with retail customers that have hourly metering equipment or are likely sources of interruptible loads.   Retail customers that voluntarily reduce their load during capacity shortages would be paid $8/kW times their customers' average hourly load reduction over the duration of the interruption. More than 900 retail customers with a total load of 253 MW signed up for this program.

Encouraged by the success of this program, ISO New England is running another load response program from November 2000 to March 2001. It is soliciting a total of 500 MW of load reduction from 10- to 15-MW industrial and commercial customers. Using the Internet-based Load Management Dispatcher (LMD), ISO New England enables to rapidly interrupt power blocks.   LMD tracks NEPOOL's market prices and provides automated notification to the energy service or load provider, or customer, when predetermined opportunities become available in the market.   Once the interruption is completed, LMD collects end-use customer usage data via a meter data recorder installed at the customer's site.   LMD then provides customer-specific hourly performance data that are matched with the hourly clearing price. This information will allow the ISO to properly credit the customer's power bill and gather metered data on the amount of interrupted MWs that can be measured against the ten-minute non-spinning reserve requirement. The ISO New England planning study reports that savings from a 500-MW load reduction could run as high as $30 million per year and reductions in air pollution could amount to 280 tons of NO_x, 200 tons of SO_2, and 230 tons of CO_2.   ISO New England plans to implement the program among a larger group if it proves to be successful.

---

[17] Sources: 1) ISO New England, "First-Ever "Click On" Load Response Program Set to Begin in New England – Pilot Program To Run Through Winter 2000-2001," *ISO New England Press Release*, October 31, 2000. 2) Market Operations ISO-NE, *Revised Concepts of the Load Response Program*, January 26, 2001. 3) Restructuring Today, *NEPOOL Votes Plan to Interrupt Big*, March 5, 2001, p. 4.

The ISO also plans to implement two types of load response program for the summer 2001. There are Type 6 (Interruptible Loads) Class 1 and Type 6 Class 2. Type 6 Class 1 is expected to respond to the ISO 30-minute advanced notice for interruption during the hours of 8.00 a.m. and 6.00 p.m. weekdays and non-holidays. Participants will receive availability payment based upon the 30-minute operating reserve market-clearing price. If the interruptible load has been called upon, Class 1 will receive credit for any energy interruption at market clearing prices. Class 2 is the purely price responsive load. When the price of power hits $100/MWh, participants can opt to cut off power and get credit at market clearing prices.

PJM Customer Load Reduction Pilot Program[18]

PJM has had a load response program termed Active Load Management (ALM) in place since 1991 – long before it became a FERC-approved ISO. Through this program, load-serving entities receive capacity credits, which reduce the utilities' costs of installed generating capacity, for load reduction that is under PJM control during an emergency. There are significant monetary penalties if load reductions requested by PJM are not implemented in full by program participants. The program was activated about six times during the summer of 1999. In addition to ALM, PJM initiated a customer load reduction pilot program during the summer of 2000. It targets existing on-site generation and load management at facilities such as hospitals, hotels, factories, and stores during emergency conditions. There are currently 35 participants with a total of 62 MW load reduction. The smallest of these generators represents 200 kW and the largest represents 15 MW. The program design permitted PJM to call on these interruptible loads during emergency conditions before implementing the ALM program. PJM pays the load-serving entities the higher of the appropriate zonal locational marginal price of energy or $500/MWh.

PJM also plans to provide the load response pilot program from June 1, 2001 through May 31, 2002. The program proposes to divide the pilot program into two options. The first option is designed to provide a method by which end-use customers may be compensated for reducing load in an emergency. Participants must be able to reduce at least 100 kW of load during 9.00 a.m. and 10.00 p.m. with one hour advanced notice. The payment will be the higher of the appropriate zonal locational marginal price of energy or $500/MWh. The second option, the PJM Economic Load Response Pilot Program, will provide a mechanism by which any qualified market participant may offer end-use customers the opportunity to reduce

---

[18] PJM Distributed Generation User Groups.

their load during PJM's high price periods. Under the second option, a web page will be created so that market participants may submit their expected amount of load reduction on a zonal basis. The PJM system operator will compile daily aggregate load reductions on a zonal basis for use in operations. Thus, it is possible that load reductions may set a price on the PJM system.

New York ISO Emergency Demand Response Program[19]

The New York ISO has proposed similar programs at the FERC: "Emergency Demand Response Program" and Day-Ahead Economic Load Curtailment Program. Both programs if approved will run from May 1, 2001 through October 31, 2002. The NY ISO will implement the program as one of its emergency procedures in response to an in-day peak hour forecast response to an operating reserve shortage. The emergency demand response program pays the higher of $500/MWh or the real-time zonal locational-based marginal price per MWh for verified load reductions for the first two hours and the real-time zonal price for the second two hours. Participants must be able to demonstrate that at least 100 kW of their loads can be curtailed within two hours advance notice. However, there will no penalties for failure to provide load reduction.

For the Day-Ahead Demand Reduction program, participants in this program can bid a specific MW curtailment (in minimum increments of 1 MW by zone) by strips of one or more hours but not exceed eight hours. The payment will be the higher of the Demand Reduction Bid or Day-Ahead locational-based marginal price. Participants will provide Customer Baseline Load (CBL) as a reference to verify its compliance with a scheduled curtailment. The CBL is based upon the five highest energy consumption levels in comparable time periods over the past ten days. Penalties may be applied for the program participants that fail to respond.

---

[19] New York ISO, Transmittal Letters to Implement Emergency Demand Response Program submitted to the Federal Energy Regulatory Commission, March 13, 2001. New York ISO, Transmittal Letters to Implement an Incentivized Day-Ahead Economic Load Curtailment Program, April 5, 2001.

The California ISO Demand Relief Program and Voluntarily Load Curtailment Program[20]

The California ISO developed two major types of Demand Response Programs for their emergency needs. One is called the Demand Relief Program and the other is called the Voluntary Load Curtailment Program.

Demand Relief Program

Like other ISOs, the California ISO operated a Demand Relief Program during summer 2000 as a pilot program, in which it sought 1,000 MW of participating loads, each of which must be greater than 1 MW. This programs pays for both capacity reserved (at about $50/kW-year) and actual energy reductions ($/MWh). Participants in the summer 2000 load response program were to be called upon when the ISO declared a Stage 1 emergency, *i.e.*, operating reserves are expected to fall below the required 7% of daily peak demand. The program was invoked 20 times during June-September 2000.

For summer 2001 (June 1 through September 30), the California ISO plans to implement its Demand Relief Program which hopes to supplement existing interruptible programs that have already been put in place by utilities. This program proposes to attract load participants that can tolerate the lower expected frequency of calls just prior to a Stage 3 emergency, *i.e.*, rotating back-outs. The program will consist of a two-tier load reduction: 1) loads without back-up generators, and 2) loads with back-up generators. The loads with back-up generators will be subject to a maximum of 7 calls during the summer with 0-3 hours for each call. The loads without back-up generators will be put into two blocks, with a specified 4-hour curtailment, of up to 24 hours per month.

Voluntary Load Curtailment Program

The California ISO plans to operate the Voluntary Load Curtailment program between June 2001 and September 2001. The program is designed to attract smaller loads and to operate through a voluntary, non-emergency program. Participants will sign up with aggregators, but can choose whether and when they are curtailed on a daily basis. Notifications will be sent to aggregators when the ISO anticipates a need to make a bilateral block energy procurement. Aggregators will have 60-90 minutes to issue their voluntary curtailment notices and to confirm the actual block of curtailment available

---

[20] California ISO, *Summer 2001 Preparedness Update and Demand Response Programs*, November 17, 2000.

to the ISO. The ISO proposes to set the curtailment payment equal to the current ISO price cap, $250/MWh.

While this kind of program is encouraging and highly useful, the results to date have been somewhat limited. For example, the California ISO solicited bids for 1,000 MW of demand reduction last summer – about 2% of their peak load. They received only a minuscule 180 MW of bids B not nearly enough to make a dent in prices or supplies B and this at a average price of about 24 cents per kilowatt-hour saved, or about 2.5 times the average price of generation in California. ISO-New England also has a pilot program going, with very sophisticated internet-based controls, but again it is for a mere 10-15 MW of load, or about .05% of the summer peak.

Chapter #11

Price-Responsive Load Among Mass-Market Customers

Daniel M. Violette
Summit Blue Consulting

Abstract: Mass-market customers are able to respond real-time pricing and other
             innovative rates. In fact, the technology available to mass-market customers
             and suppliers make these customers a unique component of a balanced
             portfolio of demand commitments and supply resources. This chapter
             illustrates the ways in which mass-market customers can respond to price
             signals and the unique attributes of these offers in helping to manage price and
             quantity risks. It then develops the case that any supplier serving these
             markets, whether it is a utility serving mass as a default provider or a
             competitive energy provider, will need to take advantage of the risk
             management attributes of demand-response offers. As a result, the relevant
             question for energy companies serving mass-market customers is not whether
             to develop offers that incorporate price-response and demand response
             attributes, but rather how to best design these offers and what portion of their
             load portfolio should be covered by these offers.

Key words: Direct Load Control (DLC), load management, retail markets, demand
             response, innovative rates

1.   INTRODUCTION

In the quest to develop competitive markets for electricity, it is important
that mass-market customers be included in the development of price-
responsive offerings. Economic efficiency and competitive considerations
call for more innovative demand response offerings to be made to mass-
market customers. These offerings will allow suppliers to manage price and

A. Faruqui and B.K. Eakin (eds.), Electricity Pricing in Transition, 165-180.
@ 2002 *Kluwer Academic Publishers.*

quantity risks associated with serving these customers and this will be a critical determinant for winning suppliers in these markets. In this chapter, mass-market customers are residential customers and small commercial customers (i.e., those commercial customers with roughly a 500 kW connected load or less).

This chapter illustrates the ways in which mass-market customers can respond to price signals and the unique attributes of these offers in helping to manage price and quantity risks. It then develops the case that any supplier serving these markets, whether it is a utility serving mass markets as a default provider or a competitive energy provider, will need to take advantage of the risk management attributes of demand-response offers. As a result, the relevant decision for energy companies serving mass-market customers is not whether to develop offers that incorporate price-response and demand-response attributes, but rather how to best design these offers and what portion of their load portfolio should be covered by these offers.

2.   BACKGROUND

A select number of states have restructured their electric markets to offer retail choice and there has been an increase in wholesale electric competition in all regions, but it can be argued that industry restructuring and the transition to competition have actually reduced the customer's ability to respond to prices. Reasons for this reduced demand response include the following:

1. The continuation of non-market, flat electricity prices as part of standard offer service available to customers, who do not choose an alternative electric provider.
2. A freeze on electricity tariffs, including a freeze on tariffed load management programs in which new participants are not allowed to enroll resulting in a decline in the number of participants.[1]
3. Facing standard offer prices that fixed, competitive electricity providers have not stepped in with innovative rate offerings with load management and/or peak period pricing.
4. Most states have adopted load profiling for mass-market customers to avoid having to install expensive interval meters.[2]

---

[1] In California, customers participating in Southern California Edison's residential direct load control program declined from over 400,000 customers in 1992 to approximately 200,000 today. PG&E actually did away with their mass-market load program entirely.

[2] A supplier serving mass-market customers has its hourly demands for wholesale market settlements calculated by using a load shape (i.e., a profile) that is fixed for a rate class of customers.   Since meters on mass-market customers only measure total monthly

It is important to understand the settlements process and the role of load profiling to understand the options and incentives for developing demand-response offers. As long as a customer remains on a set load profile, the supplier does not benefit from any change in consumption that the customer makes. For example, if a customer shifts consumption from periods with high price periods to low price periods, the supplier will still be allocated the same demand for that customer by the load profile for settlements in the wholesale market. In general, the only way to move a customer off of the load profiling is to install a revenue quality interval meter. Load profiling is used by every state that has implemented restructuring.[3]

The problem of load profiling is as acute for incumbent utility providers as it is for competitive providers. Every provider has its aggregated hourly loads for non-interval metered customers based on load profiles. As long as a customer remains on a specific load profile, no supplier will have an incentive to offer a demand-response program to its mass-market customers. Developing cost-effective solutions to this metering and measurement problem is a necessary condition for offering any type of demand-response program. The appropriate way to address this problem will vary by type of offer or program, but as each demand-response offer requires a measure of the customers' response, adjusting load profiles for customer response should not be a limiting factor.[4]

For mass-market customers, price-response options have been further limited due to a bifurcation of incentives for load management between distribution companies and energy suppliers. Distribution companies have the infrastructure to deploy mass-market demand response initiatives, but it is not clear if they have the incentive required to incur the costs. For example, suppose a distribution company invests in mass-market load management technology only to have some of the customers on the program switch to an alternate provider at some time in the future. Would the distribution company lose the demand response benefits from having these

---

consumption, the customers monthly kWh consumption is allocated to each hour of the month using this fixed profile and the resulting hourly demands are aggregated to determine each supplier's hourly demand for financial settlements in wholesale markets.

[3] Load profiling has been adopted by states to help ensure that mass-market customers will have a choice of suppliers. If a competitive electricity provider had to incur the costs of installing an interval meter for each mass-market customer that switched, it would not be economical to serve any of these customers.

[4] Approaches to address the issue of load profiles include developing estimates of load response from a sample of customers on a specific pricing offer or load management program and adjusting the load profile for these estimated changes. However, these methods must be approved for use in calculating wholesale market settlements. A discussion of the likely accuracy of these methods for mass-market customers compared to larger customer programs is presented in the next section.

customers as participants in the program? This can be addressed by having a contract between the distribution company and participating customers that continues to allow the distribution company to implement load management and obtain credit for the capacity reductions even if another company is providing energy to that customer.[5]

3.   MASS-MARKET OFFERS AND DEMAND RESPONSE

The problem of inadequate demand response is beginning to receive attention; however the focus has been on designing offerings for large customers. This focus on large customers has been due to the following:

1. A need in some regions to put demand-response programs in place quickly and a belief that large customer programs can be instituted more quickly.
2. The belief that only large customers with considerable MW curtailment potential can afford to install the metering equipment required to verify load curtailments or respond to real-time prices.
3. A concern that demand-response (e.g., curtailed load) in mass-market programs cannot be verified to the same level of accuracy as large customer programs.

Of these three reasons, only the first has merit. It may take longer to rollout a mass-market program targeting ten thousand customers than a program focused on the fifty largest customers in a region. However, mass-market programs attaining sizeable reductions (e.g., 15-20 MW) can be rolled out in a matter of months,[6] and as shown in Table 1, a number of mass-market programs put in place at utilities account for over 200 MWs of demand-responsive load.

---

[5] Baltimore Gas & Electric has addressed this issue with the Maryland PUC as of summer 2001. Baltimore Gas & Electric has approximately 300,000 residential customers on a direct load control program for air conditioners and electric water heaters. If one of these customers switches to an alternate energy provider, the peak load reduction credits will stay with Baltimore Gas & Electric. This provides the distribution company with the incentive it needs to continue to market and expand this load management offering.

[6] The Long Island Power Authority initiated a program starting up in March of this year designed to achieve 20 MW of demand reduction in a period of months (LIPA, 2001).

Table 1. Existing Mass-Market Load Management Programs

Number of Customers	MW under Control
Southern California Edison	200,000 customers (res. and ag.) with 280 MW
GPU Energy	80,000 customers with 80 MW
Baltimore Gas & Electric	300,000 customers (res. and comm.) with 350 MW
Pepco	150,000 customers with 200 MW
Florida Power & Light	600,000 customers with 700 MW
Florida Power	470,000 customers with 470 MW
ComEd	68,000 customers with 80 MW
Northern States Power	250,000 customers with 250 MW
(Source: Phone conversations with the respective companies)	

Mass-market programs have attributes that are not available from large customer programs — namely dispatchability and reliability. These attributes include the following:

– Dispatchability. Mass-market programs can provide demand relief within one minute after being called via the use of a paging communications system, which directly contacts the customers' equipment. In fact, the California ISO acknowledged seeing 200 MW of demand drop off the system within one minute of calling on Southern California Edison to implement its mass-market direct load control program.[7]
– Reliability. It can be argued that mass-market programs are more reliable due to the diversity associated with having a large number of participating customers and having load shapes that, in the aggregate, are more consistent across potential peak days.

4. VERIFYING DELIVERY

One important issue concerns whether verification/evaluation of demand response is inherently more easily addressed for large customer programs versus mass-market programs. This issue comes down to determining whether the ability to place interval meters on all participants in a large customer program necessarily makes that program more reliable and verifiable than a mass-market program.

In response to the May 17, 2000 FERC ruling requiring ISOs to develop demand response programs, the California ISO, the New York ISO, ISO New England and the PJM ISO have proposed demand response programs. At the present time, ISOs and related reliability organizations (i.e., RTOs) have been on the cutting edge of developing demand response initiatives with state regulatory agencies largely content to stay on the sidelines. Each ISO has proposed programs designed for large customer participation, and to

---

[7] Communication with Richard Cromie, Southern California Edison, October 2000.

their credit, each has developed innovative approaches for the participation of large customers. However, each ISO has also adopted verification rules that effectively eliminate the participation by mass-market customers and programs targeted at mass-market customers.

The California ISO (CAISO, 2001) Demand Relief Program calls for revenue quality meters to be installed at each participating site. The compliance calculation takes the metered MW demand during the control period and compares this to a baseline load profile that is calculated from the "immediate preceding 10 Business Days." The difference between the control period profile and the baseline load profile for each large customer is the demand response used by the CAISO in settlements. The requirement that each participant in the program have revenue-quality interval meters installed as part of the program makes mass-market programs uneconomical since mass-market programs often have many thousands of participants. Programs at other ISOs have essentially the same requirements.

The design of these verification protocols overlooks the fact that accurate values of demand response are less dependent on the accuracy with which control period loads are measured via interval meters, and more dependent on the accuracy with which the baseline load profile is estimated. The accuracy and integrity of verification is principally dependent on the assumed baseline load profile. For the CAISO demand relief program, the baseline load profile represents what each participating customer would have used if it had not controlled demand as part of the demand relief program. The difference between the estimated load profile for each customer and what that customer actually would have used on that day largely determines the accuracy of the estimated demand response for each large customer.

The variability in day-to-day and hour-to-hour load profiles will be the factor that most significantly influences the accuracy of the estimated demand response. In general, many large customers have discrete production processes that determine their daily and hourly load profiles. If a given process is not scheduled for a "control day" or for those hours subject to control, then the baseline load profile will not be an accurate indicator of what that large customer would have used, had a control day not been called.

The ability to accurately verify curtailed load will depend on the ability of the assumed baseline to accurately predict what the load would have been during a curtailment period. In contrast to large customers, the duty cycles of air conditioning and other equipment under control can be expected to exhibit less day-to-day and hour-to-hour variability on peak days. This is particularly true for small commercial customers that increasingly are becoming the focus of mass-market demand response programs since their businesses are generally open every day and maintain regular schedules. As a result, the baseline load profile for mass-market programs is likely to be estimated more accurately and the overall demand response estimated using this baseline is likely to be more accurate. This is likely to be true even if the mass-market program uses a sample of participants with interval meters to estimate current consumption as opposed to large customer programs in which every participant has an interval meter installed.

5. BUSINESS CASE FOR MASS-MARKET DEMAND RESPONSE

The business case for mass-market demand response is comprised of the following elements and are also summarized in Table 2.

1. High priced peak loads can be met more cost-effectively through mass-market load curtailment programs than by the construction of peaking power plants. Xcel Energy, in a filing before the Colorado PUC, estimated the costs of their residential load curtailment program at $250/kW and the costs of a combustion turbine at $450/kW.[8] Analysis conducted of load management programs in the ECAR region showed real levelized costs as $40/kW-yr for load management, $72/kW-yr for a combustion turbine, and $93/kW-yr for a combined cycle unit.[9]
2. The curtailable load represents a "call option" that provides a physical hedge against price spikes and supply risks related to transmission constraints or other delivery risks. Unlike financial hedges that typically require the purchase of a monthly block forward contract with a 16-hour peak period for every weekday of the month, these call options represent flexible supra-peak hedges since they can be exercised only on the days needed and for the hours in which high prices occur. This flexibility makes them more valuable than forward contracts that can be obtained in markets operated by NYMEX or by the California Power Exchange.[10]
3. Time is required to build up adequate participation in a load management program. This can be a potential hurdle, which competitors will have to overcome if they are to enter the retail market.
4. Load management can be viewed as the first step toward a mass-market demand response offering that includes distributed generation (DG). The communications infrastructure designed to switch energy-using equipment off can also be used to turn on an on-site generator. Applications of DG at large residences and small business are expected to

---

[8] Reported in an article in the Denver Post (May 29, 2001) by Steve Raabe.
[9] "Valuation of Curtailable Load Program," June 5, 2001. Report prepared for Honeywell Corporation by e-Acumen (available from author).
[10] The California Power Exchange ceased operating in spring 2001.

become cost-effective in the next five years. Working toward an integrated infrastructure that can support full distributed resources (i.e., both load management and on-site generation) is likely to provide strategic advantage as the DG technologies evolve.

5. The communications infrastructure and internet gateways being developed for new load management systems create a platform for other value-added services for upscale residential customers and small business customers.[11] These programs can include site security, appliance and equipment monitoring, and support local area wireless networks for a variety of site services.

6. Many distribution companies serving as the default provider will need to demonstrate that they are procuring electricity suppliers on behalf of their customers in an appropriate manner.[12] This is likely to include a demonstration of "best practices" in load management.

In addition to the points above, demand response offerings also offer environmental benefits by displacing generation requirements and can provide system benefits by helping to alleviate distribution system and transmission system constraints since it is an on-site resource.

Table 2. Summary of Business Case

Economics	Mass-market demand response provides kW at a lower cost than combustion turbines.
Risk Management	It provides for a cost-effective physical hedge against price and quantity risks.
Barrier to Entry	The time required to build up significant participation in a mass-market demand response program can make competitive entry into retail market more expensive for competitors.
Integration with Distributed Generation	A load curtailment demand response program is the first step toward developing an integrated demand response offering that includes on-site generation.
Strategic Infrastructure	A demand response program develops a communications and technology platform that can be used to deliver other value-added services.
Regulatory Risk	Utilities providing electricity service under tariffed rates will need to demonstrate that they are procuring supplies cost-effectively and employing best practices demand management.

---

[11] The Long Island Power Authority is installing a communicating thermostat-based load management system that utilizes an internet gateway and provides the customer with the ability to change thermostat set-points and manage energy use via the internet.

[12] For example, GPU Energy can accrue electricity supply costs that are greater than revenues collected from tariffed rates in Pennsylvania. At the end of the transition period, they would be eligible to collect these costs. However, it is likely that the State Commission would want to make sure that all these costs were prudently incurred.

6.   THE FUTURE OF MASS-MARKET DEMAND RESPONSE

The recent response to the California energy crisis of 2000 and 2001, and concerns about supply in other regions of the country has been to seek immediate and largely short-term solutions. This has resulted in a focus on large customer demand response efforts to attain sizeable, near-term results. However, as markets evolve and as entities that choose to serve the electricity needs of mass-market customers assess their future, wholesale and retail products will need to become better aligned. Successful retailers will:

1. Need to offer price differentiated and quantity differentiated service; and
2. Any flexibility in price and quantity that retailers can obtain can be used to create margins that will differentiate winners and losers in these markets.

Announced actions by electricity providers in mass markets demonstrate this trend. Recent activities include the following:

– Two programs recently announced by The New Power Company™ illustrate a commitment to working with customers to better manage energy demand. The New Power Company is a competitive energy provider that has stated its intent to become a dominant player in providing energy and related services to the mass market. Originally developed as a partnership between Enron, IBM, and AOL, The New Power Company launched a successful IPO in October 2000 raising an additional $470 million and has multi-year marketing and services agreements with AOL and IBM. The New Power Company has announced customer based demand response programs in two regions that currently allow retail choice. The programs being offered in Philadelphia and Houston combine innovative pricing with internet enabled customer energy management programs (see NewPower, 2001).

– In Houston, 500 customers are being provided with internet gateways and communicating thermostats that allow for direct load control of HVAC equipment and will also allow customers to adjust their thermostats from almost anywhere using a flexible Web interface and internet-enabled devices such as smart mobile phones and PDAs. The company is authorized to remotely adjust a customer's thermostat setting between peak hours of noon and 5 p.m. on weekdays. The New Power Company reported an "enthusiastic response from Houston area customers who are looking for innovative ways to manage their energy use."

– In Philadelphia, 300 households will participate in programs that will install time-of-use meters, programmable thermostats that provide the company with the ability to remotely change thermostat settings during high price periods to reduce consumption, and remote energy management systems for customers allowing them to adjust their HVAC energy use remotely, e.g., pre-cool the home by sending a signal from work or adjust temperature set-points while away on trips.

– Gulf Power (2001) illustrates a price-response offer. The company has developed a GoodCents SELECT™ offer for residential customers to provide customers with a choice on how much energy to use as prices vary. The offer is based on a thermostat that automatically adjusts energy consumption based on how the customer programs the thermostat to respond to price signals. The prices are tiered into four classes — low at 3.5 cents/kWh, medium 4.6 cents/kWh, high 9.3 cents/kWh, and a critical price representing extreme price spikes or supply shortages at 29.0 cents/kWh. The thermostat is tied to a communications network that informs the meter in real time regarding the applicable tiered rate and the offer is designed so that customers can purchase electricity at rates lower than the standard residential rate 80 percent of the time. "For the past 74 years we have sold electricity the same way — customers pay one flat price … now customers choose to stay with Gulf Power regular or switch to GoodCents SELECT, which lets customers shift their electricity purchases to the lower priced times of the day."

– Allegheny Power (2001) announced a real-time pricing program for its Pennsylvania residential customers that will use internet technology to reduce a customer's electricity consumption when market prices cross a pre-set threshold. This structure essentially provides Allegheny Power with a call option exercisable at a set strike price.

– Puget Sound Energy (2001) has announced a time-of-use rates for all customers. The program offers morning rates (6 to 10 AM) of 6.25 cents/kWh, mid-day rates (10 AM to 5 PM) of 5.36 cents/kWh, evening rates (5 PM to 9 PM) of 6.5 cents/kWh, and an off-peak rate for nights (9 PM to 6 AM) and weekends of 4.7 cents/kWh. The company had originally proposed rates that varied from 2.5 cents/kWh off-peak to 7.79 cents/kWh on-peak; however, state regulators opted for a more modest plan initially. Of the 300,000 customers included in the program, only 1 percent has opted to keep the old billing plan. Puget Sound is also experimenting with remote control of thermostats to curb both summer and winter peak period demand.

– Southwestern Electric Power Company has implemented innovative rate designs that allow customers to choose from alternative time-of-use options comprised of different combinations of peak and off-peak periods and prices. This creates a rate tailored to the ability of the customer to shift loads.[13]

– Salt River Project (2001) implemented time-of-use rates where customers are charged 3.89 cents evenings and mornings (8 PM through 1 PM the next day) but are charged more than four times as much per kWh during the 1 PM to 8 PM period. This compares to the standard rate plan of 8.06 cents/kWh regardless of when power is consumed. The company has been unable to keep up with requests to switch to the new rate and has met half of its annual target for participation in less than one month.

In addition to the examples above, utilities such as Xcel Energy, Long Island Power Authority, and General Public Utilities have in 2001 announced expansions to their mass market load management programs calling for recruitment of over 10,000 new mass market participants (residential, agricultural and small business customers). All of these programs are using new technology with internet gateways providing access to controls on energy using equipment.

The innovation represented in these recently announced programs illustrates four important trends for the future as summarised in Table 3.

Table 3. Future Trends in Mass-Market Demand Response

Trend One	The use of energy management as the selling point and the development of internet gateways and local area networks for delivery of energy management as a platform for other services to residential and small business customers.
Trend Two	An increased focus on small business customers for cost-effective demand response, building controls for comfort and environmental quality, and other value-added services.
Trend Three	Incorporation of on-site generation into comprehensive demand response programs.
Trend Four	A focus on mass-market demand response as a physical hedge against price and quantity risks the value of the hedge based on options models driven by price volatility and the likelihood of extreme events.

6.1 Trend One — The Use of Energy Management as an Entry Point for the Development of a Platform for Other Value-Added Home and Small Business Services

A report by the Yankee Group indicated that 21 million U.S. households want "digital remodeling" with 12 million wanting to implement home networking capabilities within the next year. The key factor in adoption of

---

[13] This is discussed in Violette (1999).

networking services is the availability of compelling applications and services. A Yankee Group survey in 2000 indicated that:

– 37% would use a network to control home heating and cooling systems;
– 36% want to control home appliances from anywhere in the home;
– 42% want to enhance home communications; and
– 39% want to network entertainment (video and music).

The desire to address home heating and cooling by those households interested in other networked services provides a key leverage point for energy companies that also needs to gain flexibility in pricing and quantity to hedge against volatile wholesale market prices. The technologies used to deliver customer choice and energy management services also serve as platforms for future home automation and monitoring services. The New Power Company is incorporating a number of new product concepts using the interactive home gateway system used to remotely communicate with the thermostat. This technology uses home networking to enable real-time monitoring and control over the internet without any new wires in the home. The Gulf Power GoodCents SELECT asks the consumer "What if your meter could talk? It would communicate with Gulf Power on how much energy you are purchasing at each price. Plus it would provide future access to other services such as cable TV and the internet. Get this communications gateway with GoodCents SELECT, too."

In addition, Honeywell Corporation's Home Systems Group, Carrier Corporation and Invensys PLC are offering internet gateway based communications systems combined with wireless local area networks (either RF or powerline carrier) that can address controls and monitors that can provide security, report on appliance performance, address indoor air quality, monitor children, provide services that help senior citizens remain in their homes longer, as well as provide entertainment, medic alert, and children alert services.

If energy providers can obtain needed flexibility in retail prices and quantities through innovative mass market offers that provide them with valuable hedges against price volatility in wholesale markets, and they can get another $20 to $30 dollars a month in home monitoring and network services, then the value proposition for serving mass-market customers with energy management technologies becomes much more attractive. The GoodCents SELECT program currently charges customers approximately $4.50 per month for participating in the program, and they are achieving customer acceptance. It is not much of a stretch to conceive of additional payments for other networked services.

6.2   Trend Two — Increased Focus on Small Business Customers

Historically, mass-market load management programs focused on the installation of switches that turned equipment on or off via a radio signal sent by the utility. Candidates for this application were residential air conditioners, water heaters, pool pumps, and select other equipment such as pumps for irrigation. The small business community was not believed to be a good candidate because they would not compromise climate control and comfort to participate in these programs. Today, however, an entirely new set of technology is available to small businesses that can allow utilities and energy suppliers to offer programs that will increase the comfort of the building everyday, improve the indoor environment, and also provide the utility with the ability to control up to 40kW of a total of 120kW connected load.

Small businesses are even more attractive for load management than residences since the load is more consistent and predictable from day to day. In addition, the movement toward the use of internet gateways as the communications system allows for loads to be managed so that building comfort is minimally impacted. As one example, Honeywell Building Systems offers a small commercial solution around an installed server in the building linked to appliances and controls via an RF-based local area network. This system offers building management internet-based remote access to control and adjust temperature settings, time clocks, air handling equipment; monitor temperature and air flows; provide custom tenant billing, indoor air quality options, equipment monitoring and predictive failure algorithms, time/event programming, alarm setting and monitoring, RTP algorithms if a rate is offered, TOU programming, and demand limit control. In return for helping to finance this building management system, the utility has the ability to call for load curtailment related to HVAC use, non-critical lighting, and other non-critical loads. Audits of buildings have shown that up to 33% of a buildings load can be shed (e.g., 40kW out of a connected 120kW load) and that this call option can be obtained at a cost of approximately $2,000 or $50/kW. This cost is roughly one-fifth the cost of obtaining a controlled kW from residential options.

The cost effectiveness of the small commercial programs and the additional services that participating buildings receive results in almost all new and expanding mass-market demand response programs to include small businesses as at least a pilot program.

6.3   Trend Three — Incorporation of Distributed Generation into Load Management Programs

This is simply a logical extension of current load curtailment programs. There are two ways to reduce the system load at customer sites during periods of high prices. The first is to reduce the use of electricity by cycling or shutting down equipment. The second is to use on-site generation to power equipment during high price periods. The new technology being used to control thermostats and cycle appliances at residences and provide energy management at small businesses can be set up to turn on distributed generation units. As 40kW to 100kW distributed generation units become available, there will be applications for these units in large residences and small businesses both as back-up generation for the customer and as another call option for suppliers, who need to manage the price and quantity risks of serving these customers.

6.4   Trend Four — The Methods for Valuing Load Management Resources Will Evolve

One of the obstacles that modern load management has to overcome in the utility setting is the archaic method of assessing the value of these resources. As a general statement, most utilities still use methods from traditional DSM approaches to assess the value of these programs. These methods are inadequate for valuing these options given current circumstances. Even in states where restructuring has not occurred, the wholesale markets for electricity have changed dramatically with increased competition, construction of merchant power plants, as well as increasingly volatile prices and increased concerns about deliverability of power across increasingly congested transmission and distribution systems. Load management options need to be viewed as call options that reduce risks associated with low probability, high consequence events. The valuation approaches need to address load management as a call option that can be exercised by the utility. This means that point estimates of avoided generation costs, or energy savings based on the estimated forward price curve will underestimate the value of these resource options. The valuation of load management in today's environment needs to be conducted using probability distributions that assess the likelihood of price spikes and deliverability constraints. These models appropriately credit load management with the value associated with its ability to deliver as a call option.

As a starting point for this approach to valuing mass-market load management, consider going to your wholesale supply and trading group and telling them that you have access to 100 MW of power that can be called upon with five minutes notice, can be called only when needed on peak days, and the call can range from two to eight hours. In addition, there are no counter party delivery or credit risks. What would they pay to have access to that type of call option?

7. SUMMARY

One can argue that electricity has been mispriced for so long that restructuring rules that continue to tie prices to tariff-based rates based on historical embedded costs will delay real competition. However, this market will work its way toward rational pricing where wholesale and retail electric products will become better aligned. Utilities and competitive energy suppliers that serve mass-market loads will need to assess the price and quantity risks associated with serving these customers, and develop partnerships through demand response offerings that provide structured hedges against these risks. Those entities that best accomplish this task; whether regulated or unregulated, will be those that will achieve success in these markets. Finally, there is no lack of opportunity for cost-effective load management offerings that provide the energy supplier with a unique resource that will have an important position in a portfolio of electricity supply facilities, contracts, and demand requirements.

REFERENCES

Allegheny Power. 2001. "Allegheny Power to Run Real-Time Pilot in Pennsylvania," reported in *The Electricity Daily*, June 18, 2001.

Braithwait, S. 2001. "Residential TOU Response in the Presence of Interactive Communication Equipment." In *Pricing in Competitive Electricity Markets*, edited by A. Faruqui and K. Eakin. Kluwer Academic Publishers: Boston, MA.

Braithwait, S. and A. Faruqui. 2001. "The Choice Not to Buy: Energy Savings and Policy Alternatives for Demand Response." *Public Utilities Fortnightly* 139 (No. 6, March 15).

CAISO. 2001. "Second Request for Bids to Provide Demand Relief (Load) for Summer 2001." California Independent System Operator Corporation, March 30.

e-Acumen. 2001. "Valuation of Curtailable Load Program" prepared for Honeywell, Inc. by S. Ringelstetter Ennis.

Hirst E. and B. Kirby. 2001. "Retail-Load Participation in Competitive Wholesale Electricity Markets," prepared for the Edison Electric Institute and the Project for Sustainable FERC Energy Policy.

Hirst, E. 2001. "Price-Responsive Retail Demand: Key to Competitive Electricity Markets," *Public Utilities Fortnightly* March 1, 2001.

Gulf Power. 2001. "GoodCents SELECT Receives Governor's New Product Award," Press Release, Gulf Power, March 22, 2001.

LIPA. 2001. Michael Marks, Applied Energy Group, Inc., "LIPA Air Conditioning Direct Load Control," presented at the Price-Responsive Load Management Conference: A New Opportunity in New York State Electricity Markets, March 22-23, 2001.

NewPower. 2001a. "Internet-Based Pilot Program Will Enable Houston Area Residents To Better Manage Energy Costs," Press Release The New Power Company, April 24, 2001.

NewPower. 2001b. "Three Innovative Energy Saving Pilot Programs in Philadelphia to Help Consumers Control their Energy Costs," Press Release, The New Power Company, May 31, 2001.

Puget Sound Energy. 2001. "Washington Regulators Okay Time-of-Use Rates," The Electricity Daily, April 30, 2001.

Salt River Project. 2001. "Arizona Utility Fails to Keep Pace with Energy Savings Requests," *The Tribune*, Mesa, Arizona, June 8, 2001.

Violette, D.M. 1999. "Conventional Pricing Wisdom Not Competitive: Innovation Creates Margins and Attracts Customers" *Electric Light and Power* (February).

Violette, D.M. 2001. *"Opportunities for Load Management in Mass Markets,"* proceedings of the EEI Retail Energy Services Conference, Chicago, March 29, 2001.

Chapter #12

RTP Customer Demand Response
Empirical Evidence on How Much Can You Expect

Steven Braithwait and Michael O'Sheasy
Laurits R. Christensen Associates, Inc.[†]

Abstract: This paper provides new evidence on customer demand response to hourly pricing from the largest and longest-running *real-time pricing* (RTP) program in the United States. RTP creates value by inducing load reductions at times of capacity constraints and high wholesale market prices, as well as load expansion when capacity is ample and wholesale prices are low. Anticipating demand response in advance can lead to lower wholesale prices and increased reliability. Results presented illustrate how customer price responsiveness varies by customer type and price level.

Key words: Real-time pricing; demand response; price elasticity; load response: customer demand; electricity demand

1.   INTRODUCTION

Prices play a fundamental role in competitive markets – they convey information about *cost* and *value* to both suppliers and consumers. Unfortunately, many restructured power markets have been designed as "one-sided markets," with competitive wholesale markets on the supply side, but with regulated retail markets and price freezes on the demand side. As a result, retail customers are generally insulated from high wholesale prices during periods of supply constraints, and thus have no incentive to reduce their usage. During episodes of demand-supply imbalance, nearly vertical demand curves (reflecting the absence of price response) intersect

---

[†] Mr. O'Sheasy was formerly Manager of Product Design at Georgia Power Company.

A. Faruqui and B.K. Eakin (eds.), Electricity Pricing in Transition, 181-190.
@ 2002 Kluwer Academic Publishers.

increasingly steep market supply curves at high demand levels, causing frequent wholesale price spikes. Several regional power markets are initiating demand response programs (*e.g.,* hourly pricing, demand-side bidding, or market-based interruptible load programs) to provide a mechanism to drive down prices during capacity-constrained episodes.

An important question for designers of demand response programs is how much load response they can expect from different types of customers and at different price levels. This paper provides new evidence on customer demand response to hourly pricing from the largest and longest-running *real-time pricing* (RTP) program in the United States. Georgia Power Company (GPC) has eight years of experience operating a large-scale RTP program, with more than 1,600 industrial and commercial customers, and 5,000 MW of subscribed demand. GPC routinely analyzes its customers' price responsiveness, and incorporates this information into load response models that are used by Southern Company in both daily dispatch operations and long-term system planning.

Experience to date with RTP illustrates the power of market pricing, even in regulated markets. RTP creates value by inducing load reduction at times of capacity constraints and high wholesale market prices, as well as load expansion when capacity is ample and wholesale prices are low. Anticipating demand response in advance (by as much as a day, or as little as an hour) can result in lower wholesale prices and higher reliability than would otherwise occur.

On a few days during the summer of 1999, GPC experienced market prices much higher than those seen in previous years, providing an important opportunity to observe how customers responded to such unprecedented price levels. This paper reports on several related analyses of customer response to RTP prices, illustrating how customer price responsiveness varies by customer type and price level.

2.   GEORGIA POWER COMPANY'S RTP PROGRAM

2.1   RTP design

Georgia Power's RTP tariff consists of a two-part design. Customers pay standard tariff prices for their typical historical usage pattern, or *customer baseline load (CBL),* and hourly RTP prices for any deviations between their actual consumption and their CBL. Viewed equivalently, but from a different perspective, customers effectively pay hourly RTP prices for all of their energy consumption, but receive a financial hedge against volatile prices in the form of a *contract for differences,* or "swap" contract, in which they are guaranteed to pay no more than their standard tariff for their CBL. Thus, to the extent that their usage remains close to their *CBL,* customers' electricity expenditures remain the same under RTP as under their standard tariff, regardless of the price level. However, if they take advantage of RTP prices by reducing usage in high-price hours, shifting usage from high-price to low-price periods, or increasing usage when prices are low, customers can see their average price of electricity fall under RTP.

RTP customers have a clear incentive to respond to hourly prices that exceed the tariff price, since load reductions below their CBL effectively represent power sales back to GPC at the high RTP prices, which show up as credits on their bill. Customers who have expanded beyond their CBL are fully exposed to RTP prices on all incremental usage. However, GPC offers optional *price protection products*, which allow customers to manage their risk from unexpectedly high RTP prices by buying forward contracts for additions to their CBL.[1]

Hourly RTP prices are designed to closely match projected marginal costs, which are calculated as Southern Company system lambdas for internal generation, and wholesale market prices for purchased power. Two forms of advance notice are offered. One is an *hour-ahead* program in which customers receive one-hour notice of the next hour's price (along with a rolling 24-hour price forecast). The other is a *day-ahead* program in which customers receive the next day's hourly prices by the afternoon of the current day.

2.2   RTP customers

In analyzing customer demand response, we typically segment GPC's RTP customers into seven broadly similar groups. One includes the *hour-ahead* (HA) customers, who are very large industrial customers with the potential to make rapid changes in load, or to substitute on-site generation (OSG) for purchased power. The *day-ahead* (DA) customers are divided into industrial and commercial categories, and then into subgroups based on the presence of OSG, or whether the customer previously participated in a supplemental energy (SE) interruptible load program. The OSG and SE segments tend to be characterized by fewer customers that are larger in size and more flexible than those in the remaining segments.

---

[1]   In a competitive retail market, customers will be able to choose the portion of their expected load to cover by purchasing forward contracts. The prices for those contracts will reflect the relevant forward prices at the time they enter the contract, this rather than a regulated tariff price.

2.3   Recent RTP price experience

In recent years, GPC's RTP customers have seen prices on a few days during the summer months that averaged between $.20 and $.40/kWh in the late afternoon hours, and in a few hours saw prices up to $.80/kWh. In 1999, however, prices averaged between $.40 and $.50/kWh on several days, and twice spiked to levels averaging in excess of $1.50/kWh. These high-price episodes provided an opportunity to develop new evidence on customer demand response at very high prices, when it is most valuable. In particular, we could examine whether customers responded differently to these very high prices than they did to more moderate price levels.

3.   RTP DEMAND RESPONSE

The following material summarizes some of the key findings from several statistical and econometric price response analyses that we conducted. First, we illustrate the overall demand response for the HA and DA customer segments, as estimated by comparing average loads on different price day-types. Second, we present price elasticity results from estimating econometric load response models for each customer segment. These models provide the ability to simulate RTP load response at alternative price levels. Finally, we characterize differences in the price responsiveness of individual customers in each segment, and show how price responsiveness differs by price level.

3.1   Overall load response

Overall load changes in response to RTP prices were estimated by calculating the difference between the *actual load* on a given price day-type, and a weather-adjusted *reference load* designed to represent the usage level that would otherwise have been expected. Figures 1 and 2 show the reference load and two particular price day-type load curves, for the HA and DA segments respectively, along with the RTP price profiles for those day-types. For clarity in viewing differences in the load patterns, each load curve has been normalized to its value in hour 4 (*i.e.*, hour ending 4 a.m.).[2]

---

[2]   Due to the extremely large differences in prices, the price profiles are shown in units of natural logarithms of the hourly prices. As an aid in interpretation, *simple differences* in logarithms represent *percentage differences* in levels. Thus, a *one-unit* difference in the logarithms of two price values reflects a *one hundred percent* difference in the prices. Thus, the approximately equal differences between the price profiles during the late

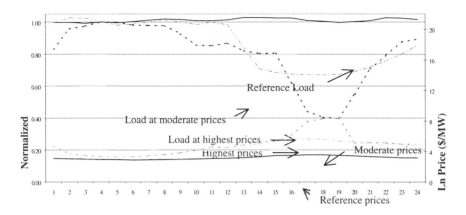

Figure 1. RTP Load Response and Prices by Price Day-type
(Loads normalized to 4 a.m. value; prices shown as logarithms)

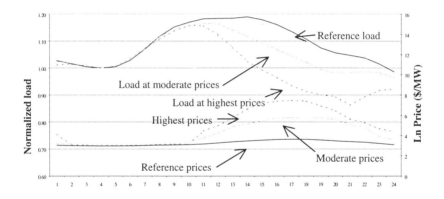

Figure 2. RTP DA Prices and Load Response by Price Day-types
(Loads normalized to 4 a.m. value)

The top load curve in each case is the normalized reference load, representing a weather-adjusted hourly average of loads on typical low-price summer weekdays. The middle curve represents the average load on a moderately high-price day (average afternoon prices between $.20 and $.35/ kWh for HA, and $.28 and $.35/ kWh for DA). The lowest curve shows the load on two of the highest price days in 1999, on which average afternoon prices averaged more than $2.00/ kWh for HA, and $1.50/ kWh for DA.

The HA reference load in Figure 1 is noticeably flat, indicative of the very large industrial customers that make up the segment. These customers

---

afternoon hours imply the same percentage difference between the reference price and the moderately high price day-type as between the latter day-type and the highest price days.

reduced their purchased energy by approximately 30 percent on the moderately high-price day type (as indicated by a normalized load of approximately .70, compared to the reference load of about 1.0). On the extremely high-price day, the load reduction increased to nearly 60 percent, providing GPC with approximately 250 MW of load relief during this high-cost, capacity-constrained episode.

The DA reference load in Figure 2 takes on a more distinctive diurnal shape that represents a mix of industrial and commercial loads. Demand response for the DA segment was approximately 10 percent on the moderately high-price day type. However, the amount of demand response more than doubled on the very high-price day to nearly 25 percent (or an hourly average of nearly 500 MW).

These load response results, combined with the price responsiveness results presented below, provide strong evidence of the substantial load relief that can be counted on from RTP customers who are exposed to market prices. Furthermore, the substantially larger load response on the extremely high-price days provides new evidence of the potential load relief that can be provided by RTP customers when it is most valuable. More detailed analyses of the load response indicated two additional facts. First, the amount of load response varied directly and consistently with the degree of price change – the higher the price-level, the greater the load response. Second, most of the additional load response on the extremely high-price days was provided by the SE and Non-SE industrial segments, whose load response nearly tripled compared to their response on the moderately high-price day-type.

3.2   Load response models

In addition to estimates of the amount of load response that has occurred historically, GPC also needs the capability to anticipate the amount of load response that it can expect in the future at different price levels. For that purpose, we developed aggregate load response models for each customer segment, using econometric analysis to estimate price elasticities that relate changes in customer loads to changes in RTP prices.

The basic load response model, shown in the logarithm form used in estimation, is given by the following:

$$ln\,(E_{dh}/\,E_h^{\,a}) = C + b_1\,ln\,(P_{dh}/\,P_h^{\,a}) + b_2\,ln\,(CDD_{dh}/\,CDD_h^{\,a}),$$

where E_{dh} is electricity usage in hour h on day d, and P_{dh} and CDD_{dh} are the RTP price and cooling degree days in the same period. The same variables with hour subscripts and a superscripts represent averages for the given hour in the reference period. The coefficients b_1 and b_2 show the effect of a change in either price or weather on electricity usage relative to its reference value.

In testing various forms of the model, we observed that the price responsiveness of the segments appeared to differ somewhat at different price levels. For example, some customers with on-site generation tended to respond strongly to relatively low prices. However, they did not continue to increase their responsiveness at the same rate at higher prices. In contrast, the two industrial customer groups without on-site generation appeared to become relatively more responsive at prices higher than those observed in previous analyses. For these reasons, we added a term to the models that allowed price responsiveness to vary with the price level (actually with the logarithm of the price level). The resulting price elasticities are not constant, but depend on both estimated coefficients and the price level. Table 1 shows the implied price elasticities for each customer segment at a range of prices.

Table 1: Price Elasticities at Different Price Levels

Price ($/kWh)	Elasticities by Customer Segment						
	HA	NSEI	NSEC	SEI	SEC	OSI	OSC
0.15	-0.180	-0.019	-0.014	-0.066	-0.039	-0.077	-0.073
0.25	-0.206	-0.029	-0.027	-0.084	-0.052	-0.073	-0.089
0.50	-0.241	-0.044	-0.045	-0.109	-0.069	-0.067	-0.110
1.00	-0.276	-0.058	-0.062	-0.134	-0.087	-0.061	-0.132

As expected, the HA, OSI and OSC, and SE-industrial segments were found to be the most price-responsive. The two Non-SE segments appear barely responsive to prices below $.25/ kWh, but increased their responsiveness substantially at the higher prices. The HA, OSC and SE segments also became more price-responsive at higher price levels. In addition, though not shown here, the commercial customer segment loads were found to be substantially more responsive to changes in weather conditions, as expected.

3.3   Customer price elasticities

To develop a more detailed picture of RTP customer response, we estimated separate price responsiveness parameters, or elasticities, at the individual customer level using a statistical approach similar to that used to estimate overall load response at the segment level. Each customer's price elasticity was calculated as the ratio of the percentage change in load between particular price day-type averages and a reference load, and the corresponding percentage difference in price.

We report average price elasticities across customers in each segment for two price day-types in 1999 – moderately high-price days, and extremely high-price days. Figure 3 compares the percentage of responders (customers whose elasticity exceeded .01 in absolute value) for each segment, calculated as a percentage of the total load in the segment, on each of the two price day-types. Two key findings are evident. First, with the exception of the hour-ahead group, the percentage of responders was greatest in the segments that contain relatively larger customers who have the technical ability to respond easily to high prices (OSG), or have shown the willingness to participate in an interruptible load program (SE). The percentage of responders was lower in the Non-SE groups, which contain relatively smaller customers. Second, when facing very high-price conditions, a substantially larger fraction of customers became price responders. The increase in the responsive portion of load at higher prices was greatest for the hour-ahead and industrial Non-SE segments.

Figure 4 shows the (load-weighted) average price elasticity (in absolute value) of the *responders* in each segment on the two price day-types. The pattern of results is similar to that for the percentage of responders, except that the values for the industrial customers are nearly uniformly larger than are those for commercial customers. Aside from the HA group, the OSG groups are typically most responsive, followed by the former SE customers and the smaller, Non-SE customers. Furthermore, for nearly all segments, the customers' average price responsiveness was substantially higher on the very high-price day, particularly for the industrial segments. The hour-ahead customers were clearly the most responsive on average, although as seen in the previous figure, only half of the load was responsive on moderately high-priced days.[3]

---

[3]   The decline in average price elasticity for the HA segment is somewhat misleading. Many of these customers tend to reduce load substantially even at relatively low prices, thus implying very high price elasticities. At higher prices, the load reductions continue to increase, but at a much lower rate, thus implying lower elasticities.

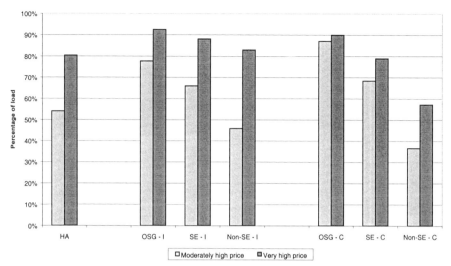

Figure 3. Pricc Responders – Percent of Load Responding
by Segment and Price Day-type

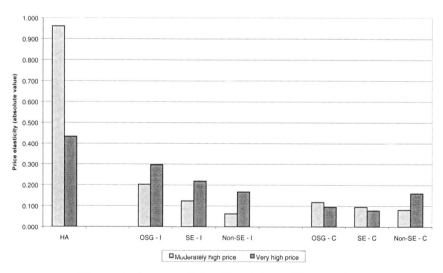

Figure 4. Average Price Elasticity of Responders,
by Segment and Price Day-type

4.   CONCLUSIONS

This paper has summarized the recent evidence on customer demand response to RTP at Georgia Power Company. Key conclusions are the following:

- RTP customers' load response to changing prices is significant and consistent; load response is consistently larger at higher prices,
- Load response to the unprecedented prices in 1999 was substantially larger than in previous years, amounting to approximately 250 MW for HA customers, and 500 MW for DA customers,
- The fraction of customers responding to RTP prices ranges from less than forty percent of smaller commercial customers, to about eighty percent of customers with on-site generation; these fractions increased substantially at very high price levels.

Chapter #13

Negawatt Pricing
Why Now is the Time in Competitive Electricity Markets

Fereidoon P. Sioshansi
Menlo Energy Economics

Abstract: Recent developments in the wholesale electricity markets have resulted in high prices and pronounced price volatility. Regional and local shortages in generation and transmission capacity are partly to blame, as are over-reliance on the spot market. This has created opportunities for making customers' demand more responsive to prices, especially during periods of high demand. Taking the concept one step further, customers with elastic demand and low-value added usage may be paid to get off the network so that others with more pressing needs can stay on. This would be highly attractive during periods when the network is capacity constrained and prices are high.

Key words: Demand response, price-sensitive load curtailment, demand-side participation, demand elasticity, real-time pricing, and peak load management

1.   INTRODUCTION

Recent developments in wholesale electricity markets have resulted in high prices and pronounced price volatility. Regional and local shortages in generation and transmission capacity are partly to blame. The increased reliance on the spot market by many *utility distribution companies* (UDCs) and *load serving entities* (LSEs) is another reason.

At the same time, the fragmentation of the industry has released the UDCs from their traditional obligation to serve. In the emerging competitive environment, independent generators build and operate plants based on economics, not because of any obligations to meet the load. Under these circumstances, it may no longer be economical to serve the infrequent peak loads, even at exorbitant prices.

A. Faruqui and B.K. Eakin (eds.), Electricity Pricing in Transition, 191-206.
@ 2002 *Kluwer Academic Publishers*

This has confronted the industry with a new business reality: the *supply side* may not be adequate to meet the peak demand. This, in turn, calls for greater participation by the *demand side,* making customers' demand more responsive to prices, especially during periods of high demand.

Taking the concept one step further, customers with elastic demand and low-value added usage may be paid to get off the network so that others with more pressing needs can stay on. This would be highly attractive during periods when the network is capacity constrained and prices are high.

This paper describes the business rationale for paying customers for *not* using power during peak demand hours. The paper illustrates how much capacity is worth when it is scarce and presents a few examples of efforts currently under way to make customer demand more responsive to prices. Obstacles to cost-effective application of these techniques, which are economic, technical, and cultural, are considerable. Recent advances in technology and the rapidly falling costs, however, continue to remove and reduce these barriers. The cost effectiveness of the suggested schemes are simply too compelling to suggest otherwise.

1.1   Half a market not as good as a full market

Most restructured electricity markets are supply-focused. Considerable effort goes towards designing well-functioning wholesale supply auctions, which – if successful – produce transparent and competitive prices. In many cases, customer demand is treated as a given. Consequently, relatively little attention is paid to ensure that the demand side of the equation plays an active role in setting the market clearing price (MCP) – and more importantly – for demand to respond to variable MCPs.

The recent experience of California and in a number of other countries has demonstrated this as a critical omission. As illustrated in **Box 1**, ignoring the elasticity of demand contributes to high prices, exasperates price volatility, and does nothing to encourage shifting usage from peak to off peak hours. The consequences are significant, as are the lost opportunities.

1.1.1   Inelastic demand exasperates California's energy crisis

One of the fundamental flaws of the California market, which is common to many other newly restructured markets around the world, is that it is virtually a half market. Only the supply side of the equation sets the market clearing price (MCP). Customer demand is treated as a *given*. It is not allowed to respond, or react, to fluctuations in price, except to a minimal degree. This major drawback is the result of a faulty assumption that demand is *inelastic*, i.e., it does not respond to high prices (Sioshansi, Energy Policy, 2001).

Consequently, customers, by and large, do not reduce consumption even during periods when prices are extremely high (**Figure 1**). The net result is that consumers continue to use electricity when it is exceedingly expensive to produce goods and services that are barely worth producing. From an economic (as well as purely logical) perspective, this is highly inefficient. According to economic theory, firms use inputs only when their use is justified by the market value of the output.

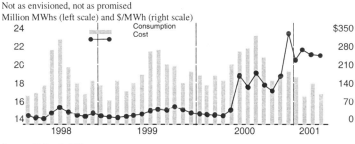

Source: Caliornia ISO
* The CA market opened in April, hence there is data for 9 months in 98

Figure 1. California load and prices, 1998-2001

For example, it would not make any sense for an aluminum smelter to produce aluminum when electricity costs 30 cents/kWh. That is why several large smelters in the Pacific Northwest have shut down operations in recent months, figuring that they can make more money by selling their energy to power hungry California.

Figure 2 illustrates how the inelasticity of demand has contributed to California's market price volatility and high prices. The explanation is simple. When demand approaches, or exceeds, available supply – which has regularly been happening in California – the relationship between a small increase in demand and price is no longer linear. Under such circumstances, a small increase in demand causes disproportionate increases in price. Conversely, it has been shown that even a small degree of demand elasticity can make a significant impact on wholesale price during peak demand episodes.

This phenomenon is exasperated by the artificial *in-elasticity* of demand, as shown in **Figure 2**. The graph on the left shows a *normal* market, with normal-looking supply and demand curves. In this case, an increase in demand (represented by an upward shift in the demand curve from D1 to D2) will result in somewhat increased price (from P1 to P2), assuming a fixed supply curve, S.

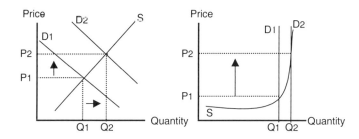

Source: EEnergy Informer, Oct 2000

Figure 2. Elementary economics - when demand is inelastic
and supply constrained, prices go through the roof

In the graph on the right, demand is shown as *perfectly inelastic* (i.e., a
vertical line), and supply with a steep upper end, representing physical
limitations of generation and/or transmission capacity. In this case, even a
small increase in demand, will result in a disproportionate price increase –
and virtually no increase in supply, since the system is running at or near full
capacity. The latter graph is a reasonable representation of the extremely
constrained California market (Menlo Energy Economics, 2001).

2.   INDUSTRY RESTRUCTURING CREATES OPPORTUNITIES FOR MARKET-DRIVEN SOLUTIONS

With restructuring new spreading across North America (**Figure 3**)
and in other parts of the world, regulated utilities are becoming open-
access *conduits* who rent their distribution wires to competing *energy
service providers* (ESPs). Moreover, in many cases, the *utility
distribution companies* (UDCs) have been forced to sell their
generation assets to independent generators. In these cases, the UDCs
and many others   serve the needs of their remaining customers by
buying from a competitive wholesale auction. Depending on the
specifics, a small to significant portion of the load may be purchased
from a spot market, where price volatility could be a problem.

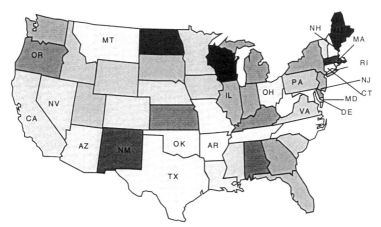

Figure 3. US States with Legislation to Introduce Retail Competition

The industry's functional unbundling has fragmented what used to be a vertically integrated supply chain. Consequently, any savings derived from implementing energy conservation or peak-load reduction at the retail level have been disconnected from those who stand to gain from it, namely generation and/or distribution companies. Moreover, the regulators now have very little leverage on many market players, notably independent generators, competitive ESPs, power traders, and re-sellers.

Other things have changed too. Historically, most customers bought electricity from monopoly utilities who offered them *stable prices* that were regulated and *low*. With deregulation, restructuring, and privatization of the electric power sector in North America and elsewhere, electricity prices are becoming *high* and highly *volatile* in many cases (**Figure 4**). Moreover, the reliability and quality of service is becoming an issue as supplies become tight in many regions of North America due to capacity shortages and/or transmission bottlenecks. These phenomena, combined with the competitive business pressures have created an atmosphere which is ripe for larger participation by the demand side of the equation (Sioshansi and Vojdani, The Electricity Journal, 2001).

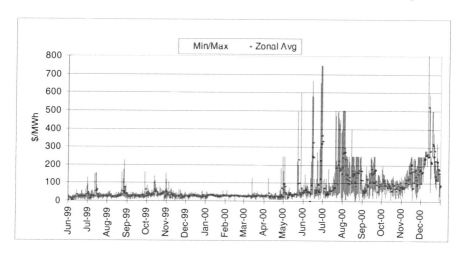

*Zonal prices for PX SP15
*Prices prior to December 2000 were capped at levels ranging from $250 - $750
Source: Gary Stern, So. California Edison Co., 2001

Figure 4. Electricity price volatility in the California market, 1999-2000

3.   DEMAND RESPONSE

Demand response DR refers to the ability to *control* and *manage* some or all of a customer's electrical load in response to variations in electricity prices. Although similar to the concept of real-time pricing (RTP), DR's main objective is different. RTP programs are primarily focused on advising customers on how to adjust their usage based on variable electricity prices. RTP programs encourage customers to:

−   Curtail load during high demand (and high-price) periods;
−   increase load during low demand (and low-price) periods; and
−   shift load from on-peak to off-peak periods.

DR programs, in contrast, are primarily focused on encouraging customers with elastic and/or discretionary demand to *forego usage* during capacity constrained periods so that others with inelastic and/or critical usage can continue to be served. DR programs:

−   maximize revenues (or yield) derived from capacity-constrained infrastructure; and
−   encourage low-value added customers to forego their usage during peak demand periods so that higher-value customers can continue to receive service.

3.1   Airline analogy

To illustrate the DR concept and to contrast it with RTP, it may be helpful to use an airline analogy. Although the two industries differ in many ways, airlines and the electric power industry have several common features. Both industries are:

- *net-worked based:* both invest in vast, capital-intensive, interconnected networks to deliver the required services;
- *capicity-constrained:* both experience bottlenecks during periods of high demand and have under-utilized capacity during all other times; and
- *high-fixed cost:* both have marginal costs which are negligible to their fixed costs.

These common features mean that both industries must design their network to serve a *finite capacity* (peak demand), which becomes over-subscribed during heavy demand periods. For example, airlines never have enough seats during heavy travel season around holidays or on popular routes during peak-demand periods. They can, of course, expand capacity to satisfy these short-duration peaks, but this would come at a high cost (e.g., more planes, more landing rights, more gates, more crew, more fuel, more maintenance, more finance costs, etc.).

Moreover, adding capacity to meet peak demand means that they will have a higher number of empty seats (and fixed costs) during low-demand periods. In short, adding capacity indiscriminately to meet the peak load may result in financial ruin. The answer, therefore, is not necessarily increased capacity, but *high utilization* of the finite capacity on the existing network. For airlines, this means high load factor. The same is true for the electric power sector.– whether we are talking about generators or distribution companies.

Since deregulation, major airlines have come to rely on highly sophisticated *yield management systems* designed to *maximize revenues* (hence profit) on finite available capacity. They do this by attempting to maximize revenues from each seat on every flight on the network by dynamically adjusting ticket prices. For example:

- seats tend to be more expensive on popular routes and/or on peak-demand periods; and
- there are fewer discount seats on popular routes and/or on peak-demand periods - and the availability of discounts is continuously adjusted as more seats are sold;
- prices on remaining unsold seats are continuously adjusted upward or downward as the departure time approaches in an attempt to fill as many

seats with as many paying passengers while extracting the maximum revenues before the plane leaves the gate.

3.2   Overbooking

Yield management programs also allow airlines to engage in another profitable practice known as *over-booking*. This practice is based on historical information about how many paying passengers may be expected to actually show up before the gate closes on each departing flight. Because a percentage of passengers with paid tickets routinely do not materialize, airlines routinely over-book flights by a corresponding percentage.

Overbooking allows airlines to sell a few more tickets and essentially ensure that every flight generates as much revenue as possible. In the absence of this practice, many flights would leave with a few empty seats because of the no-shows. Since the marginal cost of filling a few empty seats (a drink and bag of peanuts) is essentially nil, any revenue derived is pure profit.

Once in a while, every one who has bought a ticket actually shows up at the gate. In this case, the seats on the flight are over-subscribed. Since everyone with a ticket is entitled to get on board (notwithstanding the warning on overbooking on the tickets), a mechanism must be in place to get a few people to *volunteer* to get off the plane.

Airlines conduct an auction in real time to see which customers are willing to forego their "right" to fly on that flight at that time in exchange for a seat on a later fight and a travel voucher. They start the auction at low prices and raise the price until the desired number of seats are vacated.

In practice, there are bound to be a few individuals with elastic demand – i.e., people who are willing to take the next flight in exchange for a travel voucher. Incidentally, the travel vouchers offered are essentially worthless empty seats on a future flight, so the airlines suffer very little when they give away these rewards.

The logic behind the scheme is quite clever. Full fare business travelers with pressing schedules get to fly on over-booked flights. The retired couple or college student on tight travel budget wait for the next available flight – and receive what is essentially a worthless travel voucher. Financially, this is hugely lucrative. The airline collects, say, $1,400 of revenue from a business traveler on an overbooked flight, by luring a budget traveler to vacate their seat for essentially nothing.

Overbooking of flights plus the auction at the gate allows airlines two shoot two birds with one arrow by:

– filling planes with more passengers; and, more importantly;
– rationing scarce capacity on overbooked flights.

3.3 Demand response: the next-best thing to overbooking

It is instructive to examine the similarities between the airlines' practice of overbooking and the concept of DR in the electric power industry. Overbooking allows airlines to increase load factor (i.e., more filled seats) and yield (more revenues per flight) on a fixed network. In the process, they make more money – without seriously inconveniencing anybody. Moreover, the practice allows them to keep their most important business customers happy – by offering them empty seats when none are available. This miracle is produced at very little cost.

The electric power industry can do the exact same thing. It can allow the *network* (be it generation, transmission, or distribution) to be slightly over-subscribed during heavy demand periods by overbooking. Yet, by using a clever DR scheme, the network can still serve the needs of critical customers without collapsing.

The latter can be accomplished by luring customers with discretionary requirements to curtail their usage so that customers with essential needs can continue to be served. The analogy to the airlines is not perfect, but as in the case of overbooked planes, the scheme allows companies to make more money by serving customers with inelastic (and presumably high-value) demand. The increased revenues are made possible by *bribing* customers whose value for service is relatively low to *voluntarily* get off the network.

The combination of three factors, namely, high electricity prices, price volatility, and poor service reliability has created a compelling business opportunity for DR. Not surprisingly, enterprising companies are beginning to appear to offer new services to customers who are looking for ways to:

– Minimize their energy costs;
– manage exposure to price volatility; and
– actually make money by participating in load curtailment programs.

The first two they can do by participating in RTP programs and by reducing consumption during peak demand hours. The last item they can do by participating in price sensitive, voluntary load curtailment or DR programs.

The market for DR services is expected to grow in the coming years as more states open their wholesale markets to competition. The economics of such schemes will be particularly compelling in capacity-constrained markets. Rapid advancements in technology and rapidly falling prices make demand-side solutions not only feasible but affordable. The companies that stand to benefit from DR include:

– energy service companies (ESCOs);

- load aggregators, energy re-sellers, scheduling coordinators;
- energy service providers (ESPs);
- independent systems operators (ISO);
- power trading companies;
- traditional utility distribution companies and/or their non-regulated subsidiaries;
- load control and energy management service companies;
- generating companies; and
- large energy intensive customers.

3.4 Implementation details

To succeed commercially, a DR service provider will have to develop the necessary know-how and systems to handle a number of technical and business complexities including the following:

- *Volume* - Many customers' loads have to be aggregated and managed for the concept to become commercially viable.
- *Customization* - Each customer has a unique load profile and service requirement, hence the need for individualized solutions within the broad business concept.
- *Real-time* - Prices and loads must be controlled and managed in real-time, or close to it, thus requiring dynamic and instantaneous two-way communications and feedback.
- *Coordination* - Sophisticated real-time communication and coordination is required between the DR service provider and customers on the one hand, and the independent system operator (ISO) or scheduling coordinator (SC) on the other.
- *Collaboration* - Collaboration among numerous generators, utility distribution companies (UDCs), scheduling coordinators, power traders, metering and billing companies and the ISO may be necessary.
- *Monitoring* - Sophisticated metering and billing systems are required to accurately measure and report how much load has been shed by each participating client in a timely manner.
- *Settlement* - Sophisticated settlement and back office systems are required to manage the transactions efficiently, securely, and at low costs.
- *Systems* - Powerful front-, middle-, and back-office systems are needed to effectively transact business with many players, handling huge volumes of traffic in a secure uninterrupted fashion.

While these are not trivial tasks, the necessary technology and supporting systems are available or are being developed. Further details are provided in Sioshansi and Vojdani, 2001.

3.5 Electricity restructuring has brought capacity constraints to focus

Many regions of the US are short of capacity, particularly during the hot summer months when the use of air conditioning creates sharp demand peaks. This problem is aggravated by an even more acute lack of investment in transmission capacity. As a result, during critical demand periods, some load centers cannot be adequately served even though excess generation capacity may be available in a nearby region.

The effect of acute regional shortages in generation and transmission capacity combined with deregulation of wholesale electricity markets has resulted in significant increases in wholesale prices at major trading hubs in the US **(Table 1)**. The net effect of these developments has been two pronounced trends – which are likely to be amplified over time as more states open their markets to competition:

1. Wholesale electricity prices may increase in many regions of the US - notably in those regions with acute capacity shortages; and
2. Wholesale prices are likely to become more volatile as more utilities rely on competitive spot markets to secure their supplies (as opposed to generating their internal needs).

Table 1. Recent price increases at major US trading hubs

TRANSMISSION INTERCONNECTION POINT	% CHANGE 1997-2000
Texas	+293%
Louisiana-Mississippi-Arkansas	+216
Tennessee Valley Authority	+165
California-Oregon border	+162
New York-West	+138
Chicago area	+130
New England	+117
New York - east	+101
Upper Midwest	+99
Florida	+89
Mid Atlantic region	+80*

*Includes only years 1999-2000.
Source: Federal Energy Regulatory Commission (FERC)

In the short-run, there is little that can be done to address the supply-side shortages. More importantly, in competitive markets, generators may find it uneconomical to invest in peaking plants that are only seldom used. The sharp peaks in demand may only be encountered once every now and then – and only for a few hours or days. No sane investor would wish to build capacity to serve this type of load (Menlo Energy Economics, 2000).

There are, however, three broad options on the demand-side that may provide immediate results:

- increased reliance on energy conservation to make existing supplies go farther;
- increased reliance on RTP measures to reduce peak demand by shifting some load to off peak periods; and
- reliance on DR or other price response measures to *ration* limited network capacity, be it generation, transmission, or distribution capacity, during high demand periods.

Energy conservation and RTP measures are well-known and have proven track record. Concepts such as DR, however, are more surgical and highly effective against high prices – caused by sharp demand peaks and the necessity of keeping supply and demand in balance in real-time.

3.6   How would it work

The recent experience of California has demonstrated that traditional forms of load curtailment programs are unpopular, and impractical, when customers are frequently cut off with little or no advanced warning (e.g., EEnergy Informer, 2000). More flexible ways must be found to induce customers to voluntarily shed load by offering them attractive incentives. Capacity constrained networks and high wholesale prices offer the perfect incentives. Not surprisingly, a number of companies are experimenting with more flexible forms of load curtailment programs.

For example, *Idaho Power Company* in Boise, Idaho has bought roughly 250MW of load back from its agricultural customers for the summer of 2001 at a cost of $150/MWh using an Internet-based auction. Most of these customers use electricity for irrigation pumping. The rationale is simple economics. Like most utilities in the West, Idaho Power has to buy some energy from the wholesale spot market at exorbitant prices. Buying back demand from its own customers at $150/MWh is far cheaper than buying it from the spot market. The customers are happy, since getting paid $150/MWh for not pumping provides them with more income than they would from producing the crops.

Likewise, the *Bonneville Power Administration* (BPA) has proposed to shut down the aluminium smelters in the Pacific Northwest for 2 years. This would free up over 1,000 MW of capacity, which the agency does not have to buy from the high-priced wholesale spot market. Due to the current drought in the West, hydro capacity in the Pacific Northwest is limited and BPA has to buy power to meet its existing commitments. Paying customers not to use electricity during the current energy crisis is far more economical and logical. BPA has pointed out that aluminium smelters cannot economically justify operations when price of electricity rises above $30/MWh, which is likely to be the case for the next couple of years. Consequently, the agency has asked smelters to remain *off line*, allowing 1,000 negawatts to flow to customers in California and elsewhere who are willing to pay a high premium for the power.

These two examples illustrate the lazy man's version of DR. In both cases, selected customers are paid to stay off the network for extended periods when wholesale prices are known to be generally high. Such static DR schemes are easy to implement, but they are inflexible and unintelligent.

More importantly, they would not be practical for a great majority of customers. Setting aside a few big smelters and potato farmers, most customers cannot afford to forego the use of electricity for extended length of time. Many can, however, adjust or curtail their usage for short periods of time given sufficient incentives. This is where more intelligent, more flexile forms of DR programs come into play.

Sacramento Municipal Utility District (SMUD) in Sacramento, CA, for example, has a program that offers large industrial & commercial customers high incentives for curtailing usage for a few hours. A typical scheme may encourage participating customers to curtail usage, say, between 2 to 6 pm, when peak demand usually occurs in California. Although the details of these programs have not been made public, it is understood that customers who are willing to curtail partial or total usage on short notice will be offered a portion of the prevailing wholesale prices.

With wholesale prices exceeding $1,000/MWh at times, this is a powerful incentive for many customers to curtail usage totally or partially for a few hours on selected days. Looking back at there aren't that many days when such cutbacks would be needed. But on those rare occasions, having a DR program in effect would literally save the day.

In the case of SMUD, the motivation to engage in DR is more than saving money. By selectively curtailing the load of a few willing customers, SMUD can effectively avoid highly disruptive rolling blackouts, which would otherwise be necessary during emergency episodes. Rolling blackouts do not discriminate among customers with elastic and non-elastic load. DR is a much preferred alternative to rolling blackouts.

Other utilities as well as the California Independent System Operator (ISO), and the California Energy Commission (CEC) are also working to implement similar schemes in time for the summer's peak demand in 2001 and beyond. A number of private energy service companies (ESCOs) and specialty firms such as Apogee Interactive, Global Energy Partners and Silicon Energy are also actively engaged. These types of load curtailment services are sorely needed for the next couple of summers in California as well as in other regions of the US where capacity shortages and/or high wholesale prices are expected.

3.7   No pie in the sky

The DR concept would allow, in principle, *all customers*, not just big aluminium smelters and potato farmers, to decide when it is cost effective to use energy – and when it makes sense to reduce, shift usage, or shut down all operations. Presently, for most customers (i.e., everyone other than big steel mills and the like) the concept would only work if a service provider would take care of the chores. The savings may not be worth the bother for many a small to medium-sized customer. And here lies an enormous business opportunity for DR service providers.

The challenge for DR service providers is to aggregate the loads of many individual customers, each with some discretionary load. By adding these small discretionary loads, the DR service provider can accumulate a sufficient number of negawatts, which can compete against more expensive supply-side options on the wholesale spot market. The DR service provider would have to effectively manage these discretionary negawatts, and coordinate its efforts with the ISO and other critical market participants. The principles are identical to the examples cited above.

The logistical details are, of course, a bit more complicated – but not insurmountable. A lot of energy and ingenuity is at work to address these technical issues, and companies are hard at work to develop integrated systems and solutions. The opportunities to save costs and improve the efficiency of capacity-constrained networks are too large to suggest otherwise.

3.8   How much are the negawatts worth?

In principle, a megawatt saved should not be worth more than one generated. Thus, the wholesale price of energy in a competitive market sets the upper limit on how much should be paid for getting customers to curtail usage. In practice, of course, there are other costs and benefits to voluntary load curtailment that may make it worth more or less than the cost of wholesale power.

For larger and sophisticated customers, who may be active participants in the wholesale market, it may be possible to bid discretionary negawatts at the same price as the going spot market price. This would probably make sense if no middleman is involved and the size of the negawatts is considerable and firm. Not only can these customers reduce internal usage but those with emergency back-up generation can augment their negawatts with actual megawatts. The savings – i.e., MWhs not used – plus the net generation – i.e., MWhs they may be able to sell into the grid – add up and could be significant. There are many large industrial and large commercial customers who can coast along for a few hours without too much trouble – relying on stored energy, by switching fuels, or through operational flexibility's. These customers are prime candidates for participation in DR type programs.

For smaller customers with less flexibility, fewer options, and smaller discretionary loads, a middle man or DR service provider is necessary. The middleman must handle many chores, and will share in any resulting proceeds. In this case, the customers will not see the entire cost of the saved energy as incentives. Depending on the complexities of the transactions and the level of services provided, the middleman will keep a portion of the savings.

The question of how much are the negawatts worth, therefore, depends on many factors, including the following:

- the size of the negawatts;
- the firmness of the negawatts;
- the duration of the negawatts; and
- the frequency of interruptions.

Additional incentives will be offered if, in addition to load curtailment, the customer is able and willing to interject real megawatts into the grid during peak demand periods. Currently, many utilities are not particularly receptive to such buy-back arrangements. But to bring the demand-side into full play, more reliance on incentives must be introduced.

REFERENCES

EEnergy Informer. 2000. "Is There a Cure for California's Wobbly Market?" October.

EEnergy Informer. 2001. "Interruptible Load Programs Work Best When There Are No Interruptions," February.

Menlo Energy Economics. 2000. "Seeing the Light, Understanding California's Electricity market, prepared for Independent Energy Producers," June.

Menlo Energy Economics. 2001. "California's Restructured Electricity Market: How did we get into this mess, and how do we get out?" April.

Sioshansi, F. P. 2001a. "California's Dysfunctional Electricity Market: Policy Lessons on Market Restructuring." *Energy Policy* 29 (No. 9, July): 735-742.

Sioshansi, F. P. 2001b. "Demand Side Bidding: The Sequel to DSM?" ECEEE Conference, Cote d'Azur, France, June 11-16.

Sioshansi, F P. and A. Vojdani. 2001. "What could be better than RTP? Demand Response." *The Electricity Journal* (June).

Chapter #14

Innovative Retail Pricing
A Pacific Northwest Case Study

Allan Chung,[1] Jeff Lam[1] and William E. Hamilton[2]

B.C. Hydro[1] and Silicon Energy Corporation[2]

Abstract: This paper first provides a brief review of market-based pricing programs in the Pacific Northwest, a region where many customers already enjoy low cost electricity. These programs highlight the difficulty of implementing market-based pricing in a region with low cost regulated rates. The paper then provides a detailed case study of a market-based retail pilot program offered by British Columbia Hydro and Power Authority. Approximately 500 customers from its large general service rate class have subscribed to its General Rate Service TOU Rate. The case study demonstrates that a successful market-based program can be initiated in a region where regulated electricity prices are low. The approach can be compared with the market-based pricing approaches taken elsewhere in the region.

1.   BACKGROUND

The Pacific Northwest is a region with relatively low cost electricity, with hydroelectric power comprising a large portion of its electricity generation. The push for electricity deregulation has therefore been slower here than in other states with higher electricity prices.[1] The combination of low existing electricity rates and rising market electricity prices would also appear to limit the potential of market-based electricity pricing programs in this region.[2]

---

[1]   Oregon is the only state that has passed an electric restructuring law in the region (Senate Bill 1149).
[2]   Cost-based electricity rates typically reflect the low cost of producing hydroelectricity in the region. The embedded energy cost will tend to be lower than the market-based energy

A. Faruqui and B.K. Eakin (eds.), Electricity Pricing in Transition, 207-220.
@ 2002 *Kluwer Academic Publishers*

Several market-based electricity-pricing programs that have been undertaken in the Pacific Northwest are first examined in this paper. Generally, customers in this region have had access to market-based pricing either via direct access pilot programs or via market-based tariff products. The results of these pilots appear to support the view that it is very difficult to implement market-based pricing programs in regions where regulated electricity prices and costs are already low.

In contrast to the other states in the Pacific Northwest, British Columbia is served largely by a single public electricity utility, British Columbia Hydro and Power Authority (B.C. Hydro), which serves about 95% of the province's load. Although retail access pilots have not been implemented in B.C., B.C. Hydro has had some experience in developing market-based tariff products.

The main focus of this paper is to provide a detailed case study of the development and implementation of a recent market-based retail electricity pricing pilot program offered by British Columbia Hydro and Power Authority. Approximately 500 customers from its large general service rate class have subscribed to its General Rate Service TOU Rate. The case study demonstrates that a successful market-based program can be initiated in a region even with relatively low costs. The approach can be compared with the market-based pricing approaches taken elsewhere in the region.

1.1   Pacific Northwest Market-Based Pricing Programs

Oregon and Washington enjoy some of the lowest electricity prices in the U.S., as inexpensive hydroelectric power is a dominant source of electricity supply in these states. Table 1 summarises a subset of market-based pricing pilot programs that were introduced in Washington and Oregon. The state public utility commissions approved these pilot programs to provide electric utility customers with more choices when purchasing their electricity supply. It was anticipated that both electric utilities and customers would benefit from these retail choice pilots.

These programs provided subscribers with access to market-based prices either through a "portfolio" of rate options provided by the utility, or via market-based prices offered by other energy service providers (ESPs) if direct access was offered under the pilot.

---

price, as the latter is driven increasingly by the cost of more expensive generation. Thus, cost-based electricity tariffs will tend to be attractive relative to market-based electricity prices, particularly for large customers.

Table 1. [Pacific Northwest Market-Based Pricing Pilot Programs]

Pilot Program	Description	Target Market	Status
Washington			
Washington Water Power (WWP)/Avista Energy More Options for Power Service II (MOPS II) Program	Approved by PUC May, 1998. allows customer to select from a portfolio of power service options, with WWP continuing as supplier menu of options include traditional rates, rates based on month-to-month market prices, rates based on annual market prices, a standard rate offer and a green rate	Approx. 7,800 residential and commercial consumers in Washington and Idaho (or 15.9aMW)	Pilot ends April 2000.
Puget Sound Energy Power of Choice Retail Wheeling Pilot	2 year Pilot approved by PUC starting in November 1997. open access pilot program to test customer and supplier related operational issues associated with open access all participating customers allowed to meet their full load requirements through alternate supplier purchases Suppliers offered two types of pricing: fixed pricing which imply a certain amount of saving variable pricing based on market price index such as Mid-Columbia	Approx. 10% PSE customers eligible, with a maximum participation of 10,321 or 1.2 percent Suppliers chose to focus on medium commercial and industrial customers	Pilot end Dec 1999.
Oregon			
Pacific Power and Light Company "Portfolio" Pilot Program	Filed a proposal with PUC in July 1998 for a "portfolio" pilot program for residential and small commercial consumers and direct access for large commercial and industrial customers PUC approved a "shopping credit" of 1.98 cents/kWh, which reflects 12-month average peak price at COB.		7 ESPs signed to participate in direct access, but no industrials signed up 6% of eligible residential customers signed, most chose market price option

Generally, the market-based energy prices were higher than the energy charges in the standard tariffs. Examination of average 12 month strip price for COB NYMEX show that prices had risen from $25.63/MWh in June 1998 to $30.66 /MWh in June 1999. This has made it very difficult for the market-based products to compete with the traditional energy service rate. For example, under MOPS II the traditional energy rate is in the range of $24/MWh. Therefore only a small percentage of the eligible load has subscribed to the MOPS II market-based rates.

Under Puget Sound Energy's pilot program, suppliers offered two types of pricing, fixed and variable. The fixed pricing schemes were set over a defined period. The customer would pay a fixed rate which implied a certain level of savings. The variable pricing schemes were generally based on a commodity market index price, such as Mid-Columbia. As the customer's price was variable, it did not guarantee the customer any level of savings. Most customers signing up to the index prices did not have the comfort level to stay on the full term. Of the 27 accounts on the pilot, 9 accounts dropped off the pilot before it ended. Of the 9 accounts that terminated their participation, 2 had fixed prices and 7 had some form of variable indexed prices.

In summary, the pilot programs appear to be hampered by the following:

-   Cost-based electricity tariffs are lower than market-based price of electricity.
-   Customers could not handle market price volatility and price increases (indexed prices), as market prices were not fixed.
-   Market-based prices are a straight pass-through of wholesale 16-hour block energy market prices. This ignores the value that retail customers may place on shorter blocks.
-   Suppliers could not compete with cost-based tariff, as the "shopping credit" for energy was set too low. This was the complaint provided by both customers and ESP's in the Pacific Power and Light Pilot Program.
-   Open access pilots imposed new costs to distribution utility e.g., additional costs of providing information (which included load data required by alternate suppliers, software and web site development).

2. DESIGN OF B.C. HYDRO'S TIME-OF-USE (TOU) PROGRAM

2.1 Objective

B.C. Hydro developed an optional market-based TOU rate for its large commercial and light industrial customers in 1999. This class of customers takes electric service under B.C. Hydro's General Service tariff. The target market for the program consisted of approximately 22,000 secondary level commercial and industrial customers with a peak demand greater than 35 kilowatts (kW).

One of B.C. Hydro's primary objectives was to introduce more customer choice for its General Service customers, as customers in this segment had no alternative but to pay for their electricity usage at the standard flat energy and demand tariff.

Another objective was to introduce the notion of market-based pricing to this important customer class. Market-based pricing enhances efficiency as customers receive and can respond to marginal price signals. The utility also benefits as the market prices reflect its opportunity costs.

A strong motivation for offering a market-based product, even when the utility is low-cost and under regulation was that it would help any future transition towards deregulation. Valuable experience is gained by the utility in the areas of pricing, metering and billing. Moreover, information on customer behavior in response to market-based TOU prices can be collected and assessed. Finally, marketing and sales gain expertise in selling a commodity product and customers also gain experience with pricing alternatives.

2.2   Rate Design Issues

B.C. Hydro had several challenges to overcome in the development of a market-based TOU rate including:

-   Existing tariffs are primarily cost-based and they are frozen under current government legislation. A mandatory market-based TOU product could thus not be developed under the rate freeze.
-   Any optional tariff could not have negative impacts on non-participants.
-   Existing embedded costs are relatively low compared to market price.
-   A transparent hourly wholesale price index is not available for the Northwest.

The following describes the approach that was used to address these issues.

2.2.1   Rate Structure

A voluntary two-part pricing structure was chosen. The first part is based on B.C. Hydro's embedded costs, as the base tariff will apply to customers' historic load consumption. This structure ensures that historic revenue can continue to be collected on the base load.

The second part is based on marginal cost, which will apply to incremental load and will consist of an "off-peak" price and an "on-peak price." The purpose of time differentiated or TOU pricing is to encourage load growth in non-peak times, and discourage usage at peak times. The

TOU prices have been developed to take into account B.C. Hydro's situation, where load management from TOU pricing may benefit generation needs and wholesale export market opportunities during certain periods of the year. The TOU prices are therefore derived from wholesale market energy prices. Presently, there is less need to include TOU transmission pricing for incremental consumption in the BC Hydro system. This is because additional transmission capacity is generally available during the period in which the proposed TOU contracts apply.

The price structure is customer revenue neutral, which implies that each customer has to alter its consumption pattern from its historic or expected pattern to take advantage of the market-based TOU price. The fact that historic revenue continues to be collected ensures that non-participants will not be harmed by the optional TOU rate.

The current low embedded costs of the primarily hydroelectric generation system is an impediment, as it makes the standard cost-based tariff relatively attractive to the market-based price. It was determined that large industrial customers taking electric service under B.C. Hydro's Transmission Service tariff would not have as much opportunity to benefit from a market-based rate given that their average blended energy and demand rate was 3.3 cents/kWh compared to the market price of 3.5 cents/kWh.

However, there was considerably more opportunity for the General Service customers with loads greater than 35 kW to benefit as their class average blended energy and demand tariff was 5 cents/kWh. While the "on-peak" TOU price will be higher than this, reflecting the opportunity value of that power, the "off-peak" TOU price will be slightly lower than the average blended standard rate, with the aim of encouraging incremental growth at the "off-peak" times.

Value can thus be provided to these customers as they can potentially expand usage during low cost periods. If the spread between the standard rate and marginal price is small, then the benefits would only be obtained by load shifting or by conservation. Value is also provided to B.C. Hydro as greater margin can be obtained for this load than what can be gained in the export market, which means that a "win-win" situation is created.

2.2.2   Market-based TOU Prices

A major part of the rate design focused on the development of market-based TOU prices. The first issue to address was what wholesale market prices should be used, and secondly, how to convert wholesale market prices to retail prices. Currently, the closest actively traded location for wholesale electricity sales is at Mid-Columbia (Mid-C) which is a cluster of utility control areas on the Lower Columbia River at the Washington-Oregon border. Mid-Columbia was chosen as the reference location as it is the market closest to British Columbia and also least likely to be affected by transmission issues arising at the California-Oregon Border (COB). Although there is no official exchange at Mid-C, after the fact bilateral trade price survey data is available. There are also active over-the-counter trades for the physical delivery of power for future months going out up to 18 months, and price quotes are available from brokers. Although hourly prices are not available, weighted average prices for standardized 16-hour peak periods and 8 hour off-peak periods are available.

An attractive feature of the rate is that the market-based TOU prices are fixed over the course of the year. This means that customers would not have to deal with the price volatility of the spot market, which is common in real time pricing products. Customers can therefore plan their operations and energy budget ahead of time

2.2.3   Pricing Windows and Periods

The TOU rate was designed to offer customers alternative combinations of peak periods and alternative peak and off-peak prices, resulting in four different price options. The TOU peak price windows were based on the observation that the B.C. Hydro system load peaks during the winter months (November through February) during the weekday periods between 7:00 a.m. and 11:00 a.m. in the morning and between 4:00 p.m. and 8:00 p.m. during the evening. Off-peak prices apply to all other months and to the off-peak periods of the winter months. Although market prices are also high during the summer months, B.C. Hydro has in the past been transmission constrained during peak periods in these months and it has not been able to sell all generation available for export. Thus at this time there is insufficient value in encouraging load shifting away from peak times during the summer. Therefore off-peak prices apply during all periods in the summer months.

2.2.4   Conversion of Wholesale Costs to Retail Prices

The wholesale Mid-C opportunity costs over the wholesale peak period do not generally reflect the retail costs during the shorter 4 hour peak windows, as the wholesale peak price represents the average index price over a 16 hour period. Also, it is likely that retail customers place a higher value than the wholesale peak price on shorter block periods.

Derivation of Peak Price

Given that a transparent hourly index is not available for the Mid-C, the 16 hour block peak price was compressed into a 4 hour block, based on cost neutrality, i.e., the same revenue is collected from a 100% load factor load of any size. This provides a peak price of 7 cents/kWh for the 4 hour block for the four winter months for Option A. This price is also used for the 8 hour block for Option C. Finally, to provide a higher peak/off peak differential, a 10 cents/kWh peak price was chosen for Options B and D.

Derivation of Off-Peak Price

The basis for determining the off-peak price in each period is the "retail cost," defined as the average market cost for those hours that are defined as being within the off-peak period. Base off-peak prices were first derived for each price option by ensuring product revenues were equal among price options and they were then adjusted to account for risks. The "retail cost" and the difference between it and the standard tariff are key elements. If the standard rate is significantly higher than the "retail cost," then customers can have meaningful opportunities to expand usage during lower cost periods.

The risk adjustment covers the retail product's risk and return. As market prices can change, product risk includes offering fixed TOU prices over a 1-year term. Product return covers any additional costs of offering the retail product, not captured by the program charge. The risk adjustment was set at 25% of the difference between the standard rate and the off-peak retail cost.

TOU Price Options

The following graphic shows the resulting price options that customers can choose from:

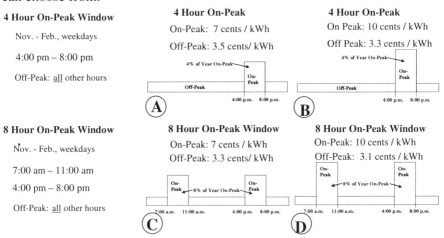

Figure 1. BC Hydro Time of Use Options

This price structure obviously does not capture the monthly price variation embodied in wholesale Mid-Columbia prices. The peak and off-peak differentials for Options B and D are also larger than what has been observed historically in hourly markets. However, it was recognized that the product would be offered on a trial basis only, with a limited number of customers and applicable only to incremental load. Options B and D would also provide information on customer response to larger price differentials, which could assist future load management initiatives.

2.2.5   The Customer's Bill

Once the subscriber selects one of the price options, the bill will comprise of three components:

- *the energy charge* - this is the summation of the peak and off-peak energy charges based on the TOU prices.
- *the delivery charge* - this is a fixed monthly assessment that the subscriber pays regardless of its current level of usage. It is set individually for each subscriber.
- *the program charge* - this covers the additional cost of serving the subscriber on the program, including metering, billing and administration.

The delivery charge is the residual revenue that is recovered from the customer after subtracting the baseline energy charge, i.e., the customer baseline load[3] priced at the chosen energy prices, from the base bill. The base bill is the customer baseline load priced at the standard tariff.

As with two-part RTP programs, the formulation ensures revenue neutrality so that if a customer's load pattern is equal to its baseline load, then the customer's bill under the TOU rate will be the same as under the standard tariff. The customer will also pay a price equal to market cost on incremental or decremental load relative to the customer baseline load.

2.2.6   Other Rate Features

Absence of Demand Charge for Incremental Changes in Consumption - the TOU prices will provide load management to assist generation needs and to take advantage of wholesale export market opportunities. Presently, there is less need to include TOU transmission or distribution pricing for incremental consumption in the B.C. Hydro system.

Bill Assurance - Under the Pilot program, bill assurance was offered to promote participation, with relatively little expected risk to B.C. Hydro. The credit is equal to the difference between the bill based on the TOU tariff and

---

[3] The customer baseline load is defined as the customer's previous year historic consumption.

the bill based on the standard tariff, with both bills calculated using actual consumption.

3.   ECONOMIC ANALYSIS OF B.C. HYDRO'S TOU RATE

An economic potential analysis was undertaken of the newly designed TOU pricing plan.[4]

Economic potential is defined to mean the gross dollar benefit that is achievable if all eligible customers switch from their existing rate and respond to the new TOU pricing plan. The estimate of economic potential includes the changes in revenue and supply costs associated with changes in consumption as participating customers respond to the new TOU prices.

EPRI's C-Valu Customer Response Model was used to calculate utility and customer benefits from offering the TOU prototype. Fifteen general rate (C&I) segments were modeled. The objective of the exercise was to determine whether B.C. Hydro should proceed with a Pilot program.

Conservative assumptions were used in the model. Average segment load shapes and average customer response elasticities were assumed. B.C. Hydro energy costs were constructed with the B.C. price based on the Mid-Columbia price adjusted for transmission constraints.

The results indicated that the customer benefits were relatively small. These estimates were generally viewed as conservative because of the assumptions that were used. The modeling revealed that mutual benefits could be achieved for both B.C. Hydro and customers, particularly those with growth opportunities. Customers with load shifting and conservation opportunities could benefit from TOU prices during peak hours. However, because peak hours are only 4% of hours in the year, the customer benefits from shifting alone are small.

On aggregate, the model results suggested that there would be economic potential benefits and that a pilot program could be supported.

4.   IMPLEMENTATION OF THE TOU PILOT

4.1   Subscription

Several tactics were used to help build subscription. These tactics were targeted at the larger customers of the General Service class.

---

[4]   William Hamilton who was with AXS Marketing LLC at the time undertook this work.

First, there was the bill assurance. This assured the customer that they would pay no more under the program than they would have paid under their standard set of prices. The bill assurance reduced the customers' perceived risk that they may pay more under the new rate.

Second, there was a pre-subscription period during which B.C. Hydro issued participation option certificates that gave customers an option to participate once the program received regulatory approval. This provided a sense of value to customers in that they were holding a place in the TOU program line up.

Third, account management staff conducted a phone blitz to solicit interest and arrange presentations with customers.

Finally, the primary means of subscription was face to face where B.C. Hydro account managers would meet with potential subscribers.

Subscription began in November 1999 and the program was fully subscribed by February 2000.

4.2   Fulfillment Process

The fulfillment process involved the acceptance of TOU agreements and initiating the meter exchange and billing change process. B.C. Hydro will have installed over 500 TOU meters in 4 months.

4.3   Value Added Solutions

It was identified early in the program design that providing customers with more than basic billing information would be required for customers to make better decisions on how to benefit from their new pricing plan.

As a value added offering, B.C. Hydro will provide customers load profile reports on a quarterly basis.

B.C. Hydro account managers will present these reports to subscribers and work with them to find better ways to take advantage of the TOU pricing plan. This may also provide B.C. Hydro with an opportunity to provide subscribers with additional energy management services.

4.4   Lessons Learned

There have been several lessons learned in the offering of this new pricing plan to customers not accustomed to having choice in prices that they pay for electricity.

It is safe to say that having choice is important to customers but choice must also come with some promise or opportunity for lower bills.

In offering choice there are some practical issues that utilities like B.C. Hydro face.

B.C. Hydro, like most other vertically integrated utilities, has revenue cycle systems that were designed and built to handle large scale, generic pricing structures and not dynamic, market-based pricing options.

We found that we had to work around the limitations of our systems. As a result we lacked the ability to track the exact status of a customer's order in the process. From the time that the customer signed up to the time that the TOU meter was installed, we did not have an automated method of tracking where the order was in the system. We can learn from the experts in the order tracking business and find a real time solution, which will allow us to manage our process better and identify and fix problem areas in the fulfillment process.

The successful introduction of optional retail pricing products requires a different set of skills, systems, and culture. The experience obtained by the regulated utility in offering pricing options is valuable and will help prepare the utility for an open market. The utility's optional pricing products are essentially "competing" against the "base" rate whatever that may be.

For some regulated utilities, the introduction of optional retail pricing products is more than just another "rate." It is a symbol of change and departure from the traditional rate making organization. We start to use words like prices instead of rates, market segments instead of rate classes, retail margin instead of rate of return/revenue requirement.

4.5   Billing Team

We chose not to modify our legacy billing system to accommodate the TOU pricing instead we used a new stand alone billing system that was being used for our largest commercial and industrial accounts. We found that different parts of the company responded differently to this change. Our billing service team embraced the change and their performance reflected that attitude. They met all targets set for them and were able to have all bills issued on time and error free. They benefited from the fact that they just completed the implementation of a new billing system for B.C. Hydro's largest business customers.

4.6   Wires Team

The meter supply process was a challenge in that we faced delays and certification problems with the supplier. When the meters did arrive there were problems with the receipt of the TOU meters into our warehousing system because they did not conform to the energy and demand meter format of most of our meters. We also chose to manage the meter exchange process from a centralized point but our work management system would not allow us to easily track the process in a centralized manner as it was built from a decentralized perspective. Once the TOU meters were issued into the field, we had no timely way of finding out whether or not they had been installed. We had to wait for the hard copy work order to be returned to our centralized point. In summary, the wire part of the company was least able to adapt to the change that the TOU program brought to the wire area. They did not have enough experience to anticipate some of the technical and process issues of meeting the special metering requirements for this TOU price offering.

4.7   Account Management Team

The subscription effort was highly successful as demonstrated by the enrolment results. We did specify a target market to help guide our account management staff in their subscription efforts. The target market was an extension of the response modeling. A market potential study based on financial benefit was completed and applied against BC Hydro's customer segments. A list of target customers was developed from that market potential study.

The final subscription objective was to reach the 500-subscription limit. The account management staff was evaluated on that basis, as it was the easiest method to motivate them. The target market analysis was only used as a guide for the subscription efforts. The end result is that we have a broad representation of all the market segments.

4.8   Customers

Based on customer feedback to date:

-    Customers understood the program;
-    Customers were not well informed about their standard prices;
     Customers liked the demand-free feature;
-    Customers had little sense of their load profile;
-    Customers liked the sense of choice;
-    Customers were looking for bottom line benefits.

Customers will adopt these new pricing options in different ways. Some will be early adopters and others will require some time to determine the benefits but you will find an untapped demand for them. Depending on the customer's organization, they will have a different process in choosing the best option. For example, owner operators are quick decision-makers compared to large multinational organizations with centralized decision making for purchasing matters.

The challenge for the utility is to find the market segments and individual customers that can best benefit from these new pricing options and can adapt to them while at the same time receiving the level of service that they require.

From a purely customer perspective in many regulated electricity markets, the notion of having choice in electricity pricing options is still a foreign one. This does not mean that customers are apathetic to electricity pricing options but instead it is an opportunity for utilities to unlock the benefit for customers and themselves.

Once you unlock the potential for these pricing options, there is no going back.

5.   SUMMARY

The following are some of the main findings:

1. Market-based pricing can be successfully developed and implemented in a low cost region.

    - The utility can increase value to customers by shaping wholesale prices and by providing a menu of options. Customers can choose the price option, which matches their ability to respond, and their operations.
    - Other attractive features of the rate include fixed TOU prices, which are set ahead of time. Customers can avoid high peak prices by shifting or conserving load, and grow during low cost off-peak periods.

2. Optional rate products provide the utility with valuable implementation experience, which will be used when markets open up.

3. In general, market-based products will be more difficult to design and implement in the Northwest if market prices continue to trend upwards and if these products have to compete against low cost-based tariffs.

    - However, market-based seasonal and peak and off-peak price differentials can still be reflected in existing cost-based tariffs.
    - When deregulation occurs, the default pricing needs to be structured so that market-based pricing can compete. This will encourage efficiency by providing customers with correct market-based price signals.

Chapter #15

Self-Designed Electric Products

Robert Camfield, David Glyer, John Kalfayan
Laurits R. Christensen Associates, Inc.

Abstract: The task of setting prices for electric service providers has evolved through at least three paradigms. These pricing modi operandi include: 1) cost-of-service class allocation, 2) marginal cost-based service options, and 3) product design packages based upon customer preferences. We suggest that a fourth approach, Self-Designed Products, resides within the realm of current capabilities.

The notion of a Self-Designed Product is the facility of offering customers the option of developing customized price and service arrangements, within profitability constraints. We envision the facility to be provided via a web site. Upon accessing the site, a customer would be presented with a menu of retail product attributes. The customer selects the level for each attribute. Given the set of selected attribute levels, an evaluation engine calculates prices by solving a constrained optimization problem, using historical loads and prices. The analysis incorporates the customer's expected load response behavior and prospective economic costs-to-serve with uncertainty.

1. INTRODUCTION

The world of electricity has changed significantly in the last five years. The dominant forces of change are service unbundling and the organization of new market arrangements for unbundled services. While service unbundling is taking place rather gradually, further evolution is anticipated. Notably, unbundling is taking place within an era of revolutionary innovation in information systems that have far-reaching impacts that can extend, in major ways, the mechanisms for market interaction.

A. Faruqui and B.K. Eakin (eds.), Electricity Pricing in Transition, 221-228.
@ 2002 *Kluwer Academic Publishers.*

Our paper explores the capability afforded by internet systems to facilitate the self-design of retail products by electricity consumers, a format referred to as simply "self-designed" products (SDP). In electricity markets where unbundling creates a highly competitive setting for the provision of power at the retail level, the interactive design of products between customers and providers through a web site can play an important role. These products are self-designed in the sense that the customer specifies customer-specific components of the pricing structure – including possibly some of the price terms themselves – rather than taking service on a standardized tariff. One form that a self-designing facility could take would allow customers to choose between several differentiated offers of power supply. A self-designing facility vastly expands the range of retail products. As a result, the value of electricity to consumers can be improved significantly.

The self-designing format can be interpreted as a manifestation of the convergence of internet-based information systems and electric power markets. This new approach to product offerings is equally applicable to incumbent and third-party providers, and there is no inherent reason for established utilities to cede this ground to their competitors.

2.   RESTRUCTURED MARKETS

Today, electricity services continue to be provided largely by vertically integrated electricity systems operating under an umbrella of franchised monopoly, with a governance structure of agency regulation. Regulated, standardized prices for retail service are set through a process of administrative procedure. This organization of electricity markets is fast giving way to a new *modus operandi*. The consensus paradigm appears to be that transport services including transmission, distribution, and connection services will continue to be regulated, while generation and power supply services will be rendered under competitive market arrangements. In place of standard utility tariffs, consumers are likely to face a wide array of pricing options.

While this vision of the future state of electricity markets is admittedly conventional, the path of change and its outcome is far from certain. Market restructuring is not proceeding at an even pace in all regions of the world, or even across the U.S. Indeed, while California and the North Eastern states have unbundled services and opened up retail to third-party access, many regions are proceeding to unwind the longstanding market arrangement of franchised monopoly rather cautiously. Arguably, a modest-paced and methodical approach to unbundling power markets is preferred insofar as restructuring inherently contains major pitfalls, particularly when errors in market design give rise to market power within wholesale generation markets. At least at this stage, restructuring can be appropriately interpreted as an experiment.

The vision of the end state of restructured markets is, moreover, rather opaque, and many dimensions and nuances of the end state cannot be adequately foreseen and anticipated. To partially fill in the blanks, and to glimpse a more in-depth view of the future, we must glean evidence from afar. In this regard, the economy-wide sweeping change ushered in by expanded information accessibility is telling. Ubiquitous internet technologies have vastly reduced the cost of gathering information, managing information, and decision-making. The internet is the natural, if not exclusive, medium for the self-design of electricity products.

3. RETAIL PRODUCT INNOVATION

In the traditional mode of agency regulation, the prices for retail electricity service are determined under the auspices of conventional embedded cost-of-service principles. These prices are established for rather delimited, class-wide, standardized services and tariffs, with occasional options of, say, time-of-use rates or curtailable service. This traditional approach is evolving, however, and many utilities today face a business environment with some degree of competitive entry. Competition often exists at the periphery of franchised monopoly, and is manifest in several forms including, but not limited to, customer demand for expanded service options, customer choice of supplier, competitive bidding of distribution line extensions, and direct franchise encroachment.

We might describe this environment of unsettled sands as *transition mode*. That is, competition is nibbling at the edges. While incumbent suppliers and regulators may be somewhat apprehensive about future developments, competitive encroachment can have positive outcomes. With more competition, incumbents may become more focused on customers, and regulators may become more open to service innovation and value-improving, incentive-compatible approaches to regulation. In response, operative strategies include cost-minimizing actions and innovations in product offerings. While we cannot make clear-cut distinctions among these various *modi operandi*, we interpret transition mode as an environment that can be very open to innovative product development. We expect that many of the innovations appearing during transition will ultimately flourish under competitive markets, as power suppliers package wholesale services into a wide variety of new retail product offerings.

The use of the internet and other communication channels is altering significantly how customers search, select, and purchase goods and services, economy wide. Standard utility tariffs, founded in the regulated world of one-way communication and reflecting little choice on the part of customers, certainly do not fit the mold of the new consumer marketplace.

The evidence – including observed selection preferences, expressions of interest, and survey research – reveals that electricity consumers have considerable variation in preferences for product attribute sets. Hence, a self-designing facility appears to be an attractive means to improve consumer value relative to traditional one-size-fits-all tariffs. Potentially, the opportunity for product innovation is vastly expanded by internet facilities for self-designed service.

4.   THE WORKINGS OF A FACILITY FOR SELF-DESIGNED PRODUCTS

While self-designed products can potentially assume several forms, our vision as follows. Imagine that customers connect to an internet web site. Access is limited by customer identification requirements and other security procedures. Once access is gained, the web site presents the customer with two menu-driven options entitled *History* and *Attributes*. *History* allows for the customer to review her recent energy and billing history. Various diagnostic capabilities can be incorporated into *History,* including, for example, the analysis and visual inspection of historical load shapes, and the calculation of average prices by month, season, and year. Such functionality implies, of course, that the web site is integrated with application software and databases, which allow individual customer load and billing history to be culled for use in the process of selection of attributes and evaluation.

The *Attributes* window allows the customer to select values or levels for service design attributes. Service attributes can be many things, of course. Possible attribute dimensions (or classes) include:

- Contract term
- Notice
- Total hours of curtailment
- Duration of any one curtailment
- Frequency of price changes
- Variation of price changes
- Rate blocks and
- Price protection options.

We envision that once an attribute class is selected, the software underlying the *Attributes* window will guide the customer through a question-and-answer style interview that helps the customer determine or select the level for the class. Once the preferred level for each attribute is selected, the service design offer is determined.

The core question at hand is "what are the price terms of the offer, given the defined attribute preferences?" The offer is developed by an evaluation engine for retail product design that is linked to the site. Software machinery culls the customer's hourly historical load shape (or inferred load shape via load profiles) and historical prices from the data bank. Along with the customer's identified attribute preferences, these historical load and price data serve as inputs into the analysis conducted by the engine. The evaluation engine develops service design offers based upon the familiar economic profit (or margin) analysis paradigm. That is, offer prices reflect expected flows of revenues minus the economic cost-to-serve. The analysis that underlies the development of offers is non-trivial because of the inherent risks regarding market and customer behavior. That is, price offers must reflect the uncertainty of economic costs, including the prospective wholesale prices for generation and transport services. In the case of the former, recent history reveals that prices for power supply can demonstrate large variation (or volatility) because of the characteristics of power supply including non-storability and network externalities. Second, customer loads change in response to price changes, and thus the supplier is exposed to both price and quantity risks. These market risks and behavioral responses should (must) be anticipated and incorporated into the analysis framework of the evaluation engine.

The service design offer obtained by the evaluation engine should be audited by a screening algorithm according to pre-defined criteria, such as profitability. Once screened, the offer is presented to the customer. That is, the web site returns a service design offer including price terms to the customer. The customer might find, for example, that the prices for some periods are somewhat high, and thus revises the allowed hours of curtailable load, or any other attribute listed in the *Attributes* window. Diagnostic tools might help the customer locate which of their attribute choices is driving the high price. The evaluation engine now performs second-round analysis – and so on. When satisfied, the customer agrees to a contract for service at the site.

In some cases, the site may be asked to return a menu of product options after the customer's initial attribute set is specified. These options would link sets of prices, or price parameters, or tie a specific price to each of several levels for a particular attribute, *when the level has not been predetermined by the customer*. For example, the monthly fixed charge of

A, B, or C is paired with usage prices of 4, 5, or 6 cents per kWh. The customer then picks from among these multiple offers.

The notion of a self-designing internet facility involves a highly integrated process that involves customer interaction, and is linked to load and billing databases, and projections of power prices or marginal costs. This process is shown in the following diagram.

Self-Designing Facility

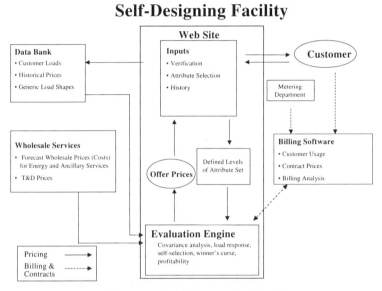

Figure 1. Self-Designing Facility

5.   CRITIQUE

While the idea is simple, implementation of a self-designing facility is challenging, and key elements appear to be:

1. A detailed analysis engine that calculates economic costs and profits. This analysis should incorporate both risk and the expected load response of the customer.
2. A mechanism to bound the attribute choices within a feasible range. The feasible range seems to be broad, however. It is essential to constrain the model solutions - i.e., price results - to profitable outcomes.

To construct these elements properly requires an understanding of the central feature of the self-designing format; its attractiveness and difficulties both lie in the fact that it allows customer-specific pricing on a broad scale. While not limited to any particular tariff group, the facility essentially brings special contracts to the mass market. A self-designing facility shapes the type of product that may be offered, and raises concerns that are worthy of our consideration.

Under certain traditional rates, notably TOU and one-part RTP, there are instant winners and losers who gain or lose because their load profiles differ from the aggregate profile on which the uniform price terms offered the group are based. When the rate is voluntary, the result is a degree of revenue erosion caused by self-selection (the selection of the rate by winners who know their patterns of usage represent windfall gains). The self-designing product facility, on the other hand, seems better able to limit such revenue exposure because the price terms offered to a customer are based on the specific history of usage for that customer.

With customer-specific pricing in the presence of reliable load data, the problem of instant winners and losers is mitigated, and the *premium* to cover the possible losses with self-selection of conventional rate options can be significantly reduced. However, to the extent that hourly data are not available, or are misrepresented, or do not accurately reflect changes to usage patterns of the customer, the problem reappears in a different guise under customer-specific pricing. There is some danger here, though modest.

The self-selection problem has a number of components or details, the importance of which vary with circumstances. Take the case of a customer with hourly load data. If the customer's hourly load data are available and verifiable, the evaluation engine can present offers that avoid instant winners and losers on the basis of load shape. The web site will use these data, in conjunction with the historical marginal cost and forecasts of future marginal costs, to calculate prices for fixed price-open quantity style offers. In order to have semi-automated calculation of such prices, the SDP software that develops the service offers must have the customer select the pricing blocks from a list of available blocks. Then, covariance analysis between costs and loads should be conducted.

The self-designing facility expands access between customer and provider in both directions. Providing fixed prices to customers facilitates customer shopping among suppliers. The supplier that provides the lowest price is most likely to get the customer. If there is a random component to the pricing offer due to uncertainty of customer usage, then the merchant that makes the biggest error in the customer's favor will most likely get the contract. This is sometimes referred to as the *winner's curse*. Use of hourly data and adequate marginal cost forecasts will diminish the potential for errors and thus limit profit losses due to the winner's curse. However, the less customer-specific information that is available, the greater the potential problem. Good load profiling is important, particularly if there is significant variation in economic costs, as in the summer.

Lack of hourly data can complicate the situation somewhat. This is particularly a problem for self-designed products that rely upon load profile data to price electricity service, in lieu of hourly loads of individual customers. In California, this situation is handled in a specific way that diminishes the problems when only aggregate usage data are available. Specifically, the pricing to the customer and the charging of costs to the merchant are based on the same basic load profile. This load profile is scaled by the customer's total usage; the shape of the profile is the key factor.

Service offered via the web site must be formulated with an awareness of sole supplier issues. Providing a customer with contractually fixed prices on open-ended quantities carries with it the necessity of being the sole supplier. If not, the customer will buy from one supplier in those hours when the supplier is cheaper, and from others at other times. This customer might take service at a fixed price during the summer, and then leave to get lower spot prices or seasonal prices for the lower-cost off-season.

6.   CONCLUSIONS

The prospective gains from self-designed pricing are many. They include overcoming *status quo* bias by giving customers "ownership" of a product that is specifically designed for them. In many cases the process can also simplify products, since the pricing does not have to include mechanisms to account for divergence of the customer from characteristics of a rate group. Customer charges tailored to specific histories and attributes can solve problems for which more complicated mechanisms are typically employed. Customer-specific pricing may also mitigate the self-selection problem attending the existence of instant winners and losers in voluntary offerings with uniform price terms across the target group. Finally, the use of a web-based self-designed products facility promises to allow providers to minimize marketing costs, while matching products closely to the customers who value them.

Section IV

The California Experience

Chapter #16

The California Electricity Manifesto
Choices Made and Opportunities Lost

Carl R. Danner, James D. Ratliff, and David J. Teece
LECG

Abstract: In January, 2001, an ad hoc group of concerned academics, former public
officials, and consultants convened at U.C. Berkeley in response to the
California Energy Crisis, issuing a "Manifesto on the California Electricity
Crisis" on January 26, 2001. The Manifesto urged government officials to
undertake a series of remedial actions to address the crisis, including raising
retail electricity rates, restoring the credit-worthiness of investor-owned
utilities, and avoiding any large or long-term expansion of government's direct
role in the industry. In the ensuing months California officials have acted in
almost direct contradiction to the Manifesto's recommendations. We assess
these actions, and attempt to foresee their likely adverse consequences for
California's electricity consumers (both households and businesses) in the
years to come.

Key words: California electricity crisis; restructuring; deregulation

1. INTRODUCTION

Starting in May 2000, the wholesale price of electricity rose sharply in
California's post-restructuring Power Exchange (PX) market. Unlike
intermittent spikes of a year earlier, PX prices rose and stayed high, through
both on and off-peak periods all summer. Nor, as had been widely expected,
did these price levels abate in the fall and winter of 2000-2001 as summer
peak electricity demands ended; instead, prices rose even higher in the face

A. Faruqui and B.K. Eakin (eds.), Electricity Pricing in Transition, 231-243.
@ 2002 *Kluwer Academic Publishers*

of sharp increases in natural-gas prices,[1] significant drought-induced declines in hydro power (especially imports from the Pacific northwest),[2] and allegations that electricity generators were exercising market power[3] (or even colluding on price) in the PX.[4] By December 2000, wholesale electricity prices were over 10 times their values of a year before.[5]

Having effectively been required to purchase their net[6] electricity requirements from the PX and the Independent System Operator (ISO),[7]

---

[1] For example, reported natural gas prices at the southern California border were $2.38-$2.45 per MMBtu for the week of December 6-10, 1999, and $15.50-$69.00 for the week of December 11-15, 2000. *California Energy Markets*, "Western Price Survey," issues of December 10, 1999 (No. 545) and December 15, 2000 (no. 597).

[2] "Spill for fish withheld because of drought," Bonneville Power Administration press release (April 3, 2001): "...'The Northwest faces one of the driest years in over 70 years of record,' said [acting BPA Administrator Steve] Wright. 'Current estimates are that natural stream flow at The Dalles will be roughly half of average. Spilling water now will worsen the shortage and drive electricity prices up even higher.'"

[3] Numerous academic analyses have sought to explain California wholesale market price dynamics, e.g. Borenstein, Severin. "The Trouble with Electricity Markets (and some solutions), University of California Energy Institute working paper (PWP-081), Berkeley, CA (January, 2001); Chandley, John D., Harvey, Scott M., and William W. Hogan. "Electricity Market Reform in California," Center for Business and Government, Kennedy School of Government, Harvard University, Cambridge, MA (November 22, 2000); Wolak, Frank A., Nordhaus, Robert, and Carl Shapiro. "An Analysis of the June, 2000 Price Spikes in the California ISO's Energy and Ancillary Services Markets," Market Surveillance Committee of the California Independent System Operator (September 6, 2000); Joskow, Paul and Edward Kahn. "A Quantitative Analysis of Pricing Behavior in California's Wholesale Electricity Market During Summer, 2000" NBER Working Paper 8157 (March, 2001); Harvey, Scott M., and William W. Hogan. "On the Exercise of Market Power Through Strategic Withholding in California," Center for Business and Government, Kennedy School of Government, Harvard University, Cambridge, MA (April, 2001).

[4] E.g., "Power Companies Agree to Turn Over Documents Demanded by Attorney General Lockyer for Investigation of Electricity Pricing Practices," press release of California Attorney General, May 22, 2001. "Grand Jury Will Probe California Markets," *California Energy Markets*, June 15, 2001 (No. 622).

[5] For example, California PX electricity prices were reported at 3-3.5 cents/kwh peak and 2.6-2.8 cents/kwh off-peak during December 6-10, 1999; and at 49.8 cents - $1.41 /kwh peak and 48.2–68.4 cents/kwh off-peak during December 11-15, 2000 (supra. note 4).

[6] Although the restructuring plan essentially required utilities to divest a large number of generating facilities, the utilities retained enough generation and power purchase contracts to serve a significant portion of their customer load (e.g., about 60 percent for PG&E). Thus, their market exposure was for the balance ("net short") of their customers' electricity requirements.

[7] Essentially, the PX ran short and medium term forward markets, while the ISO ran a residual real-time balancing market. A variety of factors led market participants to attempt to shift transactions between the PX day-head and ISO real-time markets until the PX suspended trading on January 31, 2001. The PX became insolvent in March.

major California investor-owned utilities hemorrhaged money from paying wholesale prices far higher than the regulation-capped retail prices at which California's electricity restructuring law compelled them to sell it. Pacific Gas & Electric (PG&E) reported ongoing losses of a million dollars an hour, with Southern California Edison (Edison) losing a comparable amount. San Diego Gas & Electric (SDG&E), unique in having satisfied the conditions for passing its wholesale costs directly to customers,[8] saw its retail prices soar briefly before emphatic customer protests led the California Public Utilities Commission (CPUC) and the California Legislature to re-impose retail price caps in August, 2000. As a resulting financial crisis worsened over the winter holidays, suppliers began to go unpaid, and accelerating doubt about the credit-worthiness of utilities apparently spurred some generators to make capacity unavailable to California, or to hike their bid prices to secure a risk premium. These reductions in supply and further wholesale price increases bled the utilities (and their credit-worthiness) further, creating a financial death spiral driving them toward bankruptcy.

The unprecedented rolling blackouts that ensued stunned and angered the public, creating a political crisis. Grim forecasts that the summer of 2001 would be even worse intensified the anxiety. Energy became the primary political concern of the state's voters, politicians, and media, as public officials groped for solutions while engaging in partisan attempts to assign blame elsewhere. Governor Gray Davis and a special session of the Legislature entertained numerous proposals and enacted several emergency bills while repeatedly criticizing their predecessors, utilities, electricity generators, Federal Energy Regulatory Commission (FERC) oversight of the wholesale markets, newly-elected President Bush (and Texas energy producers generally), and other targets of political opportunity.[9]

It was in this context of an electricity, financial, and political crisis that an *ad hoc* group of concerned academics, former public officials, and consultants convened—under the auspices of the Institute of Management, Innovation and Organization at the Haas School of the University of California, Berkeley—to discuss the situation and debate its possible solutions. The group decided to take a public stand to urge the State to take appropriate actions and to avoid counterproductive ones; accordingly, the "Manifesto on the California Electricity Crisis" was released at a press

---

[8] Under AB 1890 (1996), the California restructuring legislation, utility retail rates were frozen at 1996 levels (minus a customer-financed 10-percent reduction) for four years, or until a utility recovered its stranded costs of prior generation investments. Having satisfied this requirement, SDG&E began charging its customers the wholesale cost of power on July 1, 1999.

[9] Davis, Gray. "More Than California's Problem," *Washington Post* (op-ed), May 16, 2001; "Bush's Mistake In California," *New York Times* (op-ed), May 31, 2001.

conference on January 26, 2001. The more than thirty signers included two Nobel Laureates in economics, a former president of the California Public Utilities Commission, a former Governor of the Federal Reserve, well-known private energy consultants, and other distinguished professors from Berkeley, Stanford, Harvard, and other universities. Notably, this group achieved consensus even though a number of its members had previously expressed public disagreements with one another over California electricity restructuring issues.

2.   DIAGNOSIS AND PRESCRIPTIONS

The consensus expressed in the Manifesto did not question the importance or desirability of the principal goal of the 1996 restructuring - to instill competition into wholesale and retail electricity markets. Indeed, the Manifesto recognized such competition as a crucial component in solving California's energy problems, and expressed faith that the hoped-for fruits of deregulation could still be achieved.

Instead, the Manifesto was critical of the State's deregulation efforts as fundamentally flawed. In particular, restructuring was failing by standing in the way of the demand and supply pressures necessary to assure the market supply of a critical commodity. Neither a wholesale nor a retail electricity market could function for very long if prices were not permitted to bring demand and supply into balance, nor if critical intermediaries (the utilities) were to suffer insolvency as a result.

The Manifesto's analysis of these crises recognized two challenges, one near-term and one long-term. The near-term challenge was that caused by the destruction of the utilities' credit, since electricity markets could not function properly until these buyers were again seen as reliable purchasers and recipients of investment funds. The long-term challenge was to create an effective market environment to attract the investment needed to expand electricity generation capacity.

As a critical measure to address both challenges, the Manifesto called for customers to pay the full market price for at least that portion of electricity usage that utilities purchased on the wholesale market (i.e., that which the utilities did not produce themselves using their own power plants).[10] Allowing retail rates to rise would generate a triad of benefits: Higher retail prices would reduce demand, which would in turn lead to lower wholesale

---

[10] Having customers pay a market-based price on a marginal basis for at least some usage offered an opportunity to provide a broad conservation price signal without repeating the San Diego rate shock experience.

prices. Higher prices would also restore the utilities' financial viability,[11] which would both reduce the risk premium incorporated into bids and encourage additional supply into the California market. Even though any price increases obviously would be unpopular, it was critical to let the price mechanism play these traditional market roles to help stabilize these problematic markets.

The Manifesto also cautioned against harmful "solutions" the State might be tempted to pursue that might not work, or that threatened to come at an unacceptable cost to consumers and the California economy. These involved possible efforts to hide the current cost of the crisis by pushing it off into the future, or to substitute the State (and its resources and credit) in the place of private firms as generators, purchasers, or distributors of electricity. In part, these prescriptions responded to problematic calls for action made by various public figures, or contained in draft legislation.

So, the Manifesto's first caution was not to attempt to solve the immediate crisis by effectively foreclosing tomorrow's solutions. This included warnings against radical steps such as trying to create an Island California (cut off from the rest of the western electricity grid) to seek State control of the market; against the State becoming a permanent purchasing authority of electricity for consumers; and against investing State funds on large-scale energy projects.

Additionally, the Manifesto urged the State to avoid over-committing to long-term electricity contracts, since the high wholesale electricity prices of the moment were the result of market problems not yet solved. Signing long-term contracts then—before proper management of the financial crisis, thawed retail prices, and increased conservation were given a chance to lead to lower prices—would be the equivalent of buying high and selling low.

The Manifesto also urged the State not to socialize major components of California's electricity system. The drafters affirmed that it would be bad public policy for taxpayers to bear the risks of financing new generating plants, since that is a job the private sector was best equipped to perform.

3. THE STATE'S POLICY RESPONSES

In the ensuing months, Governor Davis, the Legislature, and the CPUC acted in almost direct contradiction to the recommendations offered by the Manifesto. Instead of reinvigorating the utilities as financially viable market participants, they allowed the utilities to reach virtual insolvency before

---

[11] Utility credit-worthiness could have been restored (in order to mitigate the crisis) while still leaving the ultimate financial responsibility for the utilities' debts as an issue for another day.

substituting the State's funds, credit, and hurriedly-developed electricity purchasing capability in their place to try to keep the lights on. Instead of keeping future electricity purchase contracts short-term and small, the State's new power buyers committed to over $40 billion in purchases over the coming decade. Except for a token "temporary" penny per-kilowatt-hour rise in January (later made permanent), the emergency rate increase petitions filed by the utilities in the fall were not acted upon until late March, through a CPUC process that delayed any actual customer bill impact until June — and then dedicated most of the revenue proceeds to the State to help cover its own power purchase expenses.

Fitful negotiations between utilities and the Governor about financial rescue approaches left PG&E to declare bankruptcy in April, and the Governor to announce a rushed agreement with Edison a few days later that would deliver the utility's transmission lines to State ownership. However, as of this writing only SDG&E's later-negotiated rescue plan has made significant progress towards implementation, and the accumulated power purchase debts of PG&E and Edison remain uncompensated.

A new State public power authority has been chartered with billions of dollars of borrowing authority[12] and an appointed board whose chair has endorsed construction of large State energy projects at the public's risk.[13] The State Treasurer is preparing for a $12.5 billion bond issue to help finance the State's own power purchase losses, and intends to use newly-legislated authority to recover all State power purchase and bond repayment costs through retail rates the CPUC will adopt summarily (without public review).[14] Customers, for their part, are about to be formally denied any further opportunity to shop for a retail provider whose price might undercut the State's recovery of its dedicated revenues.[15]

---

[12] The Consumer Power Conservation and Financing Authority was created by SBX1 6(2001), and authorized to borrow up to $5 billion to pursue "...augmenting electric generating facilities and to ensure a sufficient and reliable supply of electricity, financing incentives for investment in cost-effective energy-efficient appliances and energy demand reduction, achieving a specified energy capacity reserve level, providing financing for the retrofit of inefficient electric powerplants, renewable energy and conservation, and, where appropriate, developing strategies for the authority to facilitate a dependable supply of natural gas at reasonable prices to the public."

[13] "State Power Authority Takes First Step," *San Francisco Chronicle*, August 12, 2001.

[14] See CPUC, *Draft Decision of ALJ Kenney* (mailed August 28, 2001) in A.00-11-038.

[15] See CPUC, *Draft Decision of ALJ Barnett* (Mailed August 27, 2001) in A.98-07-003, and California Assembly Bill (AB) 1X, enacted February 1, 2001: "After the passage or such period of time after the effective date of this section as shall be determined by the commission, the right of retail end use customers pursuant to Article 6 (commencing with Section 360) of Chapter 2.3 of Part 1 of Division 1 of the Public Utilities Code to acquire service from other providers shall be suspended until the department [the Department of Water Resources] no longer supplies power hereunder."

At the same time, in some respects the crisis has abated positively. Widespread calls for conservation, favorable weather, and an acceleration of new power plant construction have thus far averted the summer blackouts that were forecast for California.[16] Natural gas prices are down considerably from high winter levels, and wholesale electricity market prices have recently fallen sharply to levels more typical of pre-crisis summers.[17] While characterizing the State's claim for $8.9 billion in refunds as unsupportable, FERC officials appear prepared to require wholesale power suppliers to credit California purchasers with about a billion dollars in refunds for excessive charges.[18] It will presumably come as no surprise that State officials have characterized these developments as a successful response to the crisis – for which they deserve asserted credit.[19]

To delve deeper into these developments and their implications, we will address the ongoing role of the State, and the dynamics and health of electricity markets before concluding with some observations about the future welfare of consumers and the California economy.

4. THE ONGOING ROLE OF THE STATE

For decades, Californians have taken power from both publicly and privately-owned utilities, with the latter serving the predominant number of retail customers, and the former supported by low-cost electricity from long-standing federal hydroelectric projects. Discussions about increased municipalization of power generation and delivery have persisted among

---

[16] California Independent System Operator, "CAISO Summer Assessment 2001, Version 1.0," (March 22, 2001) (describes forecasted summer electricity shortfalls); "State's Outlook Favorable for Blackout-Free Summer," *San Francisco Chronicle*, August 7, 2001.

[17] For example, August 6-10, 2001 prices were reported at $3.28-$3.72/MMBtu for natural gas (southern California border), and 5-7 cents/kwh peak, and 3-4 cents/kwh off-peak for electricity (NP15, and SP15). *California Energy Markets*, "Western Price Survey," August 10, 2001 (No. 630).

[18] "FERC Judge Washes Hands of Settlement Failure; Governor Davis Pledges Suit to Win Refunds Despite Long Odds," *California Energy Markets*, July 13, 2001 (No. 626).

[19] "State Building Natural Gas Reserve; Officials Refuse to Reveal Details of Contracts," *San Francisco Chronicle*, Thursday, August 16, 2001: "...[DWR official] Hart said the gas operation would reinforce the state's long-term power buying effort that he credits with taming the out-of-control electricity market in California"; "Electricity Crisis Lets Davis Generate an Image of Power," *San Francisco Chronicle*, July 26, 2001; "Electricity Fire Sale Reflects Success, Experts Say," *San Francisco Chronicle*, July 25, 2001: "... 'We have achieved one goal - which is to bring some normalcy to the entire process,' said Oscar Hidalgo, spokesman for the Department of Water Resources, the agency purchasing the state's power. 'The same things that happen normally to a utility are happening to us.' Hidalgo said California is now the biggest power buyer in the nation."

diehard public power advocates, but to little practical effect. Yet suddenly and dramatically, the policy responses of current State policy makers to these crises have transformed the role of government in this industry in ways that may be difficult to reverse (even if the political desire developed to do so).

During just the first eight months of 2001, various arms of the State of California: (1) became the buying agent for all net power requirements of the major investor-owned utilities (including assuming enormous associated financing obligations), (2) prepared to outlaw retail competition to create a monopoly position as a power retailer (of electricity not generated directly by the utilities); (3) made repeated attempts to acquire part or all of the state's transmission lines (while criticizing federal efforts to create a regional transmission operator that might interfere with State prerogatives as the lines' owner); and (4) created a public power authority with financing authority and an open-ended charter to built electricity facilities it considers useful. An historically-minded observer might be forgiven a confused feeling that his daily newspaper was publishing heretofore lost details of the interventionist 1930s.

Certainly the State was ill-prepared to assume this major role in the electricity industry. The State Department of Water Resources (DWR), for example, was primarily a water resource agency that was forced to hire, rapidly, a staff of energy traders and consultants to make spot electricity purchases, and help negotiate power purchase contracts. Thus, over several months a mix of contract employees and State workers experienced in other fields committed the public to over $50 billion for electricity spot purchases and future obligations, including the very long-term contracts against which the Manifesto cautioned.[20] Only time will tell whether future power purchases will be attractive by comparison to market levels,[21] although the return of natural gas prices to lower, historical levels combines with an increasing supply of power plants to suggest that ongoing electricity market prices will be far lower than those experienced during the past year — which

---

[20] "State Power Contracts Partly Revealed," *California Energy Markets*, June 22, 2001 (No. 623): ".".the State has committed to buying 600 million Mwh at a total cost of about $43 billion over the next decade"; "State Controller Calls for Nearly $6 Billion Short-Term Loan," California Energy Markets, August 10, 2001 (No. 630): .".A $5.7 billion infusion of short-term debt will be needed to cover the shortfall in state coffers caused by the billions of taxpayer dollars diverted to energy purchases, said state controller Kathleen Connell..[o]n top of the nearly $5 billion bridge loan [State Treasurer] Angelides nailed down and the $12.5 billion in long-term bonds he plans to seek in late fall to cover the net short the Department of Water Resources has bought on behalf of the three investor-owned utilities."

[21] "State Rethinks Contracts in Light of Low Prices," *California Energy Markets*, June 8, 2001 (No. 621); also supra note 22.

presumably, to at least some extent, are built into these contract prices. Reports of the recent resale of some contracted power at low market prices also do not inspire confidence in the economics of these contracts.[22] Similarly, while proposals to purchase transmission lines seemed intended primarily as a quid pro quo for State assistance in repaying utility power purchase debts,[23] the State appears to lack the institutional capability to operate the lines (which reportedly will require significant capital upgrades),[24] as well as any plan to earn an investment return on the lines above the regulated rate of return utilities have traditionally been paid on their book value. And the new public power authority is being created from scratch.[25]

At this stage, it is still largely a philosophical question as to whether this expanded State role in the electricity business will prove beneficial, or costly, to Californians. The Manifesto drew upon traditional expectations among economists (grounded in extensive experience) that such governmental roles tend to lead to higher costs and inefficiency, while risking the abuse by the state of its lawmaking authority to protect its commercial or political interests. We have already seen signs of both these concerns in the State power contracts, and the accompanying legal and regulatory efforts to insulate them against market forces by changing the rules of competition. In time, the results will be judged by the taxes and utility bills Californians will pay.

5. THE DYNAMICS AND HEALTH OF ELECTRICITY MARKETS

High wholesale electricity prices resulted from increased costs borne by power producers (e.g., increased costs for natural gas and pollution offsets), and from an imbalance between wholesale electricity market supply and demand. Of these, the State was best positioned for policy responses to the supply-demand imbalances, as the cost increases were generally beyond State control.

---

[22] "DWR Caught Long in Power Market, but It Is Not the Only One," *California Energy Markets*, July 27, 2001 (No. 628) (reporting July, 2001 market sales by DWR of 177,571 mwh of electricity originally bought for $118/mwh, and sold for $37/mwh).

[23] "Governor Davis Names Three to Negotiate California Public Interest in Utilities," press release from the Governor, January 29, 2001.

[24] "Different Ways of Coping for PG&E, Edison," *San Francisco Chronicle*, April 11, 2001.

[25] "State Power Authority to Take Energy Reins Today," *Los Angeles Times*, August 24, 2001.

Arguments persist about many of the possible causes of the wholesale market imbalances, e.g. the extent to which suppliers may have exercised market power, or whether fundamental market design flaws were largely to blame. But Manifesto authors who differed on these points agreed on two others - that high wholesale electricity prices would be reduced if (1) customers conserved electricity in response to market price signals (forcing producers to recognize that higher asking prices would lead to less electricity sold); and, if (2) California committed to paying its bills, and thereby restored the utilities to a reasonable financial condition (to reassure producers who might withhold output for fear of not being paid). These measures would also reduce expectations of future electricity market prices, thereby reducing the cost of long-term contracts that utilities ought to be allowed to enter into with few restrictions (contrary to the regulatory ban on such contracting that had left utilities so exposed to the spot market as the crisis developed).

However, the State's corresponding policy reactions were too little and too late. The rapid re-freezing of retail prices at low levels following the summer, 2000 rate shock panic in San Diego overlooked the conservation potential of the higher, market-based prices,[26] as well as opportunities to continue to send such price signals through rate designs that could have limited the pain to consumers.[27] Through the fall and into the winter of 2001, the CPUC effectively ignored the utilities' increasing financial distress by leaving retail prices frozen. Ultimately, increased retail revenues were dedicated almost entirely to the State (in its new role as power buyer) - leaving utilities nearly insolvent, and many important power producers unpaid (or only partially paid) for recent output.[28] And while significant retail rate increases were adopted at the end of March, further CPUC deliberations delayed until June any actual impact on customer bills.[29] By that time, the State had been actively signing long-term power purchase

---

[26] Bushnell, James T. and Erin Mansur. "The Impact of Retail Rate Deregulation on Electricity Consumption in San Diego," University of California Energy Institute working paper (PWP-082), Berkeley, CA (April, 2001).

[27] As the Manifesto later proposed (supra note 13).

[28] CPUC D.01-03-082 (March 27, 2001) raised retail rates by 3 cents/kwh for PG&E and Edison, and made permanent the prior 1 cent/kwn increase adopted in CPUC D.01-01-018 (January 4, 2001) – while requiring that resulting revenues be used for future power purchases only, at a time when absence of credit-worthiness precluded the utilities from purchasing their net electricity requirements on their own account. A companion order, CPUC D.01-03-081, determined an allocation of revenues to DWR for its power purchases.

[29] Specific customer rates for the three cent increase were set in CPUC D.01-05-064 (May 15, 2001), and began to be charged to customers in June.

contracts for months, seemingly without the market price-reducing benefits the retail price increases promised.

As we reach the end of summer, 2001, wholesale electricity prices have returned to much lower levels apparently influenced by significant conservation,[30] the addition of a few substantial new power plants, and sharply falling natural gas prices. Unhappily, it appears that these favorable market conditions came too late to benefit the State's extensive spot market purchases from early 2001 up to this point and its determined effort to negotiate a large portfolio of long-term electricity contracts while prices were high.

6. CONCLUSION: UNCERTAIN PROSPECTS FOR CONSUMERS AND THE ECONOMY

In fairness, California State government was dealt a difficult hand by the electricity market developments of the past year and a half. While the 1996 restructuring scheme was flawed and overly interventionist, no one foresaw the incredible run-up of wholesale electricity prices, nor its ability to persist (and worsen) throughout the fall and winter – in part due to other unusual events such as a severe drought in the Pacific northwest, and high natural gas market prices that were also unprecedented. No regulatory agency, legislature or governor could have been expected to respond immediately — and fully correctly — to such circumstances that confounded most prior expectations for these reforms.

And some governmental reactions were helpful, if belated. The public appears to have responded to extensive calls for emergency conservation, and the power plant certification process enjoyed a Prague Spring during which California's legendary approval delays were abbreviated. As a result, it may be some time before adequacy of generation capacity is again a concern for California — although, ironically, a prior surplus of capacity (spurred by a prior round of governmental error in electricity generation market management) may have been partly responsible for the attractive wholesale market prices of electricity that encouraged political pressure for California's electricity restructuring.[31]

---

[30] The California Energy Commission reported reductions in peak electricity demand (from "expected" levels adjusted for economic growth and weather) of 14.1, 10.7 and 8.9 percent respectively for June, July and August, 2001. http://www.energy.ca.gov/electricity/peak_demand_reduction.html (viewed September 4, 2001).

[31] Under the 1978 Federal Public Utility Regulatory Policy Act, the CPUC required utilities to sign standard purchase contracts for electricity from "qualifying facility" (QF) third-

However, our concern is that California businesses and consumers have been dealt an unnecessarily difficult hand by errors of crisis response by government that were avoidable. The CPUC and the Governor engaged in active denial during a critical period (from summer, 2000 nearly into spring, 2001) when any progress would have been welcomed, and likely paid large dividends in reducing the ultimate cost of the crisis.[32] For example, an earlier rate increase of perhaps half the ultimate 40 percent would have been highly useful even as late as December, 2000 to begin to influence customer demand, mitigate the credit concerns that overhung the wholesale market, and reduce wholesale prices. Yet well into the winter, the Governor continued to emphasize that no rate hike would be necessary, while publicly questioning the actions of his own CPUC appointees in adopting a modest, "temporary" increase from which the utilities were forbidden to replenish their finances. Likewise, the ad hoc creation of a multi-billion dollar State power purchasing operation — in the midst of an unprecedented, complicated crisis — represented a huge wager on the ability of government to perform such a task from almost no base of experience, or expertise. This gamble, which from the outside appeared to represent the triumph of hope and ideology over common sense, resulted in the large long-term contracts against which the Manifesto advised, and which we expect (although only time will tell) to cost the consumers of California many billions in future above-market payments for electricity. Nor do we believe the creation of the new State power authority to reflect any compelling need for taxpayer-funded electricity infrastructure, which will come with the usual concerns about the politicized (and costly) manner in which such efforts are often pursued. And while two utility financial rescue packages continue in various stages of uncertain debate and implementation,[33] it bears repeating that over

---

party generators. The CPUC's 1982 "Interim Standard Offer 4" (ISO4) contract included 10-year initial levelized prices that proved highly attractive to QFs after the economic assumptions underlying them changed (particularly when oil prices fell sharply in the early 1980s), but the required contract prices did not – until the CPUC belatedly suspended the offer in 1986. The inclusion in utility rates of the purchase cost of QF power was one factor encouraging large power customers to seek direct access to the wholesale market through restructuring, as were the low wholesale prices caused by a generation surplus to which the QF power plants contributed. Ironically, when wholesale market prices skyrocketed in 2000, many QFs sought to use a provision of AB 1890 (and later utility non-payment) to exit their ISO4 contracts and sell power for wholesale market prices instead. "QF Pricing Comparisons Lead to Wide Array of Outcomes in CPUC Case," *California Energy Markets*, May 5, 2000 (No. 565); "QFs Explore New Power Market Options," *California Energy Markets*, May 25, 2001 (No. 619).

[32] "Davis' Pledges Come Back to Haunt Him," *San Francisco Chronicle*, April 8, 2001.

[33] "What's Next for Edison? A Primer," *Los Angeles Times*, August 23, 2001; "Tentative Grid-for-Debt Deal with Sempra in the Works," *California Energy Markets*, June 22, 2001 (No. 623).

$10 billion in prior utility power purchasing losses are as yet unresolved in this patchwork of State actions.

All of this adds up to the unfortunate prediction that California's electricity consumers - both households and businesses - could be in for a rough decade of high prices that may lead some to long for the "high" utility rates of the mid-1990s that led to the restructuring experiment in the first place. Let us hope, for their sake, that our forecast is wrong.

Chapter #17

California's Electricity Crisis
What's Going On, Who's to Blame, and What to Do[*]

Jerry Taylor and Peter VanDoren
Cato Institute

1.   INTRODUCTION

Skyrocketing wholesale power prices in California and the daily threat of brownouts and blackouts have cast a pall over the merits of electricity deregulation. Liberals, led by California's governor, Gray Davis, blame the restructuring law passed in 1996 for the crisis, arguing that it left the state vulnerable to market manipulation by greedy power producers. According to Davis, the crisis is largely artificial but nonetheless a harbinger of things to come, not only in national electricity markets, but also in industries throughout the economy if we continue our mad rush toward laissez faire.

Conservatives for the most part agree that the 1996 reforms are primarily responsible for the crisis. They charge, however, that the regulations attached to those "reforms"—primarily the prohibition of long-term contracts between utilities and power generators and the imposition of a centralized daily spot market–are largely responsible for the price spike. The political right argues that California's regulations, crafted by environmental activists and anti-growth consumer groups, have long discouraged investment in new generating capacity and that the blackouts are a long-overdue flock of chickens coming home to roost.

---

[*]   The authors would like to thank Douglas Hale, Tim Brennan, Richard Gordon, Rob Bradley, and Charlotte Le Gates for their comments on earlier drafts of this paper.

A. Faruqui and B.K. Eakin (eds.), Electricity Pricing in Transition, 245-266.
@ 2002 Kluwer Academic Publishers.

While both sides are busily settling political scores, the real story of what happened in California is largely absent from most analyses. Accordingly, the important lessons that this crisis teaches about regulation and electricity are largely being overlooked: retail price controls are a recipe for disaster.

2.   THE PERFECT STORM

2.1   The Precrisis Market

California's electricity market appeared to work reasonably well from April 1998 through the spring of 2000. Wholesale electricity prices averaged $30 per megawatt-hour (MWh), or 3 cents per kilowatt-hour (kWh), in 1998 and 1999.[1] Those low prices allowed utilities to earn a return on sales and recover stranded costs even with the retail rate cap.[2] Between April 1998 and April 2000 the three incumbent utilities retired $17 billion in debt.[3] Incumbent utilities were also able to recoup some stranded costs by selling fossil-fuel power generators with a total book value of $1.8 billion for a combined $3.1 billion retail price.[4]

California's happy state of regulatory affairs changed radically in 2000-01 when two large supply shocks and one large demand shock simultaneously hit the state. None of those shocks was triggered by state policy. All of them, however, had a serious impact on wholesale electricity prices.

2.2   The Natural Gas Price Spiral

The first supply shock was a massive increase in regional wholesale natural gas prices, the fuel input for 49 percent of California's electricity

[1]   Paul Joskow and Edward Kahn, "A Quantitative Analysis of Pricing Behavior in California's Wholesale Electricity Market during Summer 2000," p. 8, Table 1, Unpublished manuscript, http://web.mit.edu/pjoskow/www/JK_PaperREVISED.pdf.
[2]   Paul Joskow, "California Can Tame Its Crisis," *New York Times*, January 13, 2001, p. A31.
[3]   Rebecca Smith and John R. Emshwiller, "California's PG & E Gropes for a Way Out of Electricity Squeeze," *Wall Street Journal,* January 4, 2001, p. A1."
[4]   Anthony   York,   "The   Deregulation   Debacle,"   January   30,   2001, http://www.salon.com/news/feature/2001/01/30/deregulation_mess/print.html.   In hindsight, the bids over book value should have been seen by analysts as a signal that industry insiders were betting with their investment dollars that future reductions in supply would make existing generators very profitable.

capacity in the first nine months of 2000[5] and nearly all its peaking capacity. In 1998-99 the average price of natural gas delivered to utilities in California was $2.70 per million British thermal units (Btu).[6] Wholesale spot gas prices at the Southern California "gate" rose to $5 per million Btu during the summer of 2000 and $25 per million Btu by December 2000[7] (the price reached $60 per million Btu on December 9, 2000).[8]

The increase in the price of natural gas was the logical consequence of the first significant cold winter after several mild winters and less-than-average amounts of natural gas available from storage.[9] The price increase was worse in California because demand was greater relative to pipeline capacity and total storage.[10] An explosion in August 2000 shut down the El Paso pipeline, which carries natural gas from Texas to Southern California. That accident reduced pipeline capacity into the state by 10 percent for several weeks.[11]

The surge in demand and subsequent price increase caught the industry by surprise. Natural gas prices had, after all, declined 25 percent in inflation-adjusted terms between 1985 and 1999.[12] Consumption from 1995 through 1999 was essentially flat.[13] Thus there were very limited incentives to increase production or hold inventory going into 2000.

[5] Edward Krapels, "Was Gas to Blame? Exploring the Cause of California's High Prices," *Public Utilities Fortnightly*, February 15, 2001, p. 29.

[6] Energy Information Administration, *Electric Power Annual 1999*, vol. 1, Table A20, p. 46, http://www.eia.doe.gov/cneaf/electricity/epav1/epav1.pdf.

[7] Krapels, p. 32.

[8] Bruce Radford, "Key to the Citygate," *Public Utilities Fortnightly*, January 1, 2001, p. 4.

[9] EIA estimated that, as of February 16, 2001, the United States had 1,038 billion cubic feet of natural gas in storage, 32.9 percent less than the five-year average. See Energy Information Administration, *Natural Gas Update*, February 22, 2001, http://www.eia.doe.gov/oil_gas/natural_gas/special/natural_gas_update/natgas_update.htm l. For a discussion of the "boom and bust cycle" that struck the natural gas market last year, see Judith Gurney, "U.S. Faces Natural Gas Price Shock," *Energy Economist* 229 (November 2000): 15-18.

[10] Storage in the West was an astonishing 55.6 percent less than the five-year average. On February 21, 2001, spot prices were $5.20 per million Btu at Henry Hub, Louisiana, $5.51 in Chicago, and $21.69 in Southern California. See Energy Information Administration, *Natural Gas Update*, February 22, 2001.

[11] Energy Information Administration, "A Look at Western Natural Gas Infrastructure during the Recent El Paso Pipeline Disruption," no date, available on the EIA *Natural Gas Update* Web site, http://www.eia.doe.gov/pub/oil_gas/natural_gas/feature_articles/2000/elpaso_disruption/el paso.pdf

[12] Alex Barrionuevo, John Fialka, and Rebecca Smith, "How Federal Policies, Industry Shifts Created the Natural Gas Crunch," *Wall Street Journal*, January 3, 2001, p. A1.

[13] Energy Information Administration, *Monthly Energy Review*, February 2001, Table 4.1, p. 73.

Given that 90 percent of a natural gas-fired generator's cost of producing electricity stems from fuel costs,[14] increased natural gas costs must increase electricity prices. A natural gas price of $20 per million Btu, for example, translates into a production cost of at least 20 cents per kWh for an average natural gas-fired plant and 32 cents per kWh for the least-efficient power plants.[15]

2.3   Why Natural Gas-Fired Electricity Determines Wholesale Prices

The least-efficient plants' costs are relevant because in commodity markets, like electricity, the costs of the most costly producer whose output is necessary to meet aggregate demand set the price for *all* the electricity sold, even power produced from other fuels. Thus, the California wholesale market cannot be understood without a full understanding of the increased cost of gas-fired electricity.

Many commentators have argued that, if utilities had signed long-term contracts with cheaper sources of power, the price spiral would have been less dramatic. Underlying that argument is the belief that the price of electricity in a free market would be a weighted average of the costs of long-term and spot prices.[16] While this belief has superficial plausibility, pricing output as a weighted average of differing prices of inputs would result in shortages.

Imagine that supermarket lettuce prices were different depending on individual farmers' costs. Once people realized that different prices existed for lettuce, shoppers would snap up the low-cost lettuce first. The supermarket would then ask the low-cost lettuce producer for more output at lower prices. But could the lettuce producer easily expand output at the same

---

[14] Personal conversation with A. Michael Schaal at Energy Ventures Analysis, Arlington, Virginia, December 19, 2000.

[15] The rate at which the energy contained in natural gas is converted into electricity is called the "heat rate." A standard heat rate for an older plant is around 10,000 Btu per kWh. The most inefficient plants require 16,000 Btu per kWh. Newer combined-cycle cogeneration plants have heat rates of around 7,000 Btu per kWh. The rate of 10,000 Btu per kWh is often used for rough calculation because natural gas prices in dollars per million Btu become electricity prices in cents per kWh. For example $10 per million Btu natural gas results in 10 cents per kWh electricity costs in a 10,000 Btu per kWh electric generating plant. See Krapels, p. 32, for an example of use of the rough calculation for an average and an inefficient plant. See J. Alan Beamon and Steven H. Wade, "Energy Equipment Choices: Fuel Costs and Other Determinants," Monthly Energy Review, April, 1996, Table 3, p. x, for a heat rate estimate on a new natural gas combined-cycle plant.

[16] For representative discussions, see Jim Yardley, "Texas Learns How Not to Deregulate," *New York Times,* January 10, 2001, p. A12; and Smith and Emshwiller, "California's PG & E Gropes for a Way Out."

low costs? He could if he didn't have to pay market prices for additional inputs, but that would be the case only if he had additional inputs under long-term contract. And if he or other producers had such spare capacity lying around under contract, market prices would already have decreased to reflect the competition among the owners of excess supply to use the units for which they had contracted.

If the low-cost producer did not have spare capacity under long-term contract and had to pay market prices for additional inputs like land and fertilizer, he would have to charge higher prices for the additional output to cover costs. In addition, the low-cost lettuce producer would realize that, rather than increase output at higher prices, he could raise prices on his existing low-cost output, eliminate the shortages, make more money than if he used weighted-average pricing, and not have to increase production.

Electricity markets are analogous to the lettuce example.[17] Even if California utilities had contracted for 99 percent of their supply at low prices, demand would exceed supply at those prices. And if the utilities attempted to expand output, they would have to pay market prices for the fuel input and would lose money on every additional sale unless the additional output was sold at market prices.

Thus, uniform prices for all sources of electricity regardless of their respective input costs is a (good) characteristic of free markets and not the result of the prohibition of long-term contracts in the California system or of the mandate that transactions take place through a centralized spot market.

2.4   The Weather

The West is more dependent on hydropower then any other region of the United States. Thus rainfall abundance and drought affect the electricity supply in the West more than any other region of the country. Just as California enacted its deregulation law in 1996, hydropower was more abundant than normal, reducing the returns that would come from investment in natural gas-fired production, just as California was switching to "deregulated" generation. Hydro output in the Canadian and U.S. West in 1997 was more than 30 percent greater than in 1992.[18]

This glut was then followed by a three-year dry spell that reduced reservoir and river-flow levels and thus reduced hydroelectric generation in

---

[17] Severin Borenstein, "The Trouble with Electricity Markets (and Some Solutions)," Program on Workable Energy Regulation Working Paper (PWP-081), January 2001, http://www.ucei.berkeley.edu/ucei/PDF/pwp081.pdf.

[18] S. A. Van Vactor and F. H. Pickle, "Money, Power and Trade: What You Never Knew about the Western Energy Crisis," *Public Utilities Fortnightly,* May 1, 2001, p. 35.

California by 20 percent from 1998 to 1999.[19] Hydropower from the Pacific Northwest further declined from an hourly average of 20,805 megawatts (MW) in 1999 to 18,075 MW in 2000. California hydropower likewise declined from an hourly average of 4,395 MW in 1999 to 2,616 MW in 2000.[20] From June through September 2000, hydro production throughout the West was, on average, 6,000 MW less than during the same months in 1999, equivalent to the output of 7 to 10 nuclear plants.[21] The practical effect of this reduction in hydroelectric generation was to leave California with little spare generating capacity during peak-demand periods.

In addition to the negative supply shocks (natural gas price increase and hydro shortage), demand increased during the summer of 2000 because of unseasonably warm temperatures (a 13 percent increase in cooling degree-days across the Pacific region from 1999 to 2000).[22] Temperatures in the Arizona subregion of the western grid averaged three to five degrees higher than normal.[23]

Accordingly, energy consumption and average daily loads during the summer of 2000 grew rapidly compared with the same period in 1999. Electricity consumption in the western states, excluding California, increased by 4.7 percent in from May 1999 to May 2000, and energy consumption in California increased by 5.8 percent over the same period. The increase in energy consumption from June 1999 to June 2000 was even greater–7.3 percent for the Western Systems Coordinating Council states, excluding California, and 13.7 percent for California. Within the ISO, average daily peak loads grew by 11 percent in May and 13 percent in June compared with the same months of 1999.[24]

Even though the increase in electricity demand may not sound dramatic, the effect on natural gas consumption was dramatic because of the hydropower reductions. During May-September 2000, natural gas consumption in California by utilities was 22.4 percent greater than for the same months in 1999.[25] In the West as a whole natural gas use for electric

---

[19] Energy Information Administration, *Electric Power Annual 1999*, vol. 1, Table A12, p. 39, http://www.eia.doe.gov/cneaf/electricity/epav1/epav1.pdf.

[20] Krapels, p. 30.

[21] Van Vactor and Pickle, p. 35.

[22] Krapels, p. 30.

[23] "Staff Report to the Federal Energy Regulatory Commission on Western Markets and the Causes of the Summer 2000 Price Abnormalities," p. 5-5, http://www3.ferc.fed.us/bulkpower/bulkpower.htm.

[24] Ibid.

[25] Consumption by utilities in May through October 1999 was 65.718 billion cubic feet (BCF). During the same period in 2000 consumption was 80. 463 BCF. See Energy Information Administration, *Natural Gas Monthly*, March 2001, Table 18, p. 43.

generation increased an astonishing 62 percent during the same period.[26] The hot summer of 2000 was then followed by a historically cold winter in 2001, ensuring that natural gas demand would remain high as heating elsewhere in the country competed with electricity production in California.[27]

We conclude from our analysis of supply and demand changes in California that the reduction in supply and increases in demand that resulted in wholesale electricity price increases are the result of natural weather variation interacting with market forces.[28]

2.4.1   Political Cloud Seeding

While we believe that the hydro shortage, natural gas price increases, weather shocks, and pipeline disruptions are the proximate causes of increased prices for wholesale electricity in California, the price increases were exacerbated by the existence of two politically created phenomena: nitrogen oxide (NO_x) emission permits and retail price controls. That "political cloud seeding" made the perfect storm even more intense and unpleasant.

2.4.2 NO_x Emission Permits

California adopted regulations in 1994 (known as the RECLAIM program) to control NO_x emissions in southern California.[29] In the winter of 1999, rights to emit NO_x were selling for about $2 per pound. By the summer of 2000, they were selling for $30 to $40 per pound.[30] Because an efficient gas-fired plant emits about a pound of NO_x per megawatt hour and an inefficient plant emits about two, NO_x credit prices of $30 to $40 per pound impose a cost of 4 to 8 cents per kilowatt-hour.[31] In January 2001,

---

[26] Van Vactor and Pickle, p. 36.

[27] The average temperature in the lower 48 states in November and December 2000 was 33.8 degrees Fahrenheit, the coldest since the start of modern nationwide record keeping in 1895. See Andrew Revkin, "Record Cold November-December in 48 Contiguous States," *New York Times,* January 6, 2001, p. A28.

[28] The events of 2000 took place in the context of an underlying trend of demand rising faster than supply. During the 1990s demand grew at 3 percent per year throughout the West, and supply increased by 1 percent per year. But the trend came about because of the very large glut in capacity in the early 1990s. See "Staff Report to the Federal Energy Regulatory Commission," p. 5-3 and introduction to section on "Are Environmentalists the Culprit?" p. 28.

[29] Gary Polakovic, "AQMD Moves to Overhaul Power Plant Emission Rules," *Los Angeles Times,* January 20, 2001, p. A23.

[30] Carl Levesque, "Emissions Standards: EPA, High Court, and Beyond," *Public Utilities Fortnightly,* January 1, 2001, pp. 46-47.

[31] Krapels, p. 32.

California regulators waived NO$_x$ permit requirements for power generators for the next three years, but the damage had been done.[32]

Although it's true that the RECLAIM program affects only those generators in the L.A. Basin–the source of only a fraction of the state's power–remember that the highest-cost source of electricity sets the price for *all* electricity sold through the western grid. Aaron Thomas, a manager at AES Pacific, points out that generators in the L.A. Basin "are setting the clearing price for everybody in California. And to the extent that that market is influencing markets in the West, all of a sudden you're getting these basin units driving costs for 50 million people in the West."[33]

2.5   The Damage from Price Controls

The wholesale electricity price increases in California also were exacerbated by the existence of retail price controls.[34] Normally, firms that increase prices experience fewer sales as a consequence. But retail price controls insulated power generators from the demand reduction consequences of their pricing decisions. They could increase prices without fear that those price increases would reduce demand and revenue.

How much would demand have been reduced if retail customers faced increased prices? Vernon Smith and his colleagues have conducted experiments to compare the behavior of auction prices under two scenarios: one in which consumers face rigid retail prices and a second in which 16 percent of customers face real prices that reflect supplier bids. Prices in the second scenario are as much as 30 percent less than prices in the first scenario.[35] Eric Hirst argues that if only 20 percent of the total retail demand faced hourly prices, and as a response to those prices reduced demand by 20

---

[32] Polakovic. In return for the waiver, power generators must install costly pollution control equipment over the next two years that would reduce emissions by 90 percent. Generators that remain in the RECLAIM market will be able to avoid buying credits by paying into a special account used to fund clean-air projects across the region.

[33] Quoted in Levesque.

[34] Even customers who switched suppliers under the retail choice program and were not officially governed by price controls were insulated from the increases in wholesale prices because the stranded-cost recovery charge varies inversely with the actual costs of electricity. High electricity prices severely reduced the stranded-cost recovery charge. See Severin Borenstein, James Bushnell, and Frank Wolak, "Diagnosing Market Power in California's Deregulated Electricity Market," Program on Workable Energy Regulation Working Paper (PWP-064), August 2000, pp. 8-9, http://www.ucei.berkeley.edu/ucei/PDF/pwp064.pdf.

[35] Stephen J. Rassenti, Vernon L. Smith, and Bart J. Wilson, "Using Experiments to Inform Privatization/Deregulation Movements in Electricity," *Cato Journal,* forthcoming; See also Steven Stoft, "The Market Flaw California Overlooked," *New York Times,* January 2, 2001, p. A19.

percent, the resulting 4 percent drop in aggregate demand could cut hourly prices by almost 50 percent.[36]

Retail price controls also reduced electricity supply because they caused the financial meltdown that led to generators fear of nonpayment. The California ISO reports that the March 2001 blackout was largely caused by generators' shutting down 2,000 MW of production because they had not been paid for the power they sold to the utilities for three or four months.[37] In light of PG&E's bankruptcy on April 6, 2001, the withholding seems very prudent.

3.   ALTERNATIVE EXPLANATIONS

Do supply and demand shocks fully explain high California electricity prices? What about market power, environmentalists' resistance to new power plants, the prohibition on long-term contracts, and the state run central auction?

3.1   Market Power

Edward Krapels believes that supply and demand shocks explain the December 2000 wholesale price in California but do not explain 5 cents per kWh of the average April through November 2000 price.[38] Paul Joskow and Edward Kahn also conclude that "high wholesale prices observed in summer 2000 cannot be explained as the natural outcome of 'market fundamentals' in competitive markets since there is a very significant gap between actual market prices and competitive benchmark prices . . . high prices experienced in the summer of 2000 reflect the withholding of supplies from the market by suppliers."[39]

3.2   Where Did All the Power Go?

Californians served by the ISO typically demand 45,000 MW of electricity during peak periods in the summer, and that electricity was available during the summer of 2000. By contrast, Californians served by the ISO demand only about 30,000 MW during peak periods in the winter. How

---

[36]   Eric Hirst, "Price-Responsive Retail Demand: Key to Competitive Electricity Markets," *Public Utilities Fortnightly*, March 1, 2001, pp. 34-41.

[37]   "California Suffers Two Days of Rolling Blackouts," *Energy Report*, March 26, 2001, p. 1.

[38]   Krapels, p. 33.

[39]   Joskow and Kahn, p. 30.

could 15,000 MW of power disappear and result in blackouts during the winter of 2000-01?

As wholesale prices increased after May 2000, the ISO enacted price controls in the market for daily backup and load-following (ancillary) reserves. On June 28, 2000, prices were limited to 50 cents per kWh and on August 7, 2000, to 25 cents per kWh.[40] And on November 1 FERC issued a "soft" price cap of 15 cents per kWh for both the California day-ahead power exchange and the real-time ancillary ISO markets.[41] As natural gas prices climbed to above $20 per million Btu and NO_x permit prices to above $30 per pound, the price caps were below the marginal costs of the least-efficient natural gas units with the greatest NO_x emissions in the Los Angeles Basin. Producers responded by stooping production or selling to other states in the West.[42]

Production also disappeared because the natural-gas units that ran all summer to replace the lost hydro output postponed needed maintenance and repairs until the winter 2000-2001.[43] Approximately 65 percent of the state's generating capacity is 35 years old or older.[44] This is particularly the case with the plants used to meet summer peak demand. Those gas-fired plants seldom operated steadily until the summer of 2000, and, because they are older and more inefficient, require additional repair and upkeep. At various times during the winter of 2000-01 power plants were out of service in unprecedented numbers.[45] During the first blackout on January 17 2001, for instance, fully 11,000 MW of in-state capacity was offline for repair and maintenance work.[46]

---

[40] Smith and Emshwiller, "California's PG & E Gropes for a Way Out"; and Joskow and Kahn, p. 29.

[41] Robert J. Michaels, "FERC's California Fix: Opportunities Lost and Found," *Public Utilities Fortnightly*, January 1, 2001, pp. 34-36. On December 15, 2001, FERC updated the order, and on March 9, 2001, FERC defined prices above 27.3 cents per kWh during stage 3 emergency periods in California in January 2001 as violating the soft price cap. See FERC press release, http://www.ferc.gov/news1/pressreleases/refunds.pdf. On March 16, FERC defined prices above 43 cents per kWh during stage 3 emergencies in California in February 2001 as violating the soft price cap. See Craig D. Rose, "Leaders Fault FERC, Call for Tougher Action," *San Diego Union Tribune*, March 17, 2001.

[42] "FERC Takes Action to Repair Calif. Market, But Few Are Pleased," *Energy Report*, December 25, 2000, p. 4.

[43] Van Vactor and Pickle, p. 36.

[44] "Reliability Picture Bleak in California," *Energy Report*, May 7, 2001.

[45] Rebecca Smith, John Emshwiller, and Mitchel Benson, "California Power Crisis: Blackouts and Lawsuits and No End in Sight," *Wall Street Journal*, January 19, 2001, p. A1.

[46] Rebecca Smith, John Emshwiller, and Mitchel Benson, "California Is Hit with Series of Blackouts," *Wall Street Journal*, January 18, 2001, p. A1.

Disbelief of the repair-and-maintenance argument is rampant. Harvey Rosenfield and Doug Heller of the Foundation for Taxpayer and Consumer Rights call for state agents to "obtain search warrants and subpoenas to enter the power plants to determine the true cause of the shortages. If necessary, the plants should be seized to protect the public health and safety."[47] But an investigation by FERC found that "generating outages in California at plants owned by Dynergy, NRG, and Reliant appeared to stem from increased demand and age of the units (boiler tube and seal leaks, turbine blade wear, valve and pump motor failures, etc.).[48]

Economists are divided about the role of market power versus necessary maintenance in explaining the extent of generator outages. Using publicly available estimates of the heat rates of electric generators and natural gas price data, Severin Borenstein, James Bushnell, and Frank Wolak generated estimates of the marginal costs of electricity. They then compared those estimates with actual prices observed on the California Power Exchange. They argue that the exercise of market power raised California electricity prices at least 16 percent above the competitive level (marginal cost) from June 1998 through September 1999.[49] Joskow and Kahn studied the summer 2000 California market and concluded that an inordinate number of plants were taken offline for maintenance when the price spikes were most intense. While that could be coincidence, it would also be perfectly consistent with the existence of market power: the withholding of some units of production to raise the price of electricity produced by other units owned by the same firm.[50]

3.3   An Alternative Explanation: Regulatory Perversity

But Bill Hogan and Scott Harvey argue that generators in California priced above marginal cost, not because they had market power, but because two characteristics of the California market created incentives for them to place high bids.[51] The California market solicited hourly bids from all generators a day in advance of production. In addition, once the day-ahead

[47] http://www.consumerwatchdog.org/ftcr/co/co000918.php3.

[48] Federal Energy Regulatory Commission, "Report on Plant Outages in the State of California," February 1, 2001, as described in *Public Utilities Fortnightly*, March 1, 2001, p. 18.

[49] Borenstein, Bushnell, and Wolak, p. 34.

[50] Joskow and Kahn, p. 20.

[51] Scott Harvey and William W. Hogan, "Issues in the Analysis of Market Power in California,"   October   27,   2000, http://ksghome.harvard.edu/~.whogan.cbg.ksg/HHMktPwr_1027.pdf. We do not discuss the California system's method for resolving transmission congestion, which Harvey and Hogan discuss on p. 8.

market cleared and produced hourly prices, the ISO asked for bids for reserve power. Under ideal conditions, arbitrage would result in similar prices for day-ahead and reserve energy in each hour, but because the ISO auction was conducted *after* the power-exchange auction, uncertainty existed about whether a generator would receive a higher price if it waited for the ISO to call on a unit as a reserve unit. This sequential feature converted the day-ahead auction from an everyone-gets-the-market-price auction (and thus generators would bid at marginal cost to ensure being selected to produce as long as costs were covered) to a pay-as-you-bid auction in which all participants, if selected, receive what they bid rather than the (potentially) higher market-clearing price.

Under such a market structure, firms lacking market power bid at expected market-clearing prices rather than at marginal cost.[52] And market-clearing prices for the reserve market during a capacity shortage in which retail consumers do not face high prices will be very high indeed because under the engineering procedures developed by the NERC, electric system operators such as the ISO must maintain generation reserves available in 10 minutes regardless of cost.[53] Producers have no reason *not* to name a stratospheric price for their power in those circumstances because the price they charge will not alter demand (remember, retail prices are capped and, even were they not, ratepayers face lagged-average rather than real-time marginal prices for electricity).

Tim Brennan argues that another characteristic of the California auction market also created incentives for high rather than low market-clearing prices.[54] The rules of the auction allowed generators to offer different amounts of electricity at different prices rather than all of their output at one price. Under those rules, generators had the incentive to offer a small amount of their output at very high prices because if the high bid was accepted they would receive that price for all their output. And if the bid were not accepted, the generators would lose only the sale of a small fraction of their possible output. Normally such bidding behavior would be unprofitable because the probability of the high bid's being accepted would be small, but, in a very tight supply situation, the probability of the bid's being accepted rises considerably, and the opportunity cost of the unsold power falls.

Even though perverse behavior by electric generators–induced either by actual withholding or by characteristics of the California auction–may have played some role in the price spike, the populist charge that the entire price

---

[52] Ibid., pp. 9-14.

[53] Ibid., pp. 4, 19; and Van Vactor and Pickle, pp. 40-41.

[54] Tim Brennan, "The California Electricity Experience," Paper presented at the 20th Annual Workshop in Advanced Regulatory Economics, Tamiment, Pa., May 24, 2001, p. 31.

spike can be explained by producer manipulation is clearly nonsensical.[55] Input costs and natural scarcities are responsible for most of the price hike.[56]

3.3.1   Are Environmentalists the Culprit?

Many observers have argued that the California electricity shortage is the result of environmentalists' and consumer activists' efforts to block new generation and transmission capacity, slowly starving the state of needed power.[57] California's reserve generating capacity decreased from 40 percent in 1990 to 15 percent at the beginning of 2000.[58] Last summer, demand for electricity was up 23 percent compared to 1992, yet generating capacity had only grown by 6 percent.[59]

3.3.2   A Murder with No Body?

Although it's certainly true that California has not seen a boom in power plant construction over the past decade, the claim that *no* new power plant has been built in California in more than 10 years is false. According to the California Energy Commission, 11 power plants (10 gas fired, 1 coal fired) with a generating capacity of 1,206 MW began operation in California in the 1990s.[60] Only two licensed projects with a generating capacity of 229 MW were dropped in the 1990s (one because of bankruptcy), and only one project was blocked by community activists: a 240 MW set of barge-mounted generators that were to supply peaking capacity to the San Francisco Bay area.

Not only were environmentalists a relative nonfactor in generator investment decisions in the early to mid-1990s, they scarcely played any role in blocking new capacity in the months leading up to the crisis. Since Governor Davis was elected in 1998, California has approved the construction of 9 power plants, and 44 plants with 22,600 MW of generating

[55] Nguyen T. Quan and Robert J. Michaels argue that, because generators have so many different opportunities to sell in the western market, the differentiation of market power from generators' simply pricing at their opportunity costs is impossible. See Nguyen T. Quan and Robert J. Michaels, "Games or Opportunities: Bidding in the California Markets," *Electricity Journal* 14, no. 1 (January-February 2001): 99-108.

[56] "Manifesto on the California Electricity Crisis," http://haas.berkeley.edu/news/california_electricity_crisis.html.

[57] See "California Messes Up," editorial, *Wall Street Journal,* December 28, 2000, p. A10.

[58] Daniel Eisenberg, "Which State Is Next?" *Time*, January 29, 2001, p. 45.

[59] Alejandro Bodipo-Memba, "Deregulation Has Been Smooth--So Far," *Detroit Free Press,* January 18, 2001.

[60] California Energy Commission, "Power Plant Projects before the California Energy Commission since 1979," January 16, 2001, http://www.energy.ca.gov/sitingcases/projects_since_1979.html.

capacity are currently under consideration by the California Energy Commission.[61] All of them will almost certainly be approved, but they will take several years to build.[62]

One could argue that investors' *fear* of environmental opposition and the costs of fighting that opposition to a successful completion explain the lack of plant construction in the 1990s, but little evidence exists to support that claim.

More obvious explanations for the lack of investment in new capacity are low prices and regulatory uncertainty. As discussed earlier, wholesale power prices in California were so low that there was little profit to be gained by increasing production. William Keese, chairman of the California Energy Commission, explains that the demand for generation that built up through the 1990s was for "needle peak demand," defined as the additional supply that would be needed to meet a 4,000 MW surge on the third consecutive day of record-high temperatures. Such peaks are rare (31 in the last 40 years), making new peaking capacity difficult to pay for. "To say people should have built power plants is not rational because they would have lost money."[63]

Moreover, regulatory uncertainty kept investors out through much of the mid-1990s. No power plant applications were filed with the California Energy Commission from 1994 to 1998 because of the investor uncertainty created by deregulation.[64]

Low prices and regulatory uncertainty also explains why states in which environmental activism are low also experienced no more investment in new power plants than California. Arizona's population grew 40 percent in the 1990s. The state has a pro-business climate, and yet its power production rose only 4 percent during the 1990s, mostly at existing plants. No one applied to build a major plant in Arizona between the late 1980s and late 1999.[65]

---

[61] From 1999 to the present, the California Energy Commission approved permits for 13 plants with a combined generation capacity of 8,464 MW. The commission is currently reviewing 13 additional projects with a combined generation capacity of 6,989 MW (none of which would be up and running until 2004); another 14 projects totaling 8,080 MW have been publicly announced but not yet submitted for state review. Two dozen additional projects totaling 12,425 MW are known to be under consideration by the investment community. "California Agency Confirms State Will Be Short Power This Summer," *Energy Report*, April 9, 2001, p. 6.

[62] John Greenwald, "The New Energy Crunch," *Time*, January 29, 2001.

[63] "CEC Chairman Says Summer 2001 Going to Be Tight," *Energy Report*, February 19, 2001, p. 5.

[64] Hal R. Varian, "Economic Scene," *New York Times*, April 5, 2001, p. C2.

[65] Peter Gosselin, "Most of West in Same Power Jam As California," *Los Angeles Times*, February 26, 2001.

If California *had* built more power generators during the 1990s, they would almost certainly have been gas-fired facilities because those were the cheapest and easiest plants to site. And because the electricity price increase is largely a reflection of the regional increase in the price of natural gas, those hypothetical plants would have reduced California prices only if enough plants had been built so that the new plants (with their lower heat rates) became the *marginal* source of electricity (rather than older natural gas plants with higher heat rates) and enough plants had been built to eliminate any scarcity rents that are currently in California prices.

In short, given the increase in natural gas prices observed in California, a massive investment in natural gas plants would still have resulted in 15 cent per kWh electricity. And if retail consumers were still paying 6.7 cents per kWh, bankruptcies and blackouts would still occur.

3.4   How Green Is the Grid?

Concern only about *California's* generating capacity, however, ignores the fact that the electric power market in the West is one large, interconnected system. There is no reason to demand that California internally generate all its power any more than to demand that Rhode Island produce all the food it consumes.

But the transportation of electricity requires transmission capacity. Even though transmission capacity costs only one-tenth as much as generation, landowners and other local residents resist new transmission capacity.[66] From 1989 through 1997 transmission capacity per MW of summer peak demand declined by 16 percent. Between 1997 and 2007, transmission capacity relative to summer peak demand is expected to decline another 13 percent.[67]

But the problem of transmission constraints exists all over the country, not just in California. While environmentalists have been known to agitate against new power lines, they're scarcely the only–or even the largest–group of NIMBY-ites to do so.

Moreover, the lack of new transmission capacity–like the lack of new generation capacity–has as much to do with economics as with politics. Incumbent utilities have little incentive to build new capacity that would make it easier for ratepayers to buy cheaper power from competitors in neighboring states.[68] And utilities do not have an incentive to invest in new

---

[66] Brendan Kirby and Eric Hirst, "Maintaining Transmission Adequacy in the Future," *Electricity Journal* 12, no. 9 (November 1999): 64.

[67] Kirby and Hirst, pp. 62-63.

[68] Peter Behr, "Shortage of Power Lines Looms," *Washington Post*, February 20, 2001, p. A1.

capacity when the profits allowed them by regulators are too low to make those investments particularly worthwhile relative to unregulated investments.[69] And with transmission rules still up in the air and unsettled at both the federal and state level, regulatory uncertainty is also damping investment.[70]

News stories also claim that natural gas pipeline capacity constraints are the product of environmentalist or NIMBY-ite opposition.[71] But there is little evidence to suggest that investors have been inhibited from increasing pipeline capacity when profit opportunities presented themselves. Extensive new pipeline capacity into northern California from Canada, for example, was built in the 1990s.[72] The Energy Information Administration observes that pipeline capacity "has grown with end-use demand, and as new supplies have developed, new pipelines have been built to bring this gas to markets."[73]

Little pipeline capacity into southern California was added during the past decade because investors found few opportunities for profit in the construction of new pipelines. The existing pipelines were not fully utilized until this year.[74] High natural gas prices, however, have revived interest in pipeline capacity expansions, and three significant projects were recently announced to take advantage of the newly discovered profit opportunities in transmission.[75] Clearly, the barriers to pipeline expansion in California are not too terribly high when profit opportunities exist.

---

[69] Ibid.; Rebecca Smith and John Emshwiller, "California Isn't the Only Place Bracing for Electrical Shocks," *Wall Street Journal*, April 26, 2001; and *High Tension: The Future of Power Transmission in North America*, Cambridge Energy Research Associates, cited in "Lack of Investment Jeopardizes Power Grid, Study Says," *Energy Report*, October 16, 2000, p. 1.

[70] Behr.

[71] Douglas Jehl, "Weighing a Demand for Gas against the Fear of Pipelines," *New York Times*, March 8, 2001, p. A1.

[72] Energy Information Administration, "A Look at Western Natural Gas Infrastructure during the Recent El Paso Pipeline Disruption," pp. 3-4, reports that capacity from Canada has had grown by more than 50 percent since 1990.

[73] See Stanley Horton, "Gas Pipelines: Solution, Not Problem," *Energy Perspective* 9, no. 2 (January 25, 2001): 3.

[74] Chip Cummins, "Natural-Gas Companies Discover California Is a Surprise Bonanza," *Wall Street Journal*, February 7, 2001, p. C1; and Energy Information Administration, "Status of Natural Gas Pipeline System Capacity Entering the 2000-2001 Heating Season," *Natural Gas Monthly*, October 2000, Table SR1, p. xiv, which reports that pipeline capacity from the southwest into California was only 55 percent utilized during 1999.

[75] "Pipe Expansions on Tap to Supply California Power Plants," *Energy Report*, September 4, 2000, p. 7; and "Herbert: Power Sector Responsible for Spike in Gas Prices," *Energy Report*, March 5, 2001, p. 5.

Accordingly, it's hard to single out the environmentalists as the "cause" of transmission constraints. While they've certainly played a role in opposing grid expansion, even states without well-organized environmentalist lobbies have found it difficult to remedy transmission congestion.

4.   LONG-TERM CONTRACTS

Long-term contracts do not offer a "better deal" than spot market purchases.[76] Long-term contracts simply reallocate the risk of price volatility from the consumer to the generator or marketer that provides the fixed-price guarantee. But the guarantee is not free. Sellers of such guarantees require a premium to accept this reallocation of risk. Generators (or marketers) do not offer fixed prices that result in lower returns than sales on the spot market. In fact, spot market prices for electric and gas utilities have historically been more favorable to consumers than contract prices.[77]

Everyone seems to have forgotten our previous love affair with long-term contracts. In the late 1970s, soaring electricity prices led Congress to pass the Public Utility Regulatory Policies Act. That law forced utilities to sign long-term contracts with independent power producers at the costs a utility would avoid because it did not have to build new supply itself. Regulators in California and New York set "avoided cost" administratively at rather high levels on the basis of the expectation of high prices for conventional fossil fuels, and utilities signed long-term contracts based on those expectations.[78] It seemed like a good deal at the time. But when electricity prices collapsed in the mid-1980s, the power companies had to keep buying this power while spot prices were around 2-3 cents per kWh. Largely because of the PURPA contracts, Californians by the mid-1990s were paying 35 percent more for their electricity than ratepayers in other states.   Thus the mindset of

---

[76] The exception arises if firms exercise market power. Under such circumstances, the greater the percentage of output the firm has sold in advance, the less the incentive for the firm to restrict output to raise the spot price. And once firms face both forward and spot sale possibilities, competition in both markets becomes more vigorous. See Borenstein, "The Trouble with Electricity Markets," p. 8.

[77] Ronald Sutherland, "Natural Gas Contracts in an Emerging Competitive Market," *Energy Policy*, December 1993, p. 1196.

[78] A 1981 Southern California Edison study forecast 1995 avoided electricity costs at 16.73 cents per kilowatt-hour, so the company willingly signed contracts to buy solar power at 15 cents per kWh even though the wholesale price of power in 1995 was actually about 2 to 3 cents per kWh. See Jeff Bailey, "Cater-Era Law Keeps Price of Electricity Up in Spite of a Surplus," *Wall Street Journal*, May 17, 1995, p. A1.

regulators going into California's deregulation was to avoid long-term contracts.

What if California's utilities had signed long-term contracts before the wholesale electricity price increases occurred? Wholesale prices would still have been sky-high from May 2000 through May 2001. That's because the causes of the increase–skyrocketing wholesale natural gas prices, a decline in regional hydroelectric power because of a three-year drought, and sharp weather-related increases in demand–have little to do with state policy. Had utilities entered into long-term contracts with generators at 6 cents per kWh, for example, before the spike hit, the discrepancy between the 6 cents per kWh contract price and the 15-50 cents per kWh cost to make that power would have forced the generators to declare bankruptcy.

5.   THE SOLUTION

Even Governor Davis understands that the repeal of retail price controls is the surest and quickest solution to the problem: "If I wanted to raise prices, I could solve this problem in 20 minutes."[79] Initially, he did not choose to do so because of the perception that demand is fixed in the short run and not significantly affected by price. Thus, freeing electricity prices would simply allow suppliers to charge whatever they wish, transferring wealth from consumers to producers.[80]

Very little empirical evidence exists on the effects of changes in electricity prices on demand because, under regulation, prices have not been allowed to vary much. But consumers in San Diego were part of a natural experiment from July 1999 through the end of August 2000. Under the terms of A.B. 1890, San Diego Gas and Electric had accumulated enough extra revenue from the start of "deregulation" in April 1998 to recover its stranded costs by the end of June 1999, so its rates were freed from controls and became a five-week moving average of wholesale prices.[81] By the time rate controls were reenacted after August 2000, retail rates had doubled. Bushnell and Mansur estimate that after controlling for weather and other sources of

---

[79]   Quoted in Smith, Benson, and Emshwiller "Major Kinks Emerge."

[80]   As we wrote this paper, the California Public Utilities Commission did increase rates 36 percent and made their temporary 10 percent hike permanent. The governor initially distanced himself from the increase but later supported it. See Todd S. Purdum, "In California, Reversal on Energy Rates," *New York Times*, April 6, 2001, p. A11.

[81]   James Bushnell and Erin Mansur, "The Impact of Retail Rate Deregulation on Electricity Consumption in San Diego," Program on Workable Energy Regulation Working Paper (PWP-082), April 2001, p. 4, http://www.ucei.berkeley.edu/ucei/PDF/pwp082.pdf.

non-price-related demand variation, a doubling of prices resulted in a demand reduction of 2.3 percent.[82]

This is an extremely low price elasticity, giving some support to the sentiments expressed by Governor Davis. But because California politicians were already discussing rate rollbacks during the summer of 2000, consumers may not have altered demand as much as they would have if they had believed that market prices were permanent. Ratepayers probably altered their investment and consumption very little because they believed (correctly) that the rate hikes would be temporary. Had they been convinced that prices would stay high for some time, even greater demand reductions would have been observed. Thus, the price responsiveness estimated by Bushnell and Mansur should be considered an extreme lower bound of the true value.

There is circumstantial evidence to back this point. Energy consultant Bill LeBlanc reported in a recent study that large commercial, industrial, and institutional consumers consider 17 percent of their aggregate demand load as "nonessential" and that, were retail electricity prices at 50 cents per kilowatt-hour (a threshold crossed by the wholesale market on dozens of occasions over the past year), 27 percent of the demand from those firms would be eliminated, resulting in "a huge, cost-effective dent in California's electricity shortfall."[83]

Price controls in petroleum markets in the 1970s were also justified on the basis of the lack of short-run price responsiveness on the part of consumers. But petroleum markets have been free of controls since the early 1980s, and in 1990-91 and 2000 price shocks occurred. How did consumers respond? In 2000 gasoline prices increased by about 50 percent. Despite the booming economy, all those SUVs, and an aggregate increase in vehicle registrations of 1.8 percent, aggregate gasoline consumption declined by about 1 percent from 1999, the first nonrecession reduction in recent history.[84] Ronald J. Planting, manager of information and analysis at the American Petroleum Institute, said, "The increase in prices was enough to spur at least some consumers to lessen lower priority travel and take other fuel-saving measures."[85] The belief that high prices do not affect demand in relatively inelastic markets and that voters will not accept price increases

---

[82] Ibid., p. 17. Steven Braithwait and Ahmad Faruqui, "The Choice Not to Buy: Energy Savings and Policy Alternatives for Demand Response," *Public Utilities Fortnightly*, March 15, 2001, p. 60, report elasticity estimates of .07 to .135 (a reduction in demand of 7 to 13.5 percent for a doubling of prices).

[83] "Experts Concerned California Setting Dangerous Precedents," *Energy Report*, March 19, 2001, p. 11.

[84] Matthew Wald, "Gasoline Use Fell Last Year, Oil Group Says," *New York Times*, January 20, 2001, p. B1.

[85] Quoted in Ibid.

does not hold in gasoline markets. Given the inelasticity of both the supply and the demand side of the electricity market, even moderate reductions in demand, as a result of freeing prices, would have a major effect on wholesale prices.

6.   INSTITUTE REAL-TIME PRICING

Although eliminating retail rate caps is a crucial component of any reform, it will fall short of fixing the fundamental problem in the electricity market. Monthly charges based on average costs do not keep up with the daily fluctuations in wholesale prices. Nor do they send the correct signals regarding marginal costs, a flaw long understood even by environmentalists who argue that average-cost pricing results in artificially low prices and, thus, excessive energy consumption.

Unfortunately, few consumers in California or elsewhere have meters that can register hourly consumption. The installation of load-sensitive meters would take time, but the technology is well established. In fact, load-sensitive meters have operated well in France for years.[86] Borenstein proposes that large commercial users be required to use real-time meters by this summer. The largest 18,000 customers account for about 30 percent of peak load in California, and the 10 million remaining customers account for the remaining 70 percent.[87] Borenstein believes that real-time pricing for the largest customers would have large benefits for the system as a whole because generators would then worry about the possible demand-reduction consequences of their pricing behavior. And the cost, time, and political difficulties of installing real-time meters for the remaining 10 million customers would be avoided.

According to EPRI users of 8,000 MW of load in California already have real-time meters in place.[88] If real-time pricing were available on a voluntary basis, EPRI estimates that peak demand would be reduced by 2.5 percent and prices by 24 percent.[89]

---

[86] Daniel McFadden, "California Needs Deregulation Done Right," *Wall Street Journal*, February 13, 2001, p. A26.

[87] Severin Borenstein, "Frequently Asked Questions about Implementing Real-Time Electricity Pricing in California for Summer 2001," March 2001, http://www.ucei.org/PDF/faq.pdf.

[88] Ahmed Faruqui et al., "California Syndrome," *Regulation* forthcoming.

[89] Ibid.

7. CONCLUSION

H. L. Mencken once said "democracy is the system that lets the people say what they want and then gives it to them, good and hard." That appears to be the case in California today. A recent Field poll asked Californians whether they would prefer a regime that capped the retail prices of electricity but produced the occasional blackout or a regime that had higher retail prices but no blackouts. Nearly two-thirds of the respondents favored the former.[90] Governor Davis is imposing an East German policy on the electricity market because most Californians prefer it.

A return to the old pre-1996 monopoly, cost-of-service, obligation-to-serve regime promises little. California regulators have demonstrated that they're not very good at overseeing that sort of enterprise. Remember, electricity rates in those "good old days" were 35 percent above the national average by the early 1990s.[91]

A complete state takeover of the system promises even less. The problem with state ownership of industries in, say, East Germany was not that those industries were run by ignorant East Germans rather than smart Californians. If the 20th century has taught us anything, it's that government is a horrible business manager and an incompetent economic planner no matter what industry we're talking about or what the nationality of the planner may be.

The simple fact is that high prices for power must be paid. Because it's politically difficult to have ratepayers pick up the tab on their monthly bill, California's politicians have decided to have taxpayers pick up the tab out of the state budget surplus. So Californians will not escape high prices. The problem with paying bills that way, however, is that the high prices will not affect electricity demand and thus will not play their intended role in allocating scarce goods as they would if they were simply passed on through the market.

That's largely because the California electricity crisis is not really a story about environmentalists gone bad, deregulatory details ignored, or unrestrained capitalists running amuck. It's a story about what happens when price controls are imposed on scarce goods.

---

[90] Rene Sanchez, "California Crisis Has Residents Seething," *Washington Post*, February 11, 2001, p. A1.

[91] John McCaughey and Kennedy Maize, "Is California Too Late for the Learning?" *Energy Perspective* 9, nos. 3-4 (February 8, 2001): 2.

Chapter #18

Empirical Evidence of Strategic Bidding in the California ISO Real-time Market

Anjali Sheffrin
California Independent System Operator

Abstract: The California deregulated wholesale electricity power market experienced tight supply conditions and unusually high prices following the summer of 2000, after a largely successful first two years of operation. This paper examines the exercise of market power in the California wholesale electricity market. Specifically, this study shows how individual suppliers' bidding behavior directly caused high market prices to be established. A small number of large suppliers were able to successfully employ bidding strategies to insure high market clearing prices, thereby increasing the price-cost mark-up by approximately 50% above competitive levels. Two main strategies were used: bidding at prices significantly above marginal cost of their generation, and withholding part of the available capacity from the market. This paper explains the methodology used to calculate the bid-cost mark-up index for each supplier, as well as the monopoly rents earned, as a result of these withholding strategies.

Key words: Strategic bidding behavior, California electricity market, Market power, Oligopoly, Bid-cost mark-up

1. INTRODUCTION

The California deregulated wholesale electricity power market experienced tight supply conditions and unusually high prices following the summer of 2000, after a largely successful first two years of operation. Previously, researchers and regulators had found evidence that the exercise of market power was causing the high price levels in the California

A. Faruqui and B.K. Eakin (eds.), Electricity Pricing in Transition, 267-281.
@ 2002 *Kluwer Academic Publishers*

Independent Operator (ISO) and California Power Exchange (PX) markets during the summer of 2000. However, they were unable to find specific evidence that any individual supplier in the market exercised market power to create those prices.[1]

This study shows how individual suppliers' bidding behavior directly caused high market prices to be established. In order to determine how individual suppliers' bids were responsible for raising prices above competitive levels, the study examined bids by individual suppliers (both in-state and importers) in the real-time imbalance energy market of the ISO. This was done by examining information on individual supplier bids into the real-time market, accounting for all bilateral and Power Exchange schedules from generation units, and utilizing generator unit specific data such as heat-rates at each range of output level and scheduled outages for each generator. Through an examination of this data, I am able to explain how individual suppliers successfully employed bidding strategies to insure high market clearing prices.

Earlier studies by the ISO examined the overall system price-cost mark-up in assessing the magnitude of the market power impacts for the combined ISO/PX markets.[2] These studies calculated how much system market clearing prices exceeded system marginal costs. System marginal costs are considered a proxy for a competitive benchmark of market clearing prices. These studies separated the causes of the overall increase in prices from May to November 2000 into component factors:

a) Increases in the cost of production,
b) increases attributable to high prices during hours of scarcity when operating reserves fell below 10%, and

---

[1]   See Appendix A of the Comments of the California ISO on the Order Proposing Remedies for California Wholesale Electric Markets in Docket Nos. EL00-95-000, et al., filed November 22, 2000. Additionally, several analyses were filed at FERC showing consistent findings on the level of price-cost mark-up in the California energy markets, including: "Diagnosing Market Power in California's Deregulated Wholesale Electricity Markets," Severin Borenstein, James Bushnell, and Frank Wolak, August 2000; "An Analysis of the June 2000 Price Spikes in the California ISO's Energy and Ancillary Services Markets," California ISO Market Surveillance Committee, September 6, 2000; "Report on California Energy Market Issues and Performance: May-June, 2000," California ISO Department of Market Analysis, August 10, 2000; and "A Quantitative Analysis of Pricing Behavior in California's Wholesale Electricity Market During Summer 2000," Paul Joskow and Edward Kahn. FERC staff, in a February 1, 2001 Report on Plant Outages, examined plant outages and concluded that it had not found evidence that the outages examined were the result of market manipulation or any form of market power abuse.

[2]   See Comments of the California ISO on Staff's Recommendation on Prospective Market Monitoring and Mitigation for the California Wholesale Electric Power Market in Docket No. EL00-95-012, March 22, 2001, Attachment B.

c) increases where market power alone resulted in prices above what would be expected in a competitive market.

The study found that increases in the cost of production accounted for only one third of the increase in hourly electricity prices (allowing for all natural gas and emission credits to be purchased at spot prices may have severely overstated the actual costs incurred). Higher loads and reduced imports accounted for an additional few percent. Price spikes represented approximately one sixth of the total increase during hours of scarcity,[3] with a generous allowance for scarcity conditions when total supply was less than total load plus 10% operating reserves. The analysis made specific allowance for scarcity conditions, because some have argued that price spikes are important market signals that facilitate reduction in demand and encourage new supply to enter the market. However, a significant level, amounting to approximately 50% of the increase in prices, could not be attributed to production cost increases or needed price signals during periods of scarcity. The analysis concluded that approximately one-half of the price increases was due to the influence of market power exercised by suppliers.

These initial study findings caused us to take steps to identify individual suppliers who caused system prices to be significantly above explainable competitive levels and examine how those suppliers were bidding to exercise this market power. This current study of individual supplier bidding behavior provides the direct link between actions of suppliers and the resulting market clearing prices which were maintained significantly above competitive levels in the California electricity markets during the period of May to November 2000.

2. METHODOLOGY

2.1 Two Forms of Withholding

I observed suppliers employed two main bidding strategies to influence market clearing prices and maximize their firms' profits. They either submitted bids at prices significantly above marginal cost of their generation unit, or withheld part of the available capacity from bidding or scheduling

---

[3] The analysis used a very generous allowance for "scarcity," defining scarcity as the condition when total supply is less than total demand, represented by total load plus 10% (for reserves). The study explicitly accounted for scarcity conditions, which may justify price spikes to attract new investment. However, many of the price spikes occurred outside these scarcity hours, giving rise to market power concerns.

into the market.[4] Both bidding strategies will prevent some economic generation capacity, which is generation with a marginal cost below the market clearing price, from being utilized to serve the load. This withholding of available and economic production resources is referred to a simply "withholding" hereafter.

I catalogued this action of withholding into two forms: economic withholding, which is bidding significantly above marginal cost of production, and physical withholding, which is not bidding or scheduling all available resources into the market. Either form of withholding prevents some lower cost resources from supplying the demand and requires the ISO to buy more costly resources to meet the load. This results in higher market clearing prices than what a competitive outcome would have produced. These two bidding strategies are illustrated in Figure 1 below.

This graph shows how a large supplier acting as an oligopolist (an oligopolist in economics is one of a small number of large suppliers able to influence prices in the market) would bid to maximize profits either by withholding capacity or bidding higher to achieve the same result. The net market outcome is higher prices than what would have resulted from a competitive market.

---

[4] Although only a fraction of actual generation to meet the load is bid into ISO real-time market, all generation serving ISO control area load must be scheduled through ISO in the day-ahead or hour-ahead time frame. Any capacity of a generation unit either scheduled into ISO or bid into ISO real-time market is considered offered into the market. Any remaining capacity of a generation unit, after deducting outages, is considered to be withheld from the market.

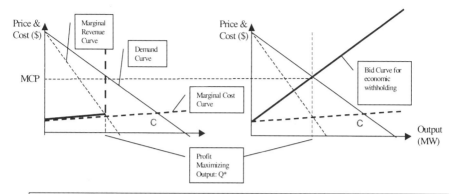

A market characterized by dominant suppliers who can be pivotal in setting the market clearing price is referred to as an oligopoly. An oligopolist in the market faces a residual demand curve, and has two bidding strategies available: physical withholding (left panel) and economic withholding (right panel). Both strategies result in the same reduced output available to the market at prices inflated above competitive levels. This outcome can be compared to expected competitive production and price levels, which are at the point marked "C."

Figure 1. Two Types of Alternative Bid Curves Allow Suppliers to Maximize Profits by Reducing Output and Inflating Prices above Competitive Levels

2.2　Determine Effective Bid Prices by Each Supplier

The study examined how a supplier's bids in the ISO's real-time market can set market clearing price. The first step of the study was to quantify the direct responsibility suppliers had in setting the market clearing price through their effective bid. The effective bid for each supplier was calculated as either the generator's highest accepted bid price or its equivalent if the full available capacity was not bid into the market.

I compared the suppliers' bids into the real-time market to their cost of supplying energy in real-time after accounting for all bilateral and PX energy schedules. If a supplier does not withhold and bids all economic resources at production costs, its effective bid price is the bid price for its highest bid accepted by the market. When a supplier withholds (economically or physically) it bids effectively at the market clearing price. In this case, the supplier's effective bid price is the market clearing price. Figure 2 is an illustration of different bidding patterns using the effective bid price and bid-cost mark-up indices.

The analysis shows that, through high bids, a supplier economically withheld its resources and set the market clearing price directly by bidding close or equal to the market clearing price. Acts of physical withholding helped set high market clearing price by making more expensive bids from other suppliers determine the market clearing price. Although actions of physical withholding set market clearing prices indirectly, they are equally as effective as economic withholding and are unquestionably as responsible for the resulting market prices.

2.3   Calculate Bid-Cost Mark-up Indices

The second step of the study was to utilize the effective bid price for each supplier by hour and calculate a bid-cost mark-up index. The bid-cost mark-up is the difference between effective bid price and the corresponding marginal cost of generation for each supplier. This calculation allowed me to identify and assess the exercise of market power for all hours during the period and analyze the ability of individual suppliers to determine prices in the market.

In calculating the suppliers cost for the bid-cost mark-up, I used characteristics of each generation unit, such as heat rates, variable costs and capacity availability. I used the California border spot market price for natural gas for all fuel purchase estimates, although this likely overestimates the price of actual fuel used. Emission costs were based on the spot price of NOx emission credits, which also likely overstated actual emission costs for production. I calculated available plant capacity by deducting all scheduled outages from a generation unit and allowing a 10% forced outage rate which is above industry standard outage rates for gas fired plants of the current age.[5]

2.4   Bid Price in Response to Higher Demand

The third step in the study was to calculate how much more individual suppliers could have increased prices had market demand been slightly higher. Each supplier had bids in the market ready to be accepted for slightly higher or lower levels of system demand (plus and minus 50 MW change in demand). The results shows, by using the typical bidding patterns identified, suppliers were not only able to effectively raise the market clearing price but also ensured it remained high within a reasonable range of variation in demand.

---

[5]   Used NERC GADS (Generation Availability Data Standards) for steam plants of average age of 40 years.

2.5 Excessive Profit Resulting from Market Power

The fourth step in the study was to calculate the monopoly rents earned as a result of these withholding strategies.[6] Monopoly rents are the amount of excess profit above competitive levels extracted by each supplier as a result of their individual bidding actions. Rent for a supplier, in a given hour, was calculated as its effective bid price minus system marginal cost (the competitive market price). The rent calculation does not use the market clearing price (MCP) but instead a supplier's bid price, thereby not attributing profit earned by price takers as monopoly rent. Actual rent earned by a supplier might be higher than the calculated monopoly rent when a higher effective bid by another supplier set the MCP.

This study defines market power as the ability of a supplier to set the market prices. It uses variable cost of production and does not consider fixed cost of production.[7] Earlier analyses have examined whether the variable profits earned during this time period were adequate after considering the fixed costs of investment. This analysis found that the amount of monopoly rent earned was excessive as measured against any standard. At prices seen in the market, investments in new plants could be paid in full in less than two years.[8]

2.6 Checking with Economic Models of Market Power

Finally, I compared the actual bidding strategies with an economic model's predictions of behavior in an oligopolistic market. This allowed us to confirm that our observations were not simple hit-and-miss pricing behavior but that the bidding behavior was consistent with what would be predicted from profit maximizing behavior in an oligopolistic market.

---

[6] The phrase monopoly rent, as defined in the text, does not imply any single firm has complete domination over the market. Here I use the term to refer to excess variable profit earned by many of the large suppliers through exercising of market power.

[7] Which is convention used in economics when examining short term operational behavior.

[8] See Comments of the California ISO on Staff's Recommendation on Prospective Market Monitoring and Mitigation for the California Wholesale Electric Power Market in Docket No. EL00-95-012, March 22, 2001, Attachment B.

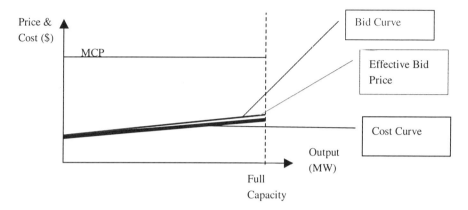

Figure 2a. **Price Taking Supplier**: The effective bid price is calculated as the suppliers' bid price at its dispatch quantity. It can be less than MCP as in the figure below or equal to MCP when the MCP line intersects the bid curve and cost curve. The bid curve shown below illustrates that this supplier's bidding has no significant bid-cost mark-up.

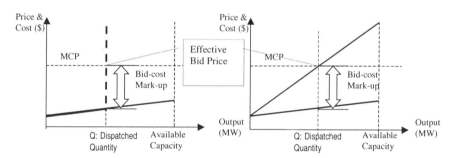

Figure 2b. **Strategic Suppliers**: Use of physical withholding or economic withholding. Effective bid prices are calculated to be equal to the Market Clearing Price (MCP) in both cases shown below. Either the supplier had a bid in at the MCP or not all capacity was bid in which is the same as bidding at or above the MCP. Under both strategies, supplier set the MCP and their calculated bid-cost mark-ups are significant as shown in the figure.

3.   RESULTS AND FINDINGS

This study reviewed bidding activities from five large in-state non-investor owned utility suppliers and 16 importers in the CAISO real-time market for each hour between May and November, 2000. In this report, I show only the results for the five large in-state generation suppliers.

Since the real-time market represents the final opportunity to sell power from a generation facility, it is easier to determine a seller's cost to produce energy because the seller has no other opportunities left to supply power. As the real-time market is the last market to sell, this means there is less concern about holding back a bid while waiting for a better opportunity in another market. This allows a direct comparison of detailed bidding data in the real-time market to the cost of supplying energy in real time after accounting for all bilateral and PX energy schedules, and calculating the level of the mark-up for each supplier. In reviewing the bids, I identified five typical bidding patterns.[9] Only one of these patterns was consistent with competitive bidding. The other four displayed some form of withholding; either physical withholding, economic withholding or a combination of both.

Withholding patterns and bid-cost mark-up indices indicate that most of the five in-state suppliers bid in patterns consistent with the exercise of market power. Many suppliers bid in excess of their marginal cost of generation either through economic or physical withholding. These bidding strategies contributed significantly to the system price spikes in the summer and fall of 2000. The observed bidding pattern shows that suppliers bid in expectation of increasing the market clearing price and maintaining high prices even if demand were to fall. As a result, many of the suppliers earned significant amounts of excess profit (or monopoly rents) under all load conditions.

3.1 Some of the Important Findings of the Study

Withholding, especially economic withholding, impacted the market for most hours from May to November, 2000. Among the 25,000 hourly bidding profiles studied (about 5000 hours for each supplier) of the five large in-state generation owners, less than 2% of these hourly profiles displayed no clear pattern of withholding or market power. The other 98% of hourly bidding

---

[9] The five typical bidding patterns are classified as follows: 1) No withholding (Pattern 0). This is the only bidding pattern that does not represent market power and did not inflate market clearing prices; 2) Full output at high mark-up (Pattern 1). Although full capacity is bid in, they are all at significantly inflated prices. The highest bid price is used in calculating the effective bid price of the supplier; 3) Physical withholding with no mark-up for bids submitted (Pattern 2). This is the pure form of physical withholding. Although there is no significant bid mark-up, there is significant unused economic generation capacity. In this case, the effective bid price is the MCP; 4) Physical withholding used in combination with significant bid mark-ups (Pattern 3). This consists of physical withholding at high loads, along with significant mark-up to ensure higher prices in case the level of system demand does not require all the bids submitted; 5) Economic withholding (Pattern 4). The only difference between Pattern 4 and Pattern 1 is that with Pattern 4 there is still available capacity that did not clear the market. Capacity is bid in at a significant mark-up. In the summary chart of bidding patterns in Figure 3, patterns 1 and 4 are combined into Economic Withholding and patterns 2 and 3 are combined into Physical Withholding.

profiles displayed various withholding patterns leading to inflated market prices. Figure 3 summarizes withholding patterns and frequencies.

Economic withholding was the dominant bidding strategy utilized to inflate the prices and had a significant impact in raising prices. Economic withholding was used more than 60% of the time if the five in-state generation owners are averaged. Two in-state suppliers used it more than 80% of the time. It was not only used consistently, but also accompanied by bid prices marked-up well above cost resulting in significant impact on the market.

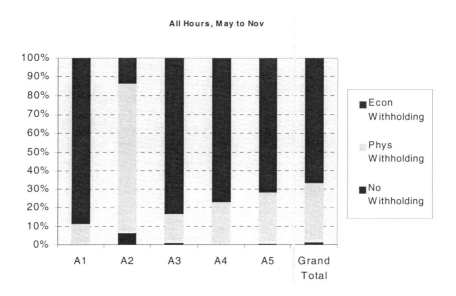

Figure 3. Frequency of Withholding by Type for 5 Large In-state Suppliers: A Withholding Strategy (both economic and physical withholding) was Utilized During May to November 2000 for More than 98% of the Hours.

Table 1. Frequency of Withholding by Types (Large In-state Suppliers)

Suppliers	A1	A2	A3	A4	A5	Grand Total
No Withholding	0%	6%	1%	0%	0%	2%
Physical Withholding	11%	80%	16%	23%	28%	31%
Economic Withholding	89%	14%	84%	77%	72%	67%
Grand Total	100%	100%	100%	100%	100%	100%

Physical withholding was less prevalent during the summer and fall of 2000. All five in-state suppliers studied, on average, used a physical withholding strategy for less than 30% of the time. Two of the suppliers used that strategy for less than 15% of the time. Physical withholding included the activity of declaring a unit on outage or otherwise unavailable or simply not scheduling or bidding available capacity into the ISO or PX market. The low frequency of observable physical withholding highlights the importance of monitoring economic withholding (bidding excessive prices above cost) and explains why the review of physical outages alone is inadequate to uncover supplier behavior producing high market prices. Economic withholding rather than physical withholding may be the most significant source of the problem in the market during the study period.

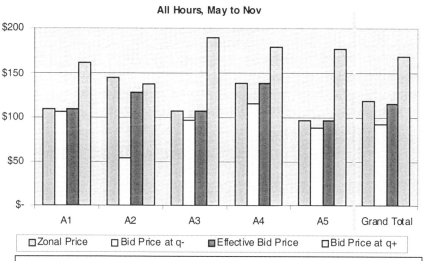

This chart compares the zonal market clearing price with the effective bid price for each supplier at their dispatched output level. This chart also reports the bid prices for a 50 MW change in the dispatch quantity (q+). It shows that if the system demand were 50 MW higher, the suppliers would have been successful at setting even higher prices. If system demand were somewhat lower, the suppliers' bids even for lesser amounts were high enough to support the high price levels and keep them from falling lower.

The average effective bid prices are close to the average zonal MCP because of the fact that for most hours, most in-state suppliers were effectively bidding and setting the zonal MCP. For any hour, when a supplier is withholding, either economically or physically, the supplier is effectively bidding at the zonal MCP, i.e., setting the zonal MCP. Only when they do not withhold, their effective bid price will be significantly below MCP.

Figure 4. High Bid Prices Guaranteed Setting High Market Clearing Prices by Five Large In-state Suppliers

Frequent economic withholding and physical withholding were accompanied by high bid prices and high bid-cost mark-up. On some summer peak months, the average bid-cost mark-up was more than $100/MW/hr. All five of the suppliers marked up their bids significantly above costs and did so in every month from May to November. Figure 4 reports the average effective bid price and bid-cost mark-up by month for each of the large suppliers.

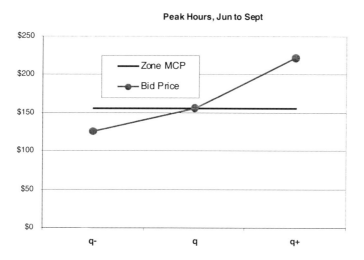

Average Bid Prices for 4 (out of 5) most aggressive large in-state suppliers:
• At q, these suppliers bid MCP most of the time. On average, it is almost equal.
• If the demand were higher by 50MW for each supplier (at q+), they had much higher bid price ready to push up the price much further.
• If the demand were lower by 50MW for each supplier (at q-), they still had fairly high bids in waiting to support high prices.
• Note the increase for higher demand will be much more significant than the drop when demand is lower. This is another indicator of well planned strategic bidding.

Figure 5. Strategic Bidding That Ensures High Market Prices for all Demand Conditions (Bidding at Output Levels Below and Above Actual Dispatch Level)

Many suppliers used well-planned strategies to ensure maximum possible prices at all load conditions. A 50 MW increase[10] in the amount demanded from a supplier's portfolio would have increased the market clearing price substantially. A review of suppliers' bid prices above and below the actual dispatch quantity revealed the strategic nature of their bid schedule. If the demand were higher, suppliers have much higher bid prices ready to push up the price much further. If the system demand were lower, suppliers still have

---

[10] The exact incremental amount used in this test is the higher of 20% of real-time dispatch quantity or 50MW. Therefore, the increment is sometimes more than 50MW. Either value is to simulate a high likelihood demand shock in the real-time market.

fairly high bids in waiting to support high prices. Figure 5 shows the impact of this bidding strategy in maintaining prices within a tight band for various output levels.

As a result of the excessive bid prices and associated bid-cost mark-ups, many large in-state suppliers and importers earned large amount of excess profit (or monopoly rent), calculated as the difference between a supplier's effective bid prices and the system marginal cost (the benchmark of a competitive market outcome).[11] Figure 6 reports the average monopoly rent per MWh for the five in-state suppliers. Figure 7 reports the total cumulative monopoly rent (sum of Rent times Dispatch Quantity) over the period of study for the suppliers that earned the largest total economic rents during the period studied.

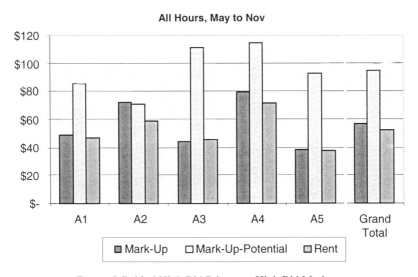

Figure 6. Behind High Bid Prices are High Bid Mark-ups

Although this total only covers the rent earned by these large suppliers, the total cost impact on the ISO real-time market as the result of their exercise of market power is larger since it must include added revenue earned by other small suppliers and generation units owned by other companies, municipal utilities, and state agencies. Even though these other

---

[11] Some additional details of the definition of monopoly rent are: Rent is defined as greater or equal to zero. Rent is calculated as bid price minus individual marginal cost if individual cost is greater than system marginal cost. Finally, the rent calculation does not use MCP but a supplier's bid price. So it does not attribute profit earned by price takers as monopoly rent, which is the indirect market impact of the strategic bidders' activity on other suppliers in the market.

suppliers may not have bid high and set the market clearing prices, they received the clearing price in a uniform price auction market.

Looking beyond the ISO real-time market, the exercise of market power had a significant impact on the PX market. Price shocks in ISO market quickly spread to PX the next day due to the expectation of high ISO prices. It is also conceivable that the suppliers used withholding as an integral part of their overall strategy to influence market prices in both ISO and PX markets. Since the trading volume is many times higher in the PX than in the ISO real-time market, the indirect impact of market power would be in multiples of that of the ISO real-time market. Further analysis of bidding data in the PX can verify the indirect and direct impact of exercise of market power.

In exploring the issue of whether an oligopoly pricing model helps to explain the observed bidding pattern in the California power market, I found evidence that the dominant bidding pattern is consistent with two characteristics of a supply function equilibrium model of oligopolist pricing. There was a positive correlation between bid-cost mark-up and system load and a positive correlation between bid-cost mark-up and the quantity dispatched from a large supplier. These confirm two of the important characteristics of a supply function equilibrium pricing strategy.

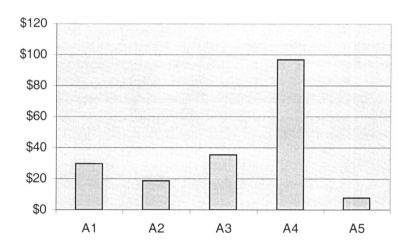

Figure 7. Estimates of Excessive Rents Earned in Real-Time Market for the 5 In-state Suppliers (Cumulative Rent for May to November 2000)

4. CONCLUSIONS

This study helps to fill an information gap left by most studies of market performance in Summer 2000. To date, most other studies have concluded that there was market power on a system-wide basis and significant price-cost mark-up indicating serious market power exercise. However, previous studies failed to identify individual exercise of market power and could not establish a link between individual strategic bidding actions and their impact on market clearing prices. This study provides the evidence that many large suppliers actively engaged in strategic bidding efforts and that their activity had a direct impact on the market prices. Moreover, their bidding strategies were not ad hoc but consistent with a certain model of oligopoly pricing behavior. This implies the systematic exercise of market power to maximize profit.

Chapter #19

Energy Modeling Forum Conference
Retail Participation in Competitive Power Markets

Elizabeth Farrow
Stanford University

Abstract: The Energy Modeling Forum held a special conference on Retail Participation in Competitive Power Markets at Stanford University on June 21-22, 2001. The conference discussion covered a range of important points on key issues, the more important of which are discussed briefly in this chapter.

Key words: Retail competition; Market-based pricing; Customer response; Load reduction schemes

1. OVERVIEW

California is now paying the price of focusing on generation to the exclusion of demand response in the design of its restructured power sector. This is not a problem peculiar to California, it applies in some degree or other to all power markets, but in California finding a way to introduce demand side participation in a more formal way offers perhaps the best hope for an accelerated solution to the immediate problem which is to reduce peak loads and change patterns of consumption.

Changing retail pricing methods can encourage conservation and load shifting. Historically, prices were averaged over large populations and over long periods of time. With a heterogeneous population, the former ensures cross-subsidization, and continuing with the latter neglects the hourly cost variation of the market. Sending price signals that accurately reflect the cost of production and transportation encourages more efficient use of resources. Customers paying full cost for what they consume must receive the full benefit for positive choices they make in modifying consumption patterns in

A. Faruqui and B.K. Eakin (eds.), Electricity Pricing in Transition, 283-293.
@ 2002 *Kluwer Academic Publishers.*

response to high prices. Real-time pricing (RTP) is one way of achieving this. RTP, in practice, involves a supplier providing hourly forward prices (day- or hour-ahead) and adding risk management. Pure spot pricing would expose the customer's entire load to the price changes. Customers could choose this option and provide their own risk management. When the supplier provides the risk management service a certain amount of load is set at a fixed price. If suppliers provide risk management they must be allowed to cover these costs with contracts.

There are other approaches to load reduction. Larger customers, by virtue of load size, have a major role to play. For example they can participate in programs offering price reductions for voluntary load shedding or they can actively participate in the wholesale market bidding in load reductions. Smaller customer too can provide valuable service in load management at an aggregate level.

The conference provided a forum for participants to share experiences, research and ideas on how to implement retail competition. The heterogeneity of the retail market was acknowledged in recognizing the need for tailor-made programs. Agreement on fundamental issues was uniform but implementation programs discussed were diverse. In discussion the role of outside influence in shaping design was highlighted, with political and regulatory barriers emerging as the most significant.

Introducing demand-side management programs to accompany the supply-side programs will create market symmetry. Solutions developed should look to the overall market and allow every customer the opportunity to respond to market conditions. Customers must be able to respond to changes in the market and must know how to respond. The former requires advanced metering and communication, while the latter is a matter of education.

Initial approaches to retail competition were based on the objective of "giving every customer a choice of supplier." Today "a choice of products and services" is being recognized as a more realistic objective. While multiple suppliers can provide a variety of products and services, given an appropriate market structure, so can an individual utility.

The biggest challenge to introducing a demand side program is gaining political acceptance. This will require a unified front to present a multi-dimensional solution that is simple to understand, easy to implement and in a short time. Above all it must be shown to work and be able to deliver results in the constrained timeframe.

2. OPENING DISCUSSION

The first session set the scene with a discussion of the factors contributing to the apparent failure of deregulation in California. Understanding what went wrong provides the platform from which the reform process can be re-started and reformulated where necessary. Key deficiencies cited were lack of transparency in pricing, no monitoring of market power, barriers to entry for new retail players, no demand-side sector and a general lack of leadership. It was suggested that regulatory not market failure led to the problem, with AB 1890 being cluttered with overlays that complicated reform. Lawmakers in seeking to protect consumers had forgotten to maintain a viable industry. Little was understood about the realities of restructuring and its implementation.

The challenge today is to "make the market capable of coming back." First and foremost decision-making has to rest with those who understand the problem and have the solutions. Additionally this group, whatever its composition, must have power to implement the proposed solutions, power that currently appears to lie with the legislature. How can the responsibility for oversight be transferred? The key is to craft solutions that are easy to understand, capable of quick, successful implementation and, above all, appeal (and can be sold) to the legislature. It was stressed that time is a limiting factor. Without results, competition will disappear very quickly and it could be within a year.

Cost-reflective pricing for customers is the cornerstone of a fully competitive retail market; it is the link between the wholesale and retail markets. Placing customers on a fixed rate tariff divorces them from the reality of the hourly wholesale market prices. Customers shielded from the volatility of prices have no incentive to respond by lowering demand and yet many could and did, as evidenced by the voluntary response seen recently. Introducing real-time pricing hinges on a number of factors not the least of which is selling the concept to politicians and customers.

3. WHY REAL-TIME PRICING?

RTP allows customers to see all the price variation. Time of use[1] (TOU) pricing does allow customers to see some of the variations in wholesale prices. However it cannot reflect all wholesale price fluctuations and it actually reflects surprisingly little. For example in June 2000 TOU pricing captured only about 14% of the day-ahead price variation, RTP captures it

---

[1] Time of use prices are fixed in advance of consumption for set blocks of hours across a day. For example, three blocks would cover off-peak, shoulder and peak periods.

all. TOU pricing doesn't create a feedback response to attempts to exercise market power, RTP does.

Some 18,000 interval meters covering 30% of peak load will be installed in California over the coming months. To supply all customers would require an incredible commitment of time and money, though ultimately it may happen. Voluntary RTP introduces a problem of adverse selection where customers with the peakiest loads are unlikely to join. While RTP need not be mandatory, it was felt it should be the default for large customers.

It was noted that RTP with forward purchase option does not compensate customers who are currently subsidized through flat (or flatter) pricing. Using energy credits based on previous year usage would solve this problem. It would be possible to have a program with voluntary RTP where RTP and non-RTP groups started with the same average price. Customers would choose which group to join. After one year the relative rates of the two groups would be reset based on their relative costs. This would result in the "lemons" problem of insurance – the RTP average will fall relative to non-RTP, making RTP more attractive. More and more customers would choose RTP as the relative rates diverged.

Efficient price regulation requires knowing the minimum cost of procurement. The regulator could determine the market price by running an energy-trading arm, but this would conflict with other regulatory responsibilities. Implementing RTP eliminates the need to know this minimum cost. It was pointed out that Californian customers will ultimately pay the equivalent of market prices for their recent consumption. However this will not appear on the bill but in future taxes. There will be no option to mitigate the effect of high prices. RTP in contrast can offer certainty through risk[2] abatement measures of the customer's choosing.

RTP forces customers to deal with price volatility, a far cry from a regulated tariff with its total certainty. Introducing a scheme, which appears to leave customers unprotected from the vagaries of the market, will not sit well politically; it will require considerable justification. Justification of RTP can be accomplished through ensuring an understanding of the difference between the effects of regulated prices and competitive prices in the retail market.

---

[2] Regulated tariffs contain a component that covers the kWh consumption cost and another that is essentially a risk premium; traditionally customers have always borne the price risk. Price volatility was managed through the expedient of building generation sufficient to meet any foreseeable load. In addition fuel adjustment charges appeared on bills

4. RETAIL TIME PRICING IN PRACTICE

4.1 Georgia

The Georgia Power voluntary demand-side program based on real-time pricing was introduced in 1992. Some 1600 large (85% industrial) customers accounting for 5GW of load have signed up for the program. It is the largest RTP program in the United States.

Georgia Power offers day- or hour-ahead notification of firm prices. The hour-ahead program has 35 customers, the largest industrials those with loads around 60-70MW. The utility uses a 2-part design based on the concept of Customer Baseline Load (CBL). The CBL is the customer-specific predetermined hourly load profile across a year. If the customer consumes electricity according to its CBL then the utility is revenue neutral with respect to the non-RTP firm load tariff. The second part of the tariff provides the customer with a risk management tool. The utility buys from (sells to) the customer when consumption is below (above) CBL at the current marginal cost of a kWh. Customers can also take advantage of additional risk management products offered by Georgia Power, including contracts for differences and caps and collars.

Customers in the program curtail load in a variety of ways consistent with their energy requirements. Some industrial customers actually close down production, others have back-up generation. A paper mill without on-site generation manages by load shifting. The large office blocks have installed energy management systems. Hi-tech companies switch the chillers off and rely on the cold-water storage system. The more energy intensive customers are more sensitive to price shifts; these customers have an elasticity of demand in the range 0.10-0.15 rather than the more commonly observed 0.05.

For customers on RTP the long run average price is less than the alternative tariff. Price-insensitive customers choose RTP over the standard tariff, saving money by self-insuring. Customer response to the program is consistent, roughly half the load on RTP responds. Customers with on-site generation tend to respond more frequently. Load reduction is consistently 10%+ but increases significantly with increasing prices. RTP was acceptable in Georgia for its perceived benefits in reducing generation requirements and it was popular with customers. It has lead to lower prices, greater reliability and improved market efficiency.

4.2   Oregon

The demand-side management program that operates in Oregon is a load-reduction program.   Known as the Energy Exchange Program, it is more flexible than the more traditional load reduction programs.[3]   A regulatory perspective was presented on this Energy Exchange program.   Customers participating in the program receive a payment in return for voluntary reduction of load.   There are requirement on size of participating loads and the amount "pledged" for reduction.   Participants benefit from reduced electricity bills as well as the payment they receive, and the ratepayers see reduced power costs.

Three utilities offer the program, Portland General Electric (PGE), PacifiCorp and Idaho Power.   All three have hourly programs, and PacifiCorp and Idaho Power offer long-term programs.   The hourly programs are designed for industrial and commercial load and notice is given the day before.   Payment for load curtailed is linked to the wholesale price. The load reduction is calculated from a comparison of actual and baseline usage (over the previous 14 days).   The long-term programs are designed for irrigation load and a fixed seasonal price applies (12.5- 15¢/kWh).   The load reduction is similarly calculated except the baseline usage is based on 5 years historical consumption.   Under the PacifiCorp program pumps must be disconnected from the distribution system (no reconnection fee applies), Idaho Power requires metering points to stay connected.   Non-compliance penalties apply in the programs.   These range from being dropped from the program to forfeiture of payments.

The total program has significant load reduction potential, 370+MW in the hourly programs and 24MW in Oregon in the long-term program, with significantly more across the system.   PGE has 23 customers participating in its short term Demand Buy Back Program, with 178MW maximum load reduction potential.   The 122 daily events (where load reduction was called) resulted in a net saving to PGE of $20+ million and a payment to customers of some $17 million.

There are some regulatory concerns with the program.   Customers can game by load shifting, inter-temporal and across sites, and by seeking double payment.   For example, customers could get paid not to irrigate when they had no water to use for irrigation.   Eligibility is another concern.   At the moment smaller customers cannot participate because of difficulties in administering the program.   Another concern is the issue of lost revenue

---

[3]   In these programs customers receive from the utility a discount in the demand charge in exchange for the right to interrupt load.   Contract are signed well in advance, specifications are all-encompassing, and failure to perform incurs stiff penalties

recovery. The utilities want the revenue back, but there is the question of entitlement.

Energy Exchange programs benefit all customers and the region increasing system reliability and reducing power costs. They are a cost-effective method of reducing load and have the potential to expand to smaller loads and can be tailored to customer loads. In many respects these programs can be viewed as an introductory step to RTP.

4.3 San Diego

Customers in Georgia and Oregon voluntarily expose themselves to controlled forms of real-time pricing. This is in sharp contrast to San Diego where all customers of San Diego Gas and Electric were summarily exposed when the retail rate freeze ended in July 1999. Most post-freeze rates were based on a 5-week moving average of PX prices plus other charges. The fixed rate of pre-July 1999 was 12.5¢/kWh and post-deregulation rates hovered around this level.[4] However in May 2000 prices in the PX began to rise dramatically and by August 2000 residential rates had more than doubled. Public outcry forced the PUC to again cap rates.

This real-life experience of exposure to market price volatility afforded a perfect opportunity to look at customer response. The research focused on the period mid-April to early September 2000 testing for a difference in electricity consumption. The first question to ask and answer - Was there a shock to retail rates? If so, was it due to retail rates? The study found a 5.2% overall average reduction in demand in August 2000. Much of the reduction came in late afternoon and late evening with over 10% reduction at 7pm, for September and August. Small customers shifted load from late evening to early morning and this behavior lasted all summer.

The constant overall increase in price had a differential impact across hours. This observation highlights the potential for more focused reduction with more refined (real-time) price movements. Without data on total consumption by class it is difficult to assess true customer response. The impact of the high prices could have been muted by refunds from the utility (rate relief from early retirement of debt) and, for some groups, a growing likelihood of a return to fixed prices.

---

[4] December 1999 through April 2000 average weekly residential rates were under 12¢/kWh.

5.   POLITICAL/REGULATORY DIMENSION

The current crisis in the Californian power industry, in large part, can be attributed to the absence of retail competition. The original restructuring approach envisioned the retail energy service providers at the center of the arrangement. The ESP on behalf of the customer using contractual relationships would procure energy, organize price hedging if required, arrange transport of the energy and arrange revenue cycle services. Unfortunately this never came to pass in any major way. There was no incentive structure to encourage new entrants into the retail market. The rate freeze (at 1996 levels) for customers using below 100kW and the additional mandated reduction in rates for small (<20kW load) customers afforded very little opportunity for new entrants.

Even with this lack of incentive at the lower end of the market, direct access did account for 15.5% of total UDC load in May 200, almost double the August 1998 figure. Customers in the main were industrial. However the energy crisis has had a devastating impact with just 2% (April 2001) of the load still direct access. Most ESPs are out of business unable to cope with negative cash flows - non-payment by UDCs and portfolios overexposure to the spot market being the major reason. The impact extends into metering services with providers going out of business and metering and communication infrastructure dismantled. Former direct access customers, who were able to return to bundled service without penalty, place an extra burden of debt on those now providing them with supply.

There are valuable lessons for those seeking solutions. The retail market must not be disconnected from the wholesale market and vice versa. The ultimate driver on a wholesale market is an efficient, healthy retail market. The absence of effective retail competition contributed to the ability of generators to run up prices to disastrous levels. End-users should not be shielded from market prices. The link between the two markets is information on energy consumption, when used and by whom. For suppliers and end-users facing market prices such information is a vital part of risk management strategy. Gathering such information and transmitting market data in a timely manner requires advanced metering and communication systems.

The policy makers of California have to decide whether they want retail competition for all or a regulated bundled utility service for all. A third option would be to have competition for industrial and large commercial users and have a regulated "market" (franchise) for the small users. Above all there should not be halfway states. Regardless of the level competition, pricing must be market-based and this will require a commitment to advanced metering.

6.   INTRODUCING REAL-TIME PRICING IN CALIFORNIA

The *potential impact* of RTP in California is the subject of EPRI research. The modeling, using a customer demand model and a wholesale price model, produces a demand response, calculates wholesale market changes and economic benefits under varying assumptions on market price, customer load shape, market shares and price elasticities.

Response was calculated for various levels of customer participation. This ranged from voluntary to mandatory. Hourly shifting (within and between days) in response to price changes is incorporated into the model through a flexibility parameter that is similar to an elasticity of substitution. Three values of the flexibility parameter were tested to cover medium, high and ultra-high responsiveness. Demand response was simulated using actual 2000 prices and ISO loads.

Demand response for medium and large commercial and industrial customers was calculated for the period May-August 2000. The model produced an average peak reduction between 2.0% - 24.4% depending on participation levels and price response. The corresponding total cost reduction for high price days ranged from 1.6% - 20.2%. Mandatory participation at all participation levels showed significant average peak reduction and cost reduction.

The need for an RTP infrastructure is a key *implementation issue.* Metering and communication systems are costly and providers tend to perceive proprietary provision as a means of ensuring maximum value to their companies. There is no one solution for metering and communication and for this reason it is imperative that open systems and standards be adopted. An open infrastructure will stimulate the development of cost-effective solutions and lowering costs will bring real-time measurement within the reach of all customers.

7.   MANAGING LOADS

Restructuring has delivered a one-sided market. With focus on bringing competition into supply, competition on the demand side was neglected. In a market with inelastic demand, high prices and price volatility, load is an untapped resource. However unlike generators, customers are many, heterogeneous and in comparison small, except for the very few large industrials. Individually small customers are probably not at all interested in providing system reliability/security services to the ISO. While load reduction programs work well with large customers there is no reason why a load aggregator working with smaller customers could not provide the same service.

It would be possible for these same load aggregators to work with the ESP on behalf of customers in designing rate structures based on the aggregated load shape. For residential customers load aggregation can be used with judicious real-time metering placement to avoid mandating RTP for all customers. Load management will have a role to play in system stability. However, before loads can participate in the market to the same extent as generators in providing services to the ISO, the operational needs and control requirements must be addressed.

Customers wanting to curtail loads may prefer interruptible contracts to the transactions cost associated with RTP. What should be the design of these interruptible contracts? Load participation in the market has a financial consequence with inter-temporal cost impacts. Interruptible supply contracts must consider notification time, duration of interruption and frequency. Optimal portfolio and scheduling of demand side resources can produce cost effective generation replacement.

8.   GAUGING CUSTOMER RESPONSE

Once demand side response programs are in place it is essential that there be some means of assessing customer response. The heterogeneity of customers cannot be over-emphasized; different customers will have different responses (if any) to different programs. Customer participation is most guaranteed by tailoring programs to load and customer characteristics. Measuring response to existing programs and analyzing this response in terms of customer classes and load characteristics can provide a valuable tool for designing programs to implement in California.

An analysis of customer response in the England and Wales market was provided. For the large industrial customers that chose to participate in the wholesale market there was significant price response. The demand responses of large customers of one particular REC[5] (Regional Electric Company) were examined. These customers showed little or no price response. This can be attributable to the nature of a market where volumes traded through the spot market are small and retailers have back-to-back contracts with generators.[6] Roughly 95% of the load was sold to end-use

---

[5] RECs are more or less equivalent to the UDCs. These held the franchise and ran the distribution network.

[6] Many RECs own or have interest in generation. The regulatory structure allowed the RECs to buy into generation with the knowledge this part of their business was unregulated. The RECs as suppliers of franchise customers were allowed to pass through

customers on fixed price contracts. These fixed price contracts can be individually tailored and quite elaborate. The data covered the period 1991-95 when market participation was limited to customers with loads >100kW. Since that time, franchises have been eliminated and changes introduced into the trading arrangements at the wholesale level.

Christensen Associates has looked at response to RTP and demand-side bidding programs. Analysis of data from programs at Georgia Power, Duke Power and GPU Energy showed that customers in aggregate do respond to hourly prices. However there was a considerable range of response across customers. Load reductions of 10% to 50% were observed with consistently larger response at higher prices. Based on existing evidence it would be possible to develop RTP load response curves for California.

In the discussion the point was made that introducing a demand-side program that used market price signals would be difficult. The pervasive fear of generators misusing market power to create a scenario of constant high prices would make these programs politically impossible to sell unless it could be demonstrated that the program would adequately protect customers from price risk and that such programs provide a natural cap on wholesale prices.

---

the procurement cost of the electricity, thus enabling them to integrate backwards into generation and pass the market risks of generation on to their franchise customers.

Section V

Markets and Regulation

Chapter #20

Market-Based U.S. Electricity Prices
*A Multi-Model Evaluation**

Hillard G. Huntington
Stanford University

Abstract: As the electricity industry moves from cost-of-service regulation to market-based pricing, market fundamentals rather than cost rules will determine electricity prices. In addition, regions will become more competitive with each other. Low-cost regions will look to sell some of its power to regions with higher generation costs, and electricity trade should grow.

Although computerized economic models of electricity markets are not needed to understand these broad trends, they help considerably in representing the main economic conditions prevailing in each region and in comparing their relative costs. These study showed how changed conditions in electricity demand growth, natural gas prices, and transmission capacity and fees could be studied in modeling systems that represented economic and technical conditions in multiple electricity regions within the United States.

Key words: Energy Modeling Forum; Price forecasting; electricity restructuring; electricity prices; forecasting

1.   INTRODUCTION

   Over the last decade many countries and regions have transformed their electricity sectors to make them more competitive.   Although the recent

---

* The author is particularly indebted to the EMF working group modelers and other participants, who developed and discussed many of the points contained in the EMF 17 report, *Prices and Emissions in a Restructured Electricity Market*, which is available from the Energy Modeling Forum, Stanford University, as well as from our website: http://www.stanford.edu/group/EMF/.

A. Faruqui and B.K. Eakin (eds.), *Electricity Pricing in Transition*, 297-313.
@ 2002 *Kluwer Academic Publishers.*

modifications of the California market design cast considerable uncertainty about how far this process will evolve,[1] competitive forces are likely to play a more influential role in the sale, transmission and purchase of electricity than they did previously. These forces are changing the industry in fundamental ways. While they are decoupling generation, transmission, distribution, and retail supply within the industry, they are also fostering much greater interdependence among regions in providing and using electricity. Moreover, these changes are occurring at a time when governments are imposing tighter controls on environmental pollutants.

New structures for organizing the industry have introduced novel ways of operating in electricity markets. They have also created alternative ways of thinking about and planning for successful business strategies. Companies can no longer ignore the potential competition from suppliers or the potential opportunity of servicing customers in other regions. In addition, governments creating market rules in one region need to be aware how their plans work with or against those rules adopted in other regions. They should develop flexible rules that produce meaningful incentives without trying to dictate the outcomes of market processes.

Although modeling competitive electricity markets is in its early stages, these frameworks are already demonstrating their value in terms of helping decision makers to anticipate and plan for widespread structural changes. Uncertainty about how restructuring will unfold and how market participants will respond to more liberalized conditions makes any single forecast of the industry's future highly suspect. But each projection contains some important elements about how competition might operate. While participants cannot predict prices and other market outcomes, they can learn to protect themselves from unexpected swings. This perspective should prepare participants to respond to unexpected developments more quickly and efficiently than otherwise, much like a person driving a car in a city neighborhood who expects the unlikely event that a child will run into the street before his vehicle.

2.   THE EMF STUDY

This chapter summarizes the recent findings of the Energy Modeling Forum's working group on electricity prices in a restructured electricity

---

[1]   Poorly designed rules have contributed significantly to California's power crisis. Implementation problems include siting delays for new plants, no long-term contracts, and fixed and subsidized retail prices that do not reflect market conditions.

industry. As in previous EMF studies,[2] the process focused partly on what could be learned from comparing the results of different models. The critical elements that were analyzed in the larger study include electricity prices, generation, capacity, interregional trade, and environmental emissions in North American electricity markets,[3] although this chapter focuses exclusively on the findings about prices. The study drew its members from leading advisors, electricity modelers, and electricity experts from government, companies, universities, and research organizations. Over the period from September 1998 through June 2000, the group met three times to discuss the key issues driving the electricity restructuring topic and how modeling results could help to develop a more comprehensive understanding of the new conditions. Participation by corporate and government advisors helped to define scenarios that would be useful for understanding the interactions among business strategies and public policy.

Participating modelers are identified along with their frameworks and organization in Table 2. While EIA's NEMS and CERI's Energy 2020 models are used primarily for developing industry outlooks and special evaluations, RFF's Haiku model was developed as a small, tractable framework for conducting risk assessment and understanding fundamental market uncertainties. MarketPoint is used mainly for such tasks as valuing electricity assets and evaluating other key business strategies. POEMS is based in part upon NEMS but has been restructured to use for electricity policy analysis and business applications. IPM is used both for evaluating policy analysis and for understanding business strategies.

---

[2] The EMF 17 working group continued the previous work initiated by the EMF 15 group on a competitive electricity industry. That particular study foresaw the problem of market power, especially when market rules prevented long-term contracts and did not allow the demand side to respond to price. Please see Energy Modeling Forum, A Competitive Electricity Industry, Stanford University, Stanford, CA, 1998. The EMF process and some earlier EMF results are described in H.G. Huntington, J.P. Weyant, and J.L. Sweeney. "Modeling for Insights, not Numbers: The Experiences of the Energy Modeling Forum," OMEGA: The International Journal of Management Science, Volume 10, No. 5, 1982, pp. 449-462.

[3] The full set of model results are discussed in Energy Modeling Forum, Prices and Emissions in a Restructured Electricity Industry, EMF Report 17, Stanford University, Stanford, California, May 2001. The group also discussed extensively emerging market design issues such as strategic behavior under different constraints, organization of the transmission system operator, and the advantages and disadvantages of considering power transmission flows rather than nodes in managing congestion. However, the group's discussions of these issues are not covered in their working group report.

Table 1. U.S. Regions Reported to EMF Working Group

NERC Subregion	Subregion Name	Geographic Area
ECAR	East Central	MI, IN, OH, WV; part of KY, VA, PA
ERCOT	Elec Reliability Counc of Texas	Most of TX
MAAC	Mid-Atlantic	MD, DC, DE, NJ; most of PA
MAIN	Mid-America	Most of IL, WI; part of MO
MAPP	Mid-Continent	MN, IA, NE, SD, ND; part of WI, IL
NE	New England	VT, NH, ME, MA, CT, RI
NY	New York	NY
FRCC	Florida	Most of FL
STV	Southeast (ex. Florida)	TN, AL, GA, SC, NC; part of VA, MS, KY, FL
SPP	Southwest	KS, MO, OK, AR, LA; most of MS, TX
NWP	Northwest	WA, OR, ID, UT, MT; part of WY, NV
RA	Rocky Mtn & Ariz-NM	AZ, NM, CO; part of WY
CNV	California & So. Nevada	CA; part of NV

Table 2. EMF 17 Modelers Submitting Results for Standardized Cases

ID in Charts	Model Name	EMF Modeler	Organization
NEMS	National Energy Modeling System	Robert Eynon Laura Martin	U.S. Energy Information Administration
POEMS	Policy Office Electricity Modeling System	John Conti Frances Wood	U.S. Department of Energy, Policy Office OnLocation, Inc.
RFF	Haiku	Dallas Burtraw Karen Palmer Ranjit Bharvirkar Anthony Paul	Resources for the Future
IPM	IPM	Elliot Liberman Boddu Venkatesh	U.S. Environmental Protection Agency ICF Consulting, Inc.
E2020*	Energy 2020	Abha Bhargava Christopher Joy	Canadian Energy Research Institute
**	MarketPoint	Dale Nesbitt Ted Forsman	Altos Management Partners

*   In this study, CERI reported results from Energy 2020 for seven Canadian regions and the United States as a whole.

** MarketPoint results are not compared geographically with the others. This model reported detailed regional results that were not easily aggregated to the core 13 regions shown in Table 2. However, discussion of their results helped to identify the basic principles of competitive markets exhibited by other models.

The working group considered the five competition scenarios: baseline or reference, high demand, low natural gas prices, expanded transmission, and a renewable portfolio standard (RPS). These cases are listed in Table 3. The baseline case adopted the economic and energy assumptions from the reference case of the U.S. Energy Information Administration's *Annual Energy Outlook (AEO)*.[4] The high demand case examined power market conditions if electricity consumption grew by 1 percent per year more rapidly than in the baseline competition case, for a total of 2.4 to 2.8 percent per year. The low gas price case kept the natural gas price paid by electric utilities in all future years at its projected 2000 inflation-adjusted (or real) level in the baseline case. The high transmission case allowed both expanded physical volume and lower transmission charges between regions of the United States. The renewables portfolio standard (RPS) assumed that the industry must meet a target share of 7.5 percent for renewables excluding hydroelectric power. As explained later, alternative versions of both the high demand and transmission cases were simulated in order to understand better the results obtained in these cases.

All of these cases assume that prices in every regional electricity market are immediately set by incremental or marginal costs in the year 2000.[5] This assumption contrasts with the AEO reference case, which assumed a mixture of deregulation and regulation in the various states. As a result, the EMF scenarios show how changing conditions influence electricity markets in a deregulated environment. The cases do not show how deregulation would affect electricity decisions relative to a regulated environment. The group did not simulate a regulated case because participants had differing opinions of what continued regulation would mean, which regions would be affected, and the degree of market deregulation in a business-as-usual scenario.

A related, important issue concerns the form of the deregulation itself. The group asked the modelers to consider a wholesale electricity market that was workably competitive. Each region had sufficient generators or access to interregional trade that muted the problems of market power being exercised on a consistent basis. This perspective also required that transmission systems within a region were sufficiently efficient in reducing congestion to allow plants to be dispatched on the basis of least cost.[6] Finally, all models, except IPM, allowed demand to respond to prices, with some, but not all,

---

[4] The Reference case for Energy 2020 was not completely aligned with the assumptions of AEO99 as provided in Table 4. Specifically, assumptions on electricity demand, heat rate improvements, transmission and distribution COS reductions, and reserve margins are not aligned with the AEO.

[5] As a result, the year 2000 values are hypothetical and do not reflect actual markets outcomes.

[6] POEMS performs the dispatch and trade at the subregional level, with representation of transmission capability among these subregions.

frameworks allowing load shifting in response to time-of-day prices. As a result, the cases are optimistic on the regulatory front by assuming that each government steers its way towards a reasonably competitive market design.

Table 3. EMF 17 Cases

Case*	Market Assumptions	Notes
Base Competition	1999 AEO** Demands and Fuel Prices See Table 4 for other assumptions.	Fully integrated so that demands respond to price.
Low Gas Prices	AEO Demands and Fuel Prices + Hold delivered gas prices constant at projected AEO price in 2000 for all classes of customers.	One case allows fuel prices to change incorporating the effects of higher natural gas prices, while the other has fixed (at baseline) natural gas prices. Electricity demands are unresponsive to electricity price changes.
High Transmission	Increase transmission capability by 50%. Reduce transmission hurdle rate to $0.10/mWh. Reduce transmission fees by 50%.	Two additional cases separate the change in transmission capacity from the change in the transmission fees and hurdle rate.
Renewable Portfolio Standard	Impose RPS goal of 7.5% non-hydropower renewables (excluding MSW) as a percent of sales by 2010. Assume the RPS requirement expires after 2015. The cost of the credit was capped at $15/mWh.	

*   All cases assume immediate deregulation in all states today.
** Although most models standardized on the 1999 Annual Energy Outlook, the NEMS system used the 2000 version.

The group did not ask to have different market designs or market imperfections simulated with these models because the systems represented in the current study do not contain the necessary institutional constraints or detailed transmission networks that allow one to consider such issues. The principal advantage of the current models lies in their detailed representation of interregional competition within the electricity industry, with appropriate links to other energy markets and the economy.

3.   THE BASELINE COMPETITION CASE

The EMF baseline competition case combines the reference case economic and energy price assumptions from the *AEO* forecast[7] with the assumption that all U.S. regional markets are immediately competitive with generation prices being set at the incremental production cost of the last unit generating in any period. Models differed in how much fixed operations and maintenance (O&M) costs they incorporated in market prices. Modelers were asked to implement the key assumptions for competition shown in Table 4, unless they had strong reasons for overriding these specifications.

Table 4. Detailed Assumptions for Baseline Competition Case

Category	Input Specification
Electricity Markets	Competitive Wholesale
	Competitive retail beginning 2000 for all states
Market Structure	Perfect competition
Electricity Demand	AEO 1999 (or 2000) Reference Case
Fuel Prices	AEO 1999 (or 2000) Reference Case
Cost of Capital	Lifetime for new plant is 20 years (17 for wind & solar). Debt/equity ratio for new builds is 60/40. The debt interest rate is 5.5% real and equity is 15% real.
Renewables	Extended wind tax credits to 2005
Generation Pricing	Marginal cost pricing as defined by each modeler
Ancillary services	Defined by each modeler
Transmission	$1/Mwh (1997$)
- Hurdle rate for trading	Postage Stamp (zonal)
- Organization	$3/Mwh (1997$) average between neighboring
- Wheeling Fees	NERC regions
O&M and G&A Costs	Savings relative to Cost-of-Service (COS):
- Generation	1.8% per year decline, 2000 to 2010
- O&M	5% per year decline, 2000 to 2010
- G&A	
Transmission Cost Reductions	0.75% per year decline, 2000 to 2010
Distribution Cost Reductions	1.5% per year decline, 2000 to 2010
Heat Rates	0.4% per year improvement for fossil plants, 2000 to 2010
Reserve Margins	Goal of 13-15% (see regional table), or endogenously derived
Stranded Cost Recovery	10 year recovery, 10% discount, 100% recovery
-Generation Assets	
Transitional Charges	
-Regulatory Assets	Recovery of existing regulatory assets
-Decommissioning Costs	Recovery of required costs
Externality Costs	None

---

[7]   NEMS used *AEO* 2000 assumptions while the other models used *AEO* 1999 assumptions.

The group was primarily interested in how changes in the economic and energy assumptions influence the competitive electricity market and emissions results. However, it is easier to understand this discussion by first emphasizing a few key characteristics of the baseline case.

3.1   National Electricity Prices

Figure 1 compares the U.S. average wholesale generation electricity price, which is earned by a generator that operates throughout the year without any downtime.[8] Generators are paying $2.93 to $3.26 per million Btu for natural gas in 2010 in this case.   All prices are in 1997 dollars. Electricity prices rise from $26 per megawatt hour (MWH) to $30 per MWH in POEMS, remain relatively steady at $29 per MWH in NEMS and at $25 per MWH in IPM, and fall from $31 to $25 per MWH in the first five years before leveling off in RFF.[9] The prices in E2020 are more cyclical, increasing from about $34 in 2000 to 36 in 2005 before dropping to about $31 in 2010. The change in E2020's price appears to correspond directly with the tightening of the market. The reserve margins in E2020 are determined endogenously and not set to those specified in AEO99. The reserve margins decrease from 25% in 2000 to 14% in 2003 and then increase to 20% in 2010. There appears to be a 1-2 year lagged price response in this model.

---

[8]   Modelers reported both "wholesale" and "delivered to consumers" competitive generation prices. The wholesale prices (reported above) are paid to generators averaged over all hours in the year, in other words paid to a generator which runs all the time ("time weighted average"). The prices to the consumer include losses and are averaged over the amounts purchased in each time period ("quantity weighted average"). These two effects lead to higher prices for the "delivered" than for the "wholesale" concept. In particular, customers buy more power at peak when prices are higher, which raises the average over the one that is time-weighted.

[9] $30 per megawatt-hour would equal 3 cents per kilowatt-hour. These are wholesale electricity prices that exclude transmission and distribution costs.

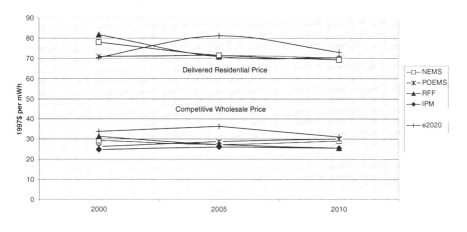

Figure 1. Electricity Prices in Baseline Case, 2000-2010

The delivered prices to consumers, also shown in Figure 1, are based on their patterns of consumption and include transmission and delivery costs. IPM results are not shown because the model does not project consumer prices. In addition, consumer prices may include transitional costs for stranded capacity where the book value of the generation assets is higher or lower than the market competitive prices. Projected delivered prices increase less quickly or decrease more quickly compared to the wholesale prices for at least two reasons. The cost of distribution was assumed to decline annually in this scenario. Moreover, the stranded cost charges decline over time in inflation-adjusted terms as well.

In NEMS and POEMS, a total value of stranded costs is determined from the net present value of net cash flows and the net book value. In POEMS positive and negative stranded costs are netted at the company level, while in NEMS they are netted across all plants in a region. This amount is recovered over 10 years in flat nominal dollars each year, which means it declines in inflation-adjusted dollars. Therefore, this component of rates declines over time. RFF calculates stranded cost on a NERC region/subregion basis in the version of the model used for this study. Under this approach, profits earned by some utilities in the region are netted against the stranded costs of others. The net result is always no stranded costs for the region as a whole and thus, there is no stranded cost recovery reflected in retail electricity prices in the RFF results. [10]

---

[10] In subsequent versions of the model, RFF has incorporated stranded costs on a utility by utility basis.

3.2   Regional Electricity Prices

Electricity prices are projected to vary considerably across the 13 regions. The wide range of regional wholesale electricity prices is displayed in Figure 2, which reports those regions in order from the most coal dependent to the least coal dependent from left to right. There is some tendency that the lowest prices are experienced in regions, such as ECAR, which have existing low cost coal and nuclear generation sources. Regions more reliant on oil and gas-fired generation and those with higher delivered fuel costs have higher prices. However, the Rocky Mountain region stands out as a coal-dependent region with relatively high wholesale electricity prices because its prices are increased by its power exports to other regions.

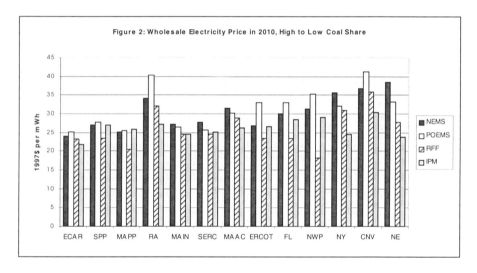

Opportunities for trading can lead to higher or lower prices than otherwise expected. For example, the Northwest region has considerable hydroelectric resources, which without trade would lead to low electricity prices. In all the models[11] except RFF, the NWP prices do not appear to be significantly lower than other regional prices, because other regions set the marginal prices. In POEMS and NEMS delivered prices in NWP are reduced by a credit or discount to account for low cost Federal preference power.

There is a tendency for the low-price regions to export power to other regions (Figure 3). Texas (ERCOT) and Florida are relatively isolated regions with little interregional trade despite lower electricity prices. New

---

[11]   The regional definition of NWP and RA are slightly different in IPM than in the other models. What has been labeled here as NWP is a subset of the full NWP used by others, and the remaining part is included with RA.

York and California export significant power during certain periods despite their high prices.

Figure 3: Export Dependence in Baseline, 2010

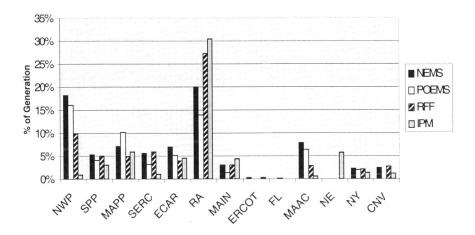

There is a tendency for the high-price regions to import power from other regions (Figure 4). New York and California imports are a relatively high percentage of their generation because trade if often more attractive than domestic electricity in these high-priced states. The southwestern region (SPP), which excludes Texas, and states like Minnesota, Iowa, Nebraska in the MAPP region import relatively small amounts of electricity due to their low generation costs. The Pacific Northwest imports electricity during the winter. Texas (ERCOT) remains relatively isolated with respect to imports as well.

Figure 4: Import Dependence in Baseline, 2010

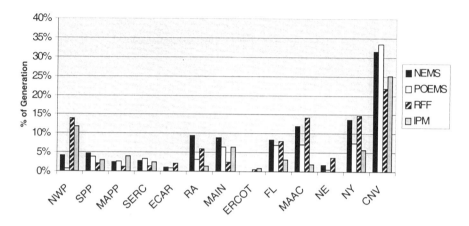

The flow of power from low-cost to high-cost states tends to equalize electricity prices across regions. Although this trade operates in these simulations, limits on transmission capacity prevent full equalization of prices across states. As a result, low-price states remain the major exporters and high-price states remain the major importers, as referenced above.

3.3   Utility Prices for Gas and Coal

Most cases follow the EIA assumptions for energy prices and indicate that the fuel prices paid by generators tend to move slightly in favor of coal over time. While inflation-adjusted (or real) coal prices remain relatively stable over the decade, natural gas prices tend to rise. In 2000, coal prices delivered to generators in most models are approximately 40-45% of the comparable gas price (Figure 5). By 2010, they decline to about 30-35% of the gas price. Despite this trend, the projections indicate a strong move towards electricity plants fired by natural gas. Past and future improvements of gas turbine technologies and higher cost of capital in competitive markets expand the use of gas in combined-cycle units over coal to meet baseload demand.

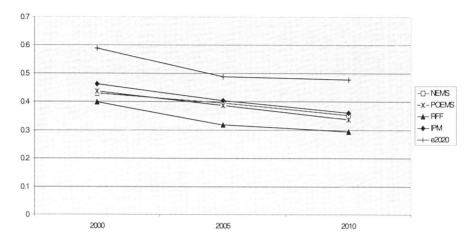

Figure 5. Coal-Gas Price Ratio for Utilities in Baseline Case, 2000-2010

4.   ALTERNATIVE SCENARIOS

Figure 6 reveals that natural gas prices and electricity demands are two important influences on the future path for electricity prices. The only other significant change at the national level occurs with the RPS case in the RFF model. Otherwise, the alternative conditions for electricity transmission and the RPS generally have very modest effects that remain below 5% of the baseline values.

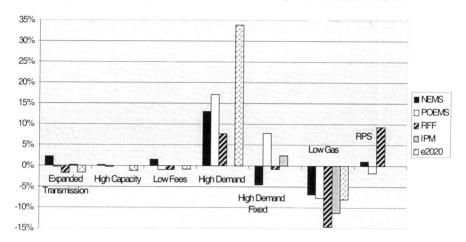

Figure 6. Percent Change in Competitive Wholesale Electricity Price from Baseline Case, 2010

Larger changes are seen for some regional electricity prices. The largest gain relative to the baseline (27%) occurs for the midwestern MAIN region in the NEMS model in the high demand case. The largest reduction relative to the baseline (22%) occurs for New York in the RFF model in the low gas price case. While the national prices change by no more than 2% from their baseline levels in the alternative transmission cases, regional prices can increase or decrease by as much as 12 or 13% in these cases.

In general, importing regions with higher prices in the baseline case will experience price reductions with more transmission access, while low cost exporting regions will see higher prices. As discussed in the section on the baseline case, these prices are the competitive wholesale levels before any adjustments for stranded costs are added.

The outlooks agree that lower natural gas prices will reduce competitive electricity prices while higher electricity demand will increase electricity prices. However, the pattern across outlooks is rather different. RFF and IPM reveal rather large reductions in electricity prices when gas prices fall; the declines in NEMS, POEMS and E2020 are more modest. By contrast, the NEMS and POEMS electricity price increases are greater when electricity demands grow more vigorously, while the RFF and IPM price increases are more modest. The increase in E2020 electricity prices is much larger than other models, because total available capacity is fixed, resulting in higher prices and declining reserve margins.

Several factors could be contributing to the higher electricity prices in the high electricity demand case. First, the industry could be pushed out further along its supply stack and forced to use more expensive units to meet the higher demand. Second, the addition of new capacity will create higher demand for fuel by generators and this increased consumption could bid fuel prices higher. Both limits on the gas resource base and gas pipeline bottlenecks could induce these higher prices. In using more natural gas, the power sector must shift gas away from other end-use sectors as well as encourage more natural gas production. These two factors will both lead to higher competitive electricity prices than under the baseline conditions.

Several other factors are also changing the average electricity price between these two cases. Even if the demand curve is shifted outward by the same proportion in all hours and days, prices could move upward more quickly during peak times and loads could be shifted away from these high-price periods. As a result, electricity use might grow more rapidly in nonpeak than in peak times, thereby decreasing the average price as consumption was shifted between periods. Another complication lies in the effect of higher demand on changing retirements and additions, both of which will influence the shape of the supply stack between the two cases.

When demands are increased and natural gas prices are fixed at their baseline values, the results show what happens to electricity prices when the increased fuel costs are ignored. POEMS shows a 7.5% increase and IPM a 2.4% increase in wholesale competitive electricity prices, which is consistent with the view that costs will rise as the industry moves further out on its supply stack. NEMS reveals a decrease of almost 5% by the end of the decade, which could be due to shifts in the supply stack attributable to retirements and additions. There is virtually no change in RFF's electricity price, and E2020 did not report results for this case. The POEMS increase is less than the 10.5% increase in total electricity demand in that model.

These results demonstrate that natural gas conditions can have a significant effect on electricity prices. In an effort to achieve greater consistency in the demand shock, the modelers in this study did not allow these higher electricity prices to reduce the use of power. In actual electricity markets and when these models are usually simulated, however, the higher power prices would offset some, but not all, of the initial growth in electricity consumption. Under these conditions, both electricity consumption and gas use by the power sector would be lower, as would the price of natural gas and electricity.

Given the important qualifications on which factors are changed in this case, the high-demand cases also reveal some information about how much electricity prices will respond relative to electricity generation when the desire for electricity expands. Table 5 shows that total generation and the competitive wholesale electricity price each rise by about 12.5-13% above baseline levels in NEMS in the high-demand case with integrated fuel prices (the last set of columns). Thus, the inferred supply response or elasticity is near unity in this model.[12] Prices increase the most in E2020; generation increases are not that high, indicating a much lower supply elasticity. Similarly, prices increase more than generation in POEMS, inferring a lower supply elasticity, although the gap between the change in price and generation is much smaller than in E2020. Prices increase less than generation in RFF, inferring a higher supply elasticity.

Table 5. Percent Change in Generation and Electricity Price in the High-Demand Cases (from the Baseline Case), 2010

	High Demand with Fixed Price		High Demand with Integrated Price	
	Generation	Price	Generation	Price
NEMS	12.54	-4.53	12.54	13.04
POEMS	10.41	7.83	10.45	17.05
RFF	10.71	-0.82	12.82	7.70
IPM	10.33	2.44		
E2020			13.21	33.81

The first set of columns in Table 5 shows the same computations when higher demands are allowed but natural gas prices are kept at their baseline levels. The results show a much lower increase in electricity prices when natural gas prices do not increase. As a result, the inferred elasticity is much greater. In fact, there is a decline in the NEMS competitive wholesale electricity price in this case, while the RFF price remains virtually unchanged. This result probably reflects the influence of retirements and additions on the supply stack in the two cases.

Competitive wholesale electricity prices fall below baseline levels in the lower gas price case as shown in Figure 6. Figure 7 shows that natural gas prices in RFF fall sharply to more than 15% below baseline values in 2005 and almost reach 20% below baseline by 2010 because they rose more quickly in their baseline case. (The figure shows decreases as negative values that increase as you move upward in the chart.) The natural gas prices relative to the baseline case fall more slowly in the other models, although the changes in gas price trends move closer to each other by the end of the decade. Moreover, the RFF coal prices also decline more than 5% below the baseline by 2010. The downward pressure on fuel prices paid by the power sector contributes to the relatively strong decline in competitive power prices in that model shown in Figure 6.

---

[12] The price elasticity of electricity supply is defined as the percentage change in electricity quantity supplied divided by the percentage change in price, holding other factors constant. Clearly, these other factors that are held constant will certainly influence the measured response. Thus, the reader is discouraged from computing implicit elasticities from the numbers in Table 5 as being a model's elasticity.

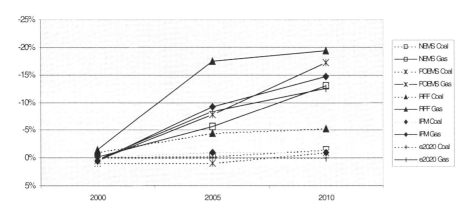

Figure 7. Percent Change in Fuel Prices for Generation in Low Gas Price Case
from Baseline Case, 2000-2010

5.   CONCLUSION

As the electricity industry moves from cost-of-service regulation to market-based pricing, market fundamentals rather than cost rules will determine electricity prices. In addition, regions will become more competitive with each other. Low-cost regions will look to sell some of its power to regions with higher generation costs, and electricity trade should grow.

Although computerized economic models of electricity markets are not needed to understand these broad trends, they help considerably in representing the main economic conditions prevailing in each region and in comparing their relative costs. These study showed how changed conditions in electricity demand growth, natural gas prices, and transmission capacity and fees could be studied in modeling systems that represented economic and technical conditions in multiple electricity regions within the United States.

Chapter #21

The Essential Role of Earnings Sharing in the Design of Successful Performance-Based Regulation Programs

Karl A. McDermott and Carl R. Peterson
National Economic Research Associates, Inc.

Abstract: Much of the literature on performance-based regulation seems to use as a cornerstone the concept that earnings sharing is antithetical to developing appropriate incentives for cost containment. However, there is a growing number of analysts who are rethinking this bias against earning sharing on both practical and theoretical grounds. The purpose of this paper is to examine the theoretical reasons and practical aspects of employing an earnings-sharing mechanism in the design of a performance-based regulation plan. In addition, we will examine some of the fundamental questions associated with the design of an appropriate earning sharing mechanism. We find that rather than being antithetical to the market process, earnings sharing is fundamental to the market process.

Key words: Earnings sharing, Performance-based regulation, market process

1. INTRODUCTION

The problem of designing a "good" performance-based regulation (PBR) program is as alive and problematic today as it was twenty years ago.[1] Even with the substantial amount of academic research devoted to the issue and years of practical experience, there still is surprisingly little consensus on the details of the programs themselves. The problem seems to lie in the inability

---

[1] McDermott (2000) defines the attributes of a "good" PBR program.

A. Faruqui and B.K. Eakin (eds.), Electricity Pricing in Transition, 315-328.
@ 2002 *Kluwer Academic Publishers.*

of academic or theoretical recommendations to coincide with the practical experience of many regulators and firms. There seems to be a disconnect between what many economists would have us believe about how markets function and how markets actually function. This disconnect has contributed to a very slow adoption rate for PBR programs outside of telecommunications.[2] Although with recent developments in energy markets and the greater reliance placed on "competitive" forces in these markets, PBR is beginning to be discussed as both a solution to restructuring problems and as an appropriate regulatory paradigm for the non-competitive portions of those markets that have been restructured.

This paper will focus on one aspect of the design of PBR programs that seems to have become one of the most difficult hurdles to implementing PBR plans—earnings sharing. After a short discussion of the reasons why this issue is still important and the basis for utilizing earnings sharing, we turn to some of the more practical reasons why earnings sharing is crucial to a successful PBR program.

1.1   Why is this issue still important?

Why is the issue of earning sharing and performance-based regulation (PBR) design still an important issue after all of these years of debate? The answer rest on three observations 1) utilities are missing some opportunities to improve profits and customer relations; 2) regulators are missing an opportunity to create more powerful incentives to lower costs and prices to customers; and 3) customers are missing an opportunity to be served by firms that have an incentive to be more innovative and service oriented. However, many analysts and regulators accept with little or no critical review the idea that the only form of incentive regulation that is appropriate and can harness the profit motive for the public good is a "high-powered" incentive structure that returns cost reduction to shareholders in a 1-to-1 ratio. It is therefore important to set the record straight on the theoretical and practical implications of utilizing only such high-powered incentive schemes.

---

[2]   The telecommunication industry may constitute a special case where the forces of technical change and competition have been more significant than in the energy industries. The introduction of, and transition to, a completely competitive marketplace implies that pure price caps might be an effective transition mechanism. However, even within telecommunications there is significant disagreement as to what role the current PBR programs have played in the industries evolution and what role they should play on an ongoing basis.

1.2 The Incentive Power of Alternative Regulatory Mechanisms

All regulatory mechanisms have inherent incentive structures. Traditional rate of return regulation (RORR) is every bit an incentive plan as the most modern price cap plan. The distinction lies in the *power* of the incentive structure.[3] The power of the incentive structures is a function of the ability to retain economic profit. In the context of regulation, economic profit will exist when revenues are in excess of the cost of service (including the cost of capital). RORR, as the name implies, allows for no economic profit to be retained by the firm. That is, the customer—or in the case of a publicly owned utility, the taxpayer—holds the "claim" on any excess revenues (i.e., the customer is the *residual claimant*). As we shall see, it is the fight for economic profit that drives investment decisions by firms and industries. Taking economic profit out of the equation must, almost by definition, distort the investment decision and in turn the incentives for efficient behavior. The characteristic implies that the incentive power of RORR is extremely low and can be referred to as a *low-powered incentive mechanism*.[4] On the other extreme, a pure price cap plan that sets prices based solely on a predetermined price index allows the firm to be the residual claimant on any excess revenues. Pure price caps provide the incentive to innovate and lower costs as any excess revenue is retained by the firm and its shareholders.[5] Such a plan is referred to as a *high-powered incentive mechanism*. Finally, an *intermediate powered incentive mechanism* combines both high and low-powered schemes in that it allows for a sharing of the residual benefits between customers and shareholders. These types of plans can range from a relatively simple bandwidth on ROE to a more complex price cap plan that incorporates earnings sharing. The sharing of the benefits makes these intermediate powered incentive schemes

---

[3] *See e.g.,* Laffont and Tirole (1993, Table 1) for a review of incentive structures of various regulatory schemes.

[4] Many observers of RORR have noted this poor incentive structure of RORR. However, there seems to be two sides of the incentive debate. One side argues that RORR is too lenient on utilities and therefore provides poor incentives for efficient behavior. Others argue that RORR is too draconian and therefore provides poor incentives for efficient behavior. Either way, RORR has been, correctly, criticized for providing poor incentives for efficient behavior.

[5] Some argue that RORR has a similar incentive structure through regulatory lag. However, regulatory lag has at least three theoretical problems with it as a high-powered incentive scheme. First, there is uncertainty concerning the length of the lag. Most regulators have the ability to cut prices nearly immediately after excess earnings is shown. Second, the length of the lag is often much shorter than the life of the new utility investment. Third, as the regulatory game is played over and over, both the utility and the regulator become adept at timing both investment and investigations.

both economically (mimics competitive markets) and politically (avoids "windfall" profits and losses) attractive.

1.3   What is the basis for utilizing an earning sharing mechanism in PBR design?

With all the discussion of pure price cap regulation and the variety of mechanisms that can be employed in designing PBR plans, it is important that we motivate the reason for emphasizing the incorporation of an earnings sharing mechanism (ESM) in PBR plans. The root cause for adopting an ESM approach rests on the fundamental philosophical point that regulation should be designed to mimic competitive markets. Bonbright (1961, p. 93) has argued that:

> Regulation, it is said, is a substitute for competition. Hence its objective should be to compel a regulated enterprise, despite its possession of complete or partial monopoly, to charge rates approximating those which it would charge if free from regulation but subject to the market forces of competition. In short, regulation should be not only a substitute for competition, but a closely imitative substitute.

Baumol and Sidak (1994, p. 5) more recently reiterated this point by noting

> … [T]hat the proper role of regulation is that of a substitute for competitive market forces where those forces are weak or absent. The regulator's task then becomes a two-part undertaking: first to determine the rules of behavior that a regulated firm could have been expected to follow if it had operated free of regulation in a market with fully effective competitive forces; second, to constrain the regulated firm to behave as it would in such a competitive market and to circumscribe it is behavior no less and no more than this.

This is a fundamental insight into the design of a proper regulatory framework. Note that Baumol and Sidak suggest that the rules of behavior should be determined *as if* an unregulated firms were operating in a market with *effective competitive forces*. A market with effectively competitive forces is neither a monopoly, nor, is it a market characterized by the perfectly competitive market characteristics emphasized in first year economics textbooks. This suggests then that the appropriate model for performance-based regulation should neither be a perfectly competitive market nor should it be a monopoly market model. Realistic competitive behavior lies somewhere in between these two extremes. The effectively competitive market model while less mathematically tractable, is none-the-less more relevant for providing the proper incentives for monopoly providers of utility services. While space considerations prohibit us from discussing these points in detail in this paper a more thorough treatment is provided in other works by the authors.[6]

2.   EARNINGS SHARING, MARKETS AND THE RESISTANCE TO PBR WITH EARNING SHARING

2.1   Using an ESM to Mimic Competitive Markets

The central role of an ESM in the design of a program that mimics competition rests on the fundamental role that profits play as the signal mechanism in competitive markets. Competitive entry and the impact this entry has on supply and prices is crucial to the proper understanding of the competitive process and in turn the process of regulating in an appropriate manner. In an effectively competitive market the dynamics of (potential) entrants center on the forecast of profitability. Firms forecast the level of sales (which may be dependant on highly complex strategies), their own costs and the expected post-entry price in order to calculate the level of profit or the net present value (NPV) of entering a marketplace.

When a firm or industry experiences an above normal level of profit[7] an *incentive* for new supply to enter the market is created.[8] The increased supply can be created by either new firms entering the market or for existing firms that expand output. Both of these processes generate a flow of resources into the market. It is this flow of resources that in turn eliminates the temporary higher profits as the new supply is marketed to customers and ultimately the new supply can lead to lower prices (or in some cases higher quality) for customers.

It is as if the consumers were initially willing to pay a "higher" price in the short-run in order to attract the resources necessary so that in the long-run firms will be enticed to enter the market and compete for the consumer's patronage in the long-run. The net benefit to customers is that the market provides them with greater choice, variety and quality of products at lower prices. Implementing an ESM provides regulators with a similar process. It

---

[6]   *See e.g.*, McDermott (2000) and McDermott and Peterson (2001a).

[7]   This is a key point. It is economic profit that spurs supply responses, not normal levels of profit that are found in RORR.

[8]   Varian (1994) discusses the impact of entry on costs and prices.

attempts to harness the same set of incentives. The difference is that third-party resources do not actually enter the market in order to discipline the competitors, although the regulated firm will alter its investment strategy to take advantage of these new incentives. Rather, the ESM itself performs the role of entry through automatically adjusting prices thereby simulating the effects of entry over time. If a utility cuts costs or provides services more efficiently through new investments, then it will experience a temporary increased return. The incremental return is the reward for the risks of the investments. Over time these returns are "shared" with customers as if entry were actually occurring.

Just as in a competitive market, the price adjustment process under ESM will over the long run "share" the profits or losses with customers as the market adapts to changing conditions while preserving the incentives to control costs. The customer, in effect, gets a reward or return for having paid the higher price to induce the new entry when the new entry results in lower prices.[9] If regulation is to mimic the competitive process appropriately then employing the profit signal through the use of an ESM mechanism makes considerable sense.

2.2   Resistance to using an ESM

The resistance to ESM adoption has come in a number of forms. First, many regulators operates under a misperception of the role of profits in a regulated environment.[10] Second, utilities and their advocates often see an ESM as reducing the incentives to cut costs. While in a very simplistic model it is certainly true that *any* sharing of cost reductions reduces the incentive to do so, however, when a more realistic model of the market process combined with the realities of the regulatory process is put forth, this concern is minimized.

2.2.1   Regulatory Resistance to PBR and ESMs

Regulators are taught that under traditional regulation we can identify *a* "fair" rate-of-return that must cap the profit level for utility. Any higher (or

---

[9]   While this discussion involves some over simplification it captures the essence of the competitive process. Other mechanisms may also be at work which benefit customers such as the improvement in the quality of the product or service, new products and services and other innovations.

[10] Most regulators fully understand the role profits play in non-regulated markets. However, the very fact that profits are regulated for utilities tends to blur the positive role profits can play in the regulated environment.

lower) return is by definition unjust and unreasonable.[11] This conception of profits is based entirely on an economic foundation that simply no longer characterizes energy and telecommunications markets (and may never have).[12] Continuing to utilize this framework places us at risk of adopting policies that are actually at cross-purposes with our ultimate goal of creating a more efficient marketplace to serve customers.[13]

Furthermore, our conception of profit must involve the understanding that in the traditional equilibrium approach to regulation profit was viewed as compensation for risk bearing. However, in a dynamic, dis-equilibrium framework, profits are about risk *taking* not just risk bearing. The distinction is this: Risk bearing is similar to an insurance function. A firm can undertake to bear risk and be compensated through "insurance premiums" for bearing this risk. The utility played this risk bearing function under traditional regulation via average pricing, the obligation to serve and various other regulatory requirements. Risk bearing institutions are generally treated with extreme caution by government because of the problems that could arise if these institutions were no longer able to bear this risk (note the example of banking regulation in the United States).

Risk taking is more of an entrepreneurial function with risk being undertaken by the firm through prospective actions and as a result they have the opportunity to earn potentially larger returns. While many authors have commented on the various gains that can be expected from the two functions, it would seem that a consensus view would argue that entrepreneurial capitalism has delivered more significant benefits to customers than a more cautious insurance like model of capitalism.[14] ESM and performance based regulation in general is designed to harness the energy of entrepreneurial capitalism while still leaving the regulator with a light handed method of regulation in case intervention is necessary.

---

[11] The concept of a zone of reasonableness is one exception. However, this concept has never been adequately integrated into the body of regulatory theory absent a theory of incentives to explain why the zone plays a positive role.

[12] Some will argue the static view that underlies rate-of-return regulation did work well in the less volatile economic conditions that characterized much of the energy and telecommunications markets from the 1920s to the 1960s. This view has some legitimacy. However, it is more likely that regulators were satisfied with the cost reductions due solely to "iron in the ground" technology changes, and were uninterested in attempting to harness gains from other sources, such as new products and services and new methods of organization that tend to bring about much larger gains.

[13] One need only look at the history of banking regulation in the United States to understand this concern.

[14] This line of thinking is often traced to Schumpeter (1942). However, other authors such as Adam Smith, Karl Marx and John Maynard Keynes have also noted the entrepreneurial nature of capitalism. A more modern approach can be found in Nelson and Winter (1982).

Therefore, under PBR profit is more than just a compensation for bearing the risk to serve a customer, it involves the prospective actions of identifying and implementing innovations and cost savings plans. These types of innovations were simply not accounted for under a traditional regulatory paradigm. As a result regulators were forced to adopt ad hoc methods to induce innovative behavior. Consider the regulatory response to Clean Coal Technology (CCT) projects in the early 1980s. In order to induce the utility to undertake these projects, regulators looked at the incremental risk that a company would bear and designed a means to offset this risk or compensate for this risk. It was difficult at best to get utilities to accept these risks.[15] In a competitive performance based world, the firm adopts a technology because of the potential reward it receives, which it weighs against the potential risks they bear.[16] In today's competitive generation market merchant plant operators are making these decisions every day with significant additions to capacity.

2.2.2   Utility Resistance to ESMs

The second reason that ESMs have been only slowly adopted is that utilities regard sharing as simply a reduction in the incentives to control costs and innovate (or more acutely, as another way for regulators to confiscate property). There is also the fear that by utilizing an ESM mechanism regulators will use it to engage in regulatory re-contracting, where the profits that the firm does earn get taken away sooner as the "new regulatory bargain" gets periodically rewritten to the companies detriment. These are reasonable concerns, however the appropriate approach to address these concerns lies in the "political economy" view of how the world operates. Most "regulatory contracts" will only be negotiated if the regulators feel that they have a mechanism to prevent the bargain from being criticized for creating windfalls for utilities. This implies that some form of earnings sharing, or re-contracting, is an essential element in any political solution. However, a known ESM adjustment mechanism is preferable to a random re-contracting approach. Moreover, by adopting an ESM we may have a solution to the concern over re-contracting in that, without an ESM there is an even greater incentive for regulators to seek to re-contract if any earnings imbalance arises. The adoption of an ESM should lead to a lower

---

[15] *See* McDermott et. al (1992).

[16] To illustrate this point, the incremental risk to undertake a CCT project might be on the order of an extra 3 percent ROE, while the incremental profit from adopting innovative technology could add as much as 10 percent to the firms ROE.

incidence of re-contracting.[17] The key to addressing re-contracting is in the design of any bargain through explicitly addressing the issue so that, violations of the contract are settled by the courts not the commissions.

There are also theoretical reasons why so-called "high-powered" incentive structures, such as pure price caps, may not yield optimal solutions. Laffont and Tirole (1993) argue that a trade-off does exist between economic profit and production efficiency, that is, if some economic profit is retained by the firm there tends to be an incentive to be efficient. If the opportunity to earn a profit is eliminated, as under RORR, customers can actually be harmed by reducing the productive efficiency of the firm. However, regulators must recognize that an opportunity cost is associated with adopting the highest powered incentive structures making such schemes potentially less efficient than lower-powered incentive plans.[18] One way of visualizing this opportunity cost is the cost of re-contracting as noted above.[19] In fact, it has been shown that if a firm attaches a positive probably on re-contracting then it has an incentive not to undertake cost saving investment and may even induce wasteful spending under a high-powered incentive scheme.[20]

3. COMPETITION AND IMPLEMENTING PBR WITH AN ESM

3.1 Regulation by PBR

Why would a regulator want to adopt a competitive market analogue as the template for regulating a utility? There are a number of good reasons starting with the fact that competition realigns the risk and reward tradeoffs facing the firm. This realignment results in improved cost minimization incentives and incentives to innovate and develop new products that are

---

[17] For example, Alabama Power has had an earnings-sharing plan that has been stable for nearly twenty years. Mississippi Power has had a similar plan for nearly a decade. In both cases customers have benefited from improved efficiency and rate stability.

[18] Burns et. al (1998) also discuss the behavior of regulated firms under alternative regulatory regimes and make an argument that sliding scale regulation has attractive economic properties.

[19] In the case were a firm is being transitioned to a truly competitive world, this opportunity cost is likely to be much lower. However, while there is much talk of moving infrastructure firms, such as electric, gas and telecom, to "competitive" markets, very few of these firms actually are "de-regulated." In addition, it is extremely unlikely, at least in the energy industry, that delivery services will be declared competitive any time soon.

[20] *See* Weisman(1993).

simply absent under traditional regulation. The benefits of the realignment of incentives are shared with customers through the ESM mechanism. The regulator is harnessing the private interest for the publics benefit. In addition under the competitive analogue firms should be allowed a certain level of price flexibility to address the increased competition in many markets. If this flexibility is combined with price caps that protect customers from monopolistic pricing then many of the characteristics of a competitive market could be captured in the regulatory design.[21]

To recap, the competitive process is essentially based on short-run incentives to beat a benchmark price that is established by the least efficient firm in the market. The reward to beating that benchmark is greater profits (or at a minimum minimizing the loss of market share). There is generally a lag in the response time of competitors when a firm innovates so that it has a cost and/or price, advantage that can be turned into increased profits in the short-run. Over time entry by new competitors or expansion by existing capacity by current competitors results in increased supplies and the incentive to cut prices to capture or keep market share and revenue flows. That is, the competitive process begins to share with customers the fruits of the earlier transfers of income to the efficient competitor (i.e. profits) that were reinvested in new capacity or served as the signal for new capacity to enter the market. In either case the customer now sees lower prices and is able to purchase more services than before which increases their economic welfare.

The ESM mechanism captures this process by incorporating lags in the time between profits and sharing. A sharing mechanism which simulates the eventual entry of competitors and if price caps are used the price cap is adjusted upwards or downward to reflect the trend in the benchmark firm's actual costs of meeting demand. Furthermore, by incorporating the ESM the focus can be shifted to the design of incentives while avoiding the problems of "windfall profits."

3.2   Focusing on the Details

The central issues that regulators should be focused on is the nature of the sharing mechanism an its parameters. For example, the nature of any dead bands in the ROE mechanism, the actual sharing rate, the symmetry of the mechanism and length of lag between sharing adjustments.

---

[21] In effect the competitive market has a price cap that is set by the *least efficient firm* capable of meeting customers demands. The incentive is form more efficient firms to price under this level to gain customers and profits. Thus price "caps" and flexibility are two common elements of a competitive process. *See* McDermott and Peterson (2001b), (uses the concept of the inefficient firm to set access prices for telephone networks.)

3.2.1 Sharing parameters and the lag time

The balance between sharing and the lag between sharing adjustments should recognize the long-term investment life of capital. For example, it may be that selecting a 75 percent sharing with customers could imply a simple payback period that is unacceptable to utility managers. Yet on societal grounds, the investment should go forward as it may have a positive net present value. The result of such a system may be that utilities never invest but only try to adjust variable costs over time. This will limit the ultimate gains that customers will experience. In attempting to capture immediate gains for the customers, regulators could actually limit the benefits customers receive over time. Just as in traditional regulation performance based regulation involves balancing a variety of tradeoffs.

While a 50/50 sharing rate has certain intuitive appeal, it presumes that the risks of innovation should be borne on an equal basis. If the reward is tipped toward the firm who has the ability to control the innovative process, then the customer could gain just as much under a 60/40 sharing if the total profits under this approach are greater than under the 50/50 approach.[22] If the mechanism is also symmetric then by moving toward greater shares for the firm the company also bears the risk of greater losses and this should force the company to make reasoned decisions regarding the risks to engage in and the expected payoffs of different innovations to explore. By adopting a smaller share for the customers an Aristotelian form of ethics is being employed that suggests that those who put in the greatest effort receive the greatest share. In the past, traditional regulation often forced the customers to share the greatest burden through cost-plus regulation even though the customer had little or no control over the paths taken by the firm.[23]

3.2.2 Benefits of varying lag times

The adoption of a significant lag period has a number of aspects to it in the final analysis. First, if performance regulation has benefits some of those benefits come through reduced regulatory intervention. Second, with longer periods the firm has greater incentives to maximize the ultimate cost reductions shared with customers because they get the full use of those savings during the lag. From a theoretical standpoint entry does not take

---

[22] The choice between 50 percent of $1 or 40 percent of $2 seems to be straightforward.

[23] Even where a commission would "punish" a company for imprudent decisions the ultimate responsibility still rested with the consumer as the perception of risk increased for the firm and future rate cases would involve higher ROE calculations. Very often the Commission would never address the long-run issue of which is cheaper in a present value sense, disallowing "imprudent costs" and thereby increasing regulatory risk or allowing the cost and lowering regulatory risk.

place instantaneously if it did there would be little or no incentive to innovate and cut costs. During this lag firms recover the costs of investing in innovative investments.[24] Balancing this is a concern that the more frequently the rates are adjusted the less concern there will be regarding windfall profits. A reasonable time frame would seem to exist between three and five years. Of course, the company always has the ability to reduce prices under the price cap approach before the adjustment period occurs. They may do this because of competition or to build good will with their customer base.

3.2.3   Deadbands

Under ESM there is sometimes employed a concept known as the "dead band" this refers to a range around a base ROE that will not trigger a rate adjustment.[25] Deadband helps insulate both customers and the companies from small changes in earnings due to the natural ebb and flow of general business conditions and serve an important role in keeping the plan simple and easy to administer.

Dynamic deadbands provide an even more innovative approach to utilizing incentives and earnings sharing.[26] Dynamic deadbands move the target with the performance of the company under the plan. For example, if a quality factor is included in the plan, the ROE deadband could be linked to how well the company does relative to the quality standard.[27] Such a proposal bridges a gap between pure price cap plans and pure earnings sharing plans. A dynamic ROE bandwidth has two main functions. First, it sends the signal that *superior* performance will be rewarded. RORR provides the incentive for minimally acceptable performance. Second, and more importantly, it adds some of the positive features of a high-powered incentive plan (i.e., the company is the residual claimant) without increasing the likelihood of re-contracting. The result is that the firm will strive to meet the performance standards *and* do so in a least-cost manner so it can retain as much revenue as possible. This benefits customers through a more efficient utility that provides service that meets customer expectations.

---

[24] One of the most successful government policies to utilize this concept is patent protection. Patent protection is a kind of social compact that allows monopoly profits in return for sharing the information with society and elimination of the monopoly after a lag period.

[25] Deadbands can also be used for other measures that are often found in PBR plans such as quality of service measures.

[26] A more detailed discussion of dynamic deadbands can be found in McDermott (2000).

[27] For example, if the target ROE was 10 percent plus or minus 100 basis points, a dynamic deadband could move the target to 10.5 percent if certain criteria are met or to 9.5 percent if they are not met. Such a program mimics competitive markets much better than a set ROE that RORR uses.

4. CONCLUSIONS

Earnings sharing mechanisms help alleviate the fundamental tension between profit regulation and the promotion of a more entrepreneurial utility sector. Market competition works through the entry and exit of firms and the increasing or decreasing of supply in response to profit opportunities. Traditional regulation abstracted from these market realities in the hope that a simple system (i.e., RORR)[28] would be the best system. Unfortunately, this "simple" system broke the link between risk taking and profits, truncating the possible gains from this link that is so pervasive in non-regulated industries. While much research and experimentation with PBR or alternative forms of regulation has been undertaken in the past twenty years, there still does not seem to be a consensus on the appropriate design of these programs. Earnings sharing is the mechanism that can bring these two sides together and promote a more efficient utility sector in the 21st century.

REFERENCES

Baumol W. and G. Sidak,. 1994. *Toward Competition in Local Telephony.* MIT Press: Cambridge, MA.

Bonbright, James. 1961. *Principles of Public Utility Rates.* Columbia University Press: New York.

Burns, Philip, Ralph Turvey, and Thomas G. Weyman-Jones. 1998. "The Behaviour of the Firm Under Alternative Regulatory Constraints." *Scottish Journal of Political Economy* 45-2: 133-157.

Fraser, Rob. 1998. "Modifying the RPI-X Regulatory System to Include Profit-Sharing." *Australian Economic Papers* (December): 372-382.

Laffont, J.J and J. Tirole. 1993. *A Theory of Incentives in Procurement and Regulation.* MIT Press: Cambridge, MA.

McDermott, Karl A., Koby A. Bailey, and David. W. South. 1992. *Examination of Incentive Mechanisms for Innovative Technology Applicable to Utility and Non-Utility Power Generators,* Argonne National Laboratory, Report No. ANL/EAIS/TM-102 (revised August 1993).

McDermott, Karl A. 1994). Designing the New Regulatory Compact," working paper, Illinois Commerce Commission. (revised 1996).

McDermott, Karl A. 2000. "Prepared Direct Testimony," before the North Dakota Public Service Commission, *In the Matter of the Application of Northern States Power Company d/b/a Xcel Energy for Authority to Operate Under Performance-Based Regulation in*

---

[28] While rate of return regulation is simple in that it abstracts from the realities of the economic world, it is often extremely complicated to implement in practice. The difficulty does not lie in the complications of the system, but in the extremely poor level of information regulators have to make decisions. This poor level of information requires either ever more sophisticated techniques of analysis or simplistic assumptions that often do not coincide with reality.

North Dakota, Case No. PU-400-00-195; and *In the Matter of the Application of Otter Tail Power Company for Authority to Operate Under Performance Based Regulation in North Dakota* Case No. PU-401-00-36, Bismarck, North Dakota.

McDermott, Karl A., and Carl R. Peterson. 2001a. "Designing the New Regulatory Compact: The Role of Market Processes in the Design of Dynamic Incentives," Working Paper, National Economic Research Associates, Chicago, IL. (Preliminary Draft presented at *Advanced Workshop in Regulation and Competition, Incentive Regulation: Making it Work,* CRRI, Rutgers University, January 19, 2001.

McDermott, Karl A. and Carl R. Peterson. 2001b. "The Efficiency of the Inefficient Firm Standard in Setting Network Access Charges," paper presented at 20th Annual Eastern Conference, CRRI, Rutgers University, May 25, 2001.

Nelson, Richard R., and Sidney G. Winter. 1982. *An Evolutionary Theory of Economic Change.* Harvard University Press: Cambridge, MA.

Schumpeter, J.A. 1942. *Capitalism, Socialism and Democracy.* Harper: New York.

Varian, Hal R. 1994. "Entry and Cost Reduction." Working Paper, Department of Economics, University of Michigan.

Weisman, D. 1993. "Superior Regulatory Regimes in Theory and Practice." *Journal of Regulatory Economics* 5: 355-366.

Chapter #22

How Transmission Affects Market Power in Reserve Services

Laurence D. Kirsch
Laurits R. Christensen Associates, Inc.

Abstract: In electric power markets, transmission constraints facilitate the exercise of market power by limiting the ability of some competing suppliers to serve loads in particular locations. Such constraints can facilitate the exercise of market power in markets for both electrical *energy* services and electrical *reserve* services. This chapter is concerned with the interaction of transmission constraints with market power in electricity reserve markets. It finds that market power in reserve markets arises from the inability of new entrants to compete due to either: a) lack of sufficiently inexpensive opportunities to cite new generation; or b) lack of transmission facilities to transport reserve services to high-priced markets. The existence of reserve markets changes the ways in which electricity suppliers can exercise market power. If energy were the only electricity services, suppliers could exercise market power only if they withheld capacity. With both energy and reserve services, however, suppliers can also exercise market power by shifting capacity among services, generally away from reserve services and toward energy service.

Key words: transmission pricing, transmission congestion, operating reserve, ancillary services, market power

1. INTRODUCTION

In electric power markets, transmission constraints facilitate the exercise of market power by limiting the ability of some competing suppliers to serve loads in particular locations. Analyses of electricity market power therefore

A. Faruqui and B.K. Eakin (eds.), Electricity Pricing in Transition, 329-343.
@ 2002 *Kluwer Academic Publishers.*

need to focus on markets that are geographically defined according to frequently constrained transmission facilities. Such analyses are almost always concerned with the markets for electrical *energy* services. Transmission constraints can, however, also facilitate the exercise of market power in markets for electrical *reserve* services. Thus, a complete analysis of market power may need to consider how transmission affects markets in both energy and reserve services.

This chapter is concerned with the interaction of transmission constraints with market power in electricity reserve markets. The first two sections define market power and explain the general relationship between energy and reserve markets in the absence of market power. The second two sections describe the effects of transmission constraints in fostering market power in energy and reserve markets, respectively. The next two sections present methods for evaluating and mitigating market power in reserve markets. The final section summarizes our conclusions.

2.   DEFINING MARKET POWER

Market power is the ability of a market participant to influence price. It is best measured by the welfare losses that consumers sustain because market prices exceed the market's marginal cost of supply; but it is also meaningfully measured by the amount by which the price exceeds marginal cost.[1] In a market with a single service, the supplier can exercise market power *only* if it withholds some portion of its supply from the market. In a market with multiple services, the supplier can *also* exercise market power by shifting capacity away from services with the least elastic residual demand.[2]

Figure 1 presents a simple analysis of how suppliers exercise and profit from market power, and of how consumers lose. It shows an electrical energy market for a single hour. Demand is represented by the curve D, which is downward-sloping because loads increase as prices fall. Generators' marginal costs are represented by the curve MC, which is upward-sloping because the least expensive generators supply power before the more expensive generators do.

If all firms are price takers, the market price will be P* while the quantity supplied and consumed will be Q*. This is the outcome that maximizes society's net benefits of electricity. It would be inefficient for output to be *higher* than Q* because the marginal cost of higher output would exceed the

---

[1] A classic measure of market power is the Lerner index, which is the percentage by which the market's marginal cost is less than the market price.

[2] "Residual demand" is market demand minus the supply of other suppliers.

marginal value that consumers would attach to such output. It would also be inefficient for output to be *lower* than Q* because the marginal value that consumers would lose from an output reduction would exceed the marginal cost savings from producing less output.

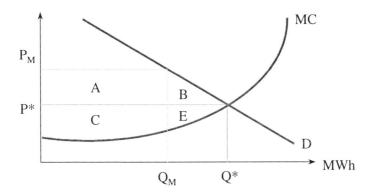

Figure 1. Welfare effects of exercising market power

To exercise market power, suppliers would reduce supply below the efficient level Q*. If they reduced output to Q_M, for example, they could raise price up to P_M. Suppliers would want to reduce output because the profit that they would lose on the output reduction (Q* - Q_M) would equal the area E, while the profit that they would gain on continuing sales Q_M would equal the much larger area A. In other words, suppliers want to reduce output to the extent that their increased profits on continuing sales exceed their lost profits on reduced sales.

The exercise of market power has two bad effects on consumers. First, it induces consumers to reduce consumption and thereby lose the net benefits of this consumption. In Figure 1, area B represents these lost net benefits. Second, it forces consumers to pay higher prices P_M on continuing consumption Q_M. In Figure 1, area A represents consumers increased payments on this consumption. Thus, in Figure 1, the exercise of market power causes consumers to lose an amount of net benefits equal to the sum of areas A+B, where area A is a redistribution of economic value from consumers to suppliers while area B is a net loss of economic value to society.

3.   RELATIONSHIPS AMONG ENERGY AND RESERVE MARKETS

The different types of operating reserves are distinguished by the speed with which they become available and the length of time that they remain available.   The different regions of the U.S. define reserves services somewhat differently, sometimes giving them different names and sometimes recognizing only a subset of the services that are provided in other regions.   The Federal Energy Regulatory Commission's Order No. 888 gave some uniformity to these definitions by naming the following reserve services:[3]

- *Regulation service* is available instantaneously to respond to load fluctuations and to provide an initial response to the loss of power system facilities.   Such responses must be sustainable for up to ten minutes.
- *Spinning reserve service* is available almost immediately, but in not more than ten minutes, to provide a temporary response to load changes and the loss of power system facilities.   Such responses must be sustainable for up to 30 minutes (up to 120 minutes in some regions).
- *Supplemental reserve service* is available within ten minutes to provide a more sustained response to power system changes that are more durable than those addressed by spinning reserves.   Such responses must be sustainable for up to 30 minutes (up to 120 minutes in some regions).
- *Backup reserve service* is available within 30 minutes (60 minutes in some regions) to provide a sustained response to power system changes.   Depending upon the region, such responses must be sustained for hours or days.

Order No. 888 requires jurisdictional utilities to provide the first three services to all transmission system users on a non-discriminatory basis.

Under competitive conditions, each generator will choose to sell the profit-maximizing combination of energy and reserve services.   Figure 2 illustrates how a generator makes this choice under the relatively simple circumstances wherein: a) there is a single reserve service; and b) the generator has a reserve capability equal to two times its minimum load, an

---

[3]   Federal Energy Regulatory Commission, Promoting Wholesale Competition Through Open Access Non-Discriminatory Transmission Services by Public Utilities and Recovery of Stranded Costs by Public Utilities and Transmitting Utilities, Order No. 888, 75 FERC 61,080 (1996).

incremental energy cost of $50 per MWh, and no start-up or shut-down costs.

- When the energy price plus two times the reserve price is less than the generator's incremental energy cost of $50, the generator does not operate. The factor of two reflects the assumption that 2 MW of reserve capability are available for each 1 MW of minimum load. In this case, the revenues from each 0.5 MW of reserve capability are not sufficient to offset the loss on the sale of each 1.0 MW of energy.
- When the energy price is below $50 but the energy price plus twice the reserve price exceeds $50, the generator will operate at its minimum load and provide its maximum quantity of reserves. In this case, the revenues from each 0.5 MW of reserve capability are more than sufficient to offset the loss on the sale of each 1.0 MW of energy.
- When the energy price exceeds $50 but exceeds the reserve price by *less* than $50, the generator will provide its maximum possible quantity of reserves because reserve sales will be more profitable than energy sales. Because energy sales are nonetheless profitable, the generator will use the remainder of its capacity to produce energy.
- When the energy price exceeds the reserve price by *more* than $50, the generator will provide its maximum possible quantity of energy because energy sales will be more profitable than reserve sales. The generator will not provide reserves.

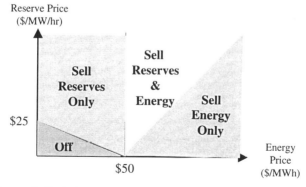

Figure 2. Relationship among energy and reserve markets

4.   TRANSMISSION EFFECTS ON MARKET POWER IN ENERGY MARKETS

By dividing power systems into separate regional markets, transmission constraints create opportunities for generators to exercise market power in those regions that import power from other regions.   In the absence of transmission constraints, all generators could serve all loads.   But in the presence of transmission constraints, generators that are located in exporting regions are limited in their ability to serve importing regions, while generators located in importing regions need fear only limited competition from generators in other regions.

These conclusions can be illustrated by a numerical example.   In this example, all suppliers are price takers except for Firm X, which may or may not exercise market power depending upon circumstances.   We consider a simple power system with a Western region and an Eastern region.   Firm X is located in the Eastern region.   As shown in Figure 3, demand in the Western region is 1 GW lower, at all prices, than demand in the Eastern region; and, excluding Firm X's supply, the Western region has cheaper supply than does the Eastern region.   A reasonable expectation is that cheap power produced in the West will flow eastward to serve the relatively high load in the East.

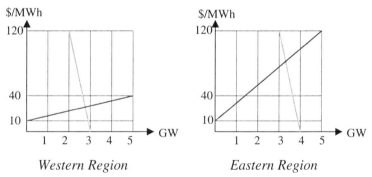

Figure 3. Market supply and demand without firm x and without interregional trade

Firm X has 2 GW (2,000 MW) of capacity located in the Eastern region. This capacity is able to produce power at an incremental cost of $20 per MWh, which, as Figure 3 implies, is cheap relative to its competitors.

In the absence of transmission constraints, Firm X's most profitable strategy is to provide 2,000 MW of energy.   Figure 4 shows that this leads to market energy prices of $31.13 per MWh in both the West and East, with 780 MW flowing eastward.

```
Gen   = 3,521 MW                        Gen   = 2,960 MW
Load = 2,741 MW    Flow = 780 MW        Load = 3,741 MW
Price = $31.13                          Price = $31.13
```

Western Region   *Eastern Region*

Figure 4. Solution with no transmission constraint

If transmission capability were instead limited to 600 MW, society would still be best served if Firm X provided 2,000 MW of energy. Because the unconstrained flow is 780 MW, the 600 MW flow limit would be binding. Figure 5 shows that this 600 MW limit would cause prices to differ among regions, being \$30.10 per MWh in the West and \$34.48 per MWh in the East.

```
Gen   = 3,349 MW                        Gen   = 3,113 MW
Load = 2,749 MW    Flow = 600 MW        Load = 3,713 MW
Price = $30.10                          Price = $34.48
```

Western Region   *Eastern Region*

Figure 5. Solution with a 600 MW transmission constraint: social optimum

But the 600 MW flow limit would provide Firm X with an opportunity to exercise market power. Given the demand and competitive supply shown in Figure 3, Firm X would maximize its profits by providing only 1,389 MW of energy. As shown in , this would drive the Eastern price up to \$45.84 per MWh, but would have no effect on the Western price.

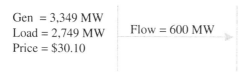

```
Gen   = 3,349 MW
Load = 2,749 MW    Flow = 600 MW
Price = $30.10
```

Western Region   *Eastern Region*

Figure 6. Solution with a 600 MW transmission constraint: firm X exercises market power

Table 1 summarizes the welfare impacts of Firm X's exercise of market power. Firm X increases its own profits by \$6,932. Its Eastern competitors enjoy a \$15,572 profit gain. The reason this occurs is that, at the same time that Firm X reduces its sales from 2,000 MW to 1,389 MW in order to drive up prices, its Eastern competitors increase their sales from 1,113 MW up to

1,629 MW. The competitors not only enjoy the benefits of a price increase but also the benefits of picking up some of the sales that Firm X foregoes.

Eastern consumers, on the other hand, lose benefits of $41,647, mostly because they pay higher prices, partly because they reduce consumption. These lost benefits exceed, by $19,133, the increased profits that suppliers enjoy because of market power. This illustrates the general result that market power confers benefits on suppliers that are always smaller than the losses that it imposes on consumers.

Table 1. Welfare effects of firm X exercising market power (without reserve markets)

	Competitive Case	Market Power Case	Net Welfare Effect
Firm X	$28,958	$35,889	$6,932
Other Eastern suppliers	$13,618	$29,191	$15,572
Eastern consumers			$(41,637)
Total			$(19,133)

5.   TRANSMISSION EFFECTS ON MARKET POWER IN RESERVE MARKETS

Transmission constraints limit generators' access to reserve markets in the same way as they limit access to energy markets. In particular, transmission constraints divide power systems into separate regional markets that are the same for reserves as for energy. A key difference between energy and reserve markets, however, is that only some generators can provide reserve services. Furthermore, the interactions among energy and reserve markets can influence the ways that suppliers exercise market power.

Consider, again, the same power system of the preceding examples, including the demand and supply conditions shown in Figure 3. Suppose that there is no transmission constraint, but that there is now a requirement that the power system must have 500 MW of spinning reserves.[4] In this situation, society's net benefits from electricity are maximized if Firm X provides 1,833 MW of energy and 167 MW of reserves, thus offering all

---

[4]   For simplicity, the examples have fixed reserve requirements that do not depend upon reserve prices. This is in fact the practice in some control areas. Other control areas, by contrast, have reserve requirements that depend upon loads, thus making the requirement indirectly dependent upon energy price. In principle, however, reserve requirements should depend upon reserve price. Because reserves are insurance against generation-related outages, a control area's willingness to pay for reserves should be limited by the value of this insurance. In practice, some independent system operators limit their payments for reserves through price caps that crudely reflect willingness to pay.

2,000 MW of its capacity to the market. Figure 6 shows that this would result in systemwide energy prices of $33.07 per MWh and systemwide reserve prices of $13.07 per MW per hour. The difference between the energy and reserve prices – $20.00 – is the incremental energy cost of Firm X, who is (in effect) the marginal supplier of reserves.

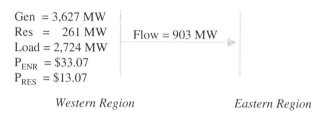

Gen = 3,627 MW
Res = 261 MW   Flow = 903 MW
Load = 2,724 MW
P_{ENR} = $33.07
P_{RES} = $13.07

Western Region   *Eastern Region*

Figure 6. Solution with no transmission constraint: social optimum

Unfortunately, what is best for society is not the same as what is best for Firm X. Firm X can increase its profits by shifting its capacity from reserves to energy, providing 1,898 MW of energy and 102 MW of reserves. As shown in Figure 7, this drives systemwide energy prices down to $33.02 per MWh and while driving up system reserve prices to $15.64 per MW per hour. Table 2 shows that this shift in service offerings causes Firm X's profits to increase, even though Firm X is not withholding any of its capacity from the market. In other words, where there are both energy and reserve markets, it is possible for suppliers to exercise market power even without withholding any capacity. Suppliers can achieve this by shifting capacity from markets with relatively inelastic supply (reserves, in this example) to markets with relatively elastic supply (energy, in this example).

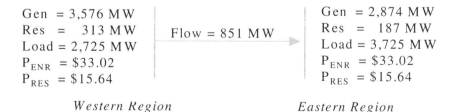

Gen = 3,576 MW   Gen = 2,874 MW
Res = 313 MW   Flow = 851 MW   Res = 187 MW
Load = 2,725 MW   Load = 3,725 MW
P_{ENR} = $33.02 P_{ENR} = $33.02
P_{RES} = $15.64 P_{RES} = $15.64

Western Region   *Eastern Region*

Figure 7. Solution with no transmission constraint:
Firm x exercises market power

If there is a 600 MW flow constraint, with 200 MW of reserves required in the West and 300 MW of reserves required in the East, society would be best off if Firm X provided 1,808 MW of energy and 192 MW of reserves. This would result in the production and prices shown in Figure 8. In the

West, the energy prices and reserve prices are $31.05 and $10.00, respectively. In the East, these respective prices are $39.72 and $19.72, which again differ by the $20.00 incremental energy cost of Firm X, who is the marginal supplier of reserves.[5] The transmission constraint causes the Eastern prices to be higher than the Western prices.

Table 2. Effects of firm x exercising market power
(with reserve markets, with no transmission constraint)

	Competitive Case	**Market Power Case**	**Difference**
Firm X Energy Sales (MW)	1,833	1,898	3.57%
Energy Price ($/MWh)	33.07	33.02	-0.14%
Firm X Reserve Sales (MW)	167	102	-39.05%
Reserve Price ($/MW/hr)	13.07	15.64	19.64%
Firm X Energy Profits	$23,951	$24,714	$764
Firm X Reserve Profits	$2,187	$1,595	$(592)
Firm X Total Profits	$26,138	$26,309	$172

Once again, however, Firm X's private interests diverge from the social optimum. Firm X would maximize its profits by providing 1,587 MW of energy and 78 MW of reserves, thus withholding 335 MW of capacity from the market. As shown in Figure 9, this would result in a moderate increase in the Eastern energy price, to $44.83 per MWh, and a substantial increase in the Eastern reserve price, to $85.87 per MW per hour.[6] Firm X pursues this

---

[5] We ignore the fact that Eastern reserves can be more cheaply provided by using the constrained transmission to provide reserves. Specifically, if the eastward flow were limited to 599 MW, the last MW of transmission capacity could be used to provide reserves for the East. Holding back that last MW of transmission would cost $8.67 per MWh, the excess of the Eastern energy price over the Western energy price. The incremental cost of the 1 MW of transmission-facilitated reserves would be $18.67 per MW per hour, the $8.67 congestion cost plus the $10.00 price of reserves in the West. This $18.67 incremental cost is lower than the $19.72 Eastern reserve price shown in Figure 8. These considerations get directly at the issue of evaluating "capacity benefit margin", which is the use of transmission to provide reserves; but they would unduly complicate the exposition of the text.

[6] In the results of Figures 6, 7, and 8, reserves are provided by units that are already committed to provide energy; so the reserve prices all equal the energy price minus the incremental energy cost of the marginal source of reserves. In the result of Figure 9, all units that are committed to provide energy are providing the maximum quantities of reserves of which they are capable; so the marginal source of reserves is a generator that is committed especially to provide reserves. The accompanying commitment costs results in a reserve price that exceeds the energy price.

strategy because, as shown in Table 3, the increased reserve price yields a profit that more than compensates Firm X for its lost profit on reduced energy sales.

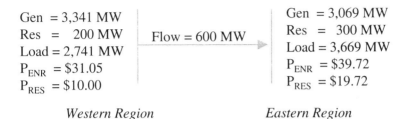

Gen = 3,341 MW		Gen = 3,069 MW
Res = 200 MW	Flow = 600 MW	Res = 300 MW
Load = 2,741 MW		Load = 3,669 MW
P_{ENR} = \$31.05		P_{ENR} = \$39.72
P_{RES} = \$10.00		P_{RES} = \$19.72
Western Region		*Eastern Region*

Figure 8. Solution with a 600 MW transmission constraint: social optimum

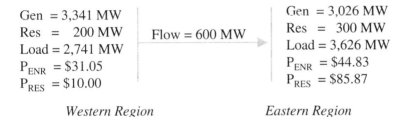

Gen = 3,341 MW		Gen = 3,026 MW
Res = 200 MW	Flow = 600 MW	Res = 300 MW
Load = 2,741 MW		Load = 3,626 MW
P_{ENR} = \$31.05		P_{ENR} = \$44.83
P_{RES} = \$10.00		P_{RES} = \$85.87
Western Region		*Eastern Region*

Figure 9. Solution with 600 MW transmission constraint (firm X exercises market power)

Table 3. Effects of firm exercising market power

	Competitive Case	Market Power Case	Difference
Firm X Energy Sales (MW)	1,808	1,587	-12.20%
Energy Price (\$/MWh)	39.72	44.83	12.86%
Firm X Reserve Sales (MW)	192	78	-59.46%
Reserve Price (\$/MW/hr)	19.72	85.87	335.38%
Firm X Energy Profits	\$71,803	\$71,150	\$(652)
Firm X Reserve Profits	\$3,795	\$6,698	\$2,903
Firm X Total Profits	\$75,598	\$77,848	\$2,251

6.   EVALUATING MARKET POWER IN RESERVE MARKETS

Market power is generally evaluated according to two different types of measures: *structural measures* that infer the *ability* of firms to exercise market power according to the numbers or sizes of competing firms; and *behavioral measures* that examine the *actions* of suppliers. Structural measures include:

- Number of competitors
- Market shares of competitors (concentration measures)

Behavioral measures include:

- "Pivotal bid" frequency by which a particular supplier's bids determine the market clearing price
- Ratios of bid volumes to required volumes
- Ratios of actual prices to competitive prices
- Ratios of bid prices to incremental costs
- Capacity withheld from the market

These measures suffer from a variety of limitations. First, it is not always easy to define the relevant geographical markets. Although transmission constraints are key to defining market regions, loop flows can make regional definitions ambiguous; and changing power system conditions assure that the relevant regions will change over time, sometimes rapidly and unpredictably.

Second, these measures tend to be indicators rather than definitive guides. In some situations, a market with six suppliers with certain market shares can be very competitive, while in other situations; a market characterized by the same market share statistics may be oligopolistic.

Third, the underlying data may be ambiguous or difficult to obtain. For example, incremental costs are difficult to measure. This is especially true for input-constrained resources (such as energy-limited hydroelectric power) that have opportunity costs that depend upon uncertain future market conditions. Ambiguous incremental costs can allow suppliers to mask their exercise of market power, as can strategies such as switching capacity among energy and reserve markets rather than withholding capacity.

Finally, market power is limited, at least in principle, by the potential for entry by new competitors. If the incumbent suppliers get too greedy and raise prices too high, they will attract competition that will bring prices down. The fact that a market has only three suppliers may not indicate the existence of a market power problem if those suppliers have a sufficient fear of new entry. The evaluation of market power should therefore consider the factors that tend to limit the entry of new competitors.

These limitations in the structural and behavioral measures should not be regarded as fatal. They do indicate, however, that the evaluation of market power requires consideration of multiple factors and perspectives to determine whether firms' actions are aimed at manipulating market prices rather than merely responding to those prices.

7. MITIGATING MARKET POWER IN RESERVE MARKETS

There are four general strategies for mitigating the potential or actual exercise of market power in reserve services. These strategies may be pursued simultaneously.

First, new competition should be encouraged. To the extent consistent with environmental protection, this can include reduced regulatory obstacles to siting of new generation and transmission facilities and to upgrading of existing facilities. It can also include institutional reforms that allow wider geographical participation in markets.

Second, trading processes should be more flexible. In particular, the development of forward markets should be encouraged. Because incumbent regulated utilities retain such large shares of retail loads, rules should be developed that will allow incumbent utilities to trade in forward markets without fear of second-guessing by regulators. Without participation by incumbent utilities, the growth of forward markets will be stultified.

Third, reserve requirements should be flexible. Reserve requirements have traditionally been based upon rules of thumb like "spinning reserves must equal X% of load" or "spinning reserves plus supplemental reserves must be sufficient to cover the largest system contingency". These rules of thumb do not take advantage of recent advances in engineering and computer analysis of power systems, nor do they consider market conditions. Consequently, reserve requirements should be reviewed to consider: a) their relationship to the probabilities and costs of customer outages; b) whether reserve requirements might depend upon the current market prices of energy and reserve services; and c) the circumstances under which one type of reserves might substitute for another type.

Fourth, as a last resort, supplier bids for reserve services may be subject to administrative limitations. These can include price caps and minimum MW requirements from certain generators. Such remedies can create more problems than they solve.

8.   CONCLUSIONS

Market power results from barriers to competition. In reserve markets, market power arises from the inability of new entrants to compete due to either: a) lack of sufficiently inexpensive opportunities to cite new generation; or b) lack of transmission facilities to transport reserve services to high-priced markets. These two barriers to competition can be regarded as arising from restrictions on entry (new generation) and restrictions on trade (new transmission). Because of construction lead times, these barriers are always present in the short term (up to a few years). Incumbent firms' pricing strategies may consider the possibility that their short-term pricing decisions can affect longer-term entry by new competitors.

The existence of reserve markets changes the ways in which electricity suppliers can exercise market power. If energy were the only electricity service, suppliers could exercise market power only if they withheld capacity. With both energy and reserve services, however, suppliers can also exercise market power by shifting capacity among services, generally away from reserve services and toward energy service.

Inflexible reserve rules increase opportunities for exercising market power. By making reserve requirements insensitive to reserve prices, system operators can inadvertently create situations in which reserve suppliers are able to tremendously increase reserve prices. A better approach is to recognize that consumers are not willing to pay infinite amounts of money for reliability, but are willing to accept lower reliability as costs rise. Thus, it would be both reasonable and efficient for reserve requirements to fall as reserve prices rise. The appropriate trade-offs between requirements and price can be determined through a probability analysis that treats reserves as a form of insurance transmission pricing, transmission congestion, operating reserve, ancillary services, and market power. In electric power markets, transmission constraints facilitate the exercise of market power by limiting the ability of some competing suppliers to serve loads in particular locations. Such constraints can facilitate the exercise of market power in markets for both electrical *energy* services and electrical *reserve* services. This chapter is concerned with the interaction of transmission constraints with market power in electricity reserve markets. It finds that market power in reserve markets arises from the inability of new entrants to compete due to either: a) lack of sufficiently inexpensive opportunities to cite new generation; or b) lack of transmission facilities to transport reserve services to high-priced markets. The existence of reserve markets changes the ways in which electricity suppliers can exercise market power. If energy were the only electricity services, suppliers could exercise market power only if they withheld capacity. With both energy and reserve services, however, suppliers can also exercise market power by shifting capacity among services, generally away from reserve services and toward energy service.

Chapter #23

Real-Time Pricing and Demand Side Participation in Restructured Electricity Markets

Robert H. Patrick & Frank A. Wolak
Rutgers University and Stanford University

Abstract: This chapter presents the results of an econometric analysis of the customer-level demand for electricity of large and medium-sized industrial and commercial customers purchasing electricity under half-hourly spot prices and demand charges coincident with system peak in the England and Wales (E&W) electricity market. These customer-level demand estimates can be used to forecast customer-level price responsiveness, and design alternative time-of-use and fixed retail pricing options. In particular, we focus on how customers respond to real-time prices and how this response can be used in designing demand side bids. First, we present some background on the E&W electricity market and describe the pool-price contract under which customers purchase electricity according to the half-hourly spot price of electricity.

Key words: real-time pricing; pool-price contracts; England and Wales; spot pricing

1.   INTRODUCTION

All of the wholesale electricity markets operating in the US and abroad lack the price-responsive wholesale demand necessary to achieve the full benefits of a competitive market. Without a significant fraction of final demand responding to the half-hourly or hourly wholesale price, load-serving entities cannot credibly submit demand-side bids into the half-hourly or hourly spot market for wholesale electricity. These demand-side bids can increase the incentive generators have to bid willingness to supply curves

A. Faruqui and B.K. Eakin (eds.), Electricity Pricing in Transition, 345-360.
@ 2002 *Kluwer Academic Publishers.*

closer to their true marginal cost curve, and therefore set wholesale electricity prices closer to the perfectly competitive market price.

Throughout the US, very few retail consumers face prices that vary with the half-hourly or hourly wholesale price of electricity. For this reason, these customers have no incentive to reduce their consumption during hours when the wholesale price of electricity is extremely high and substitute into hours when the price is very low. Because they pay according to the same fixed price regardless of the current wholesale price of electricity, any incentive is to reduce consumption is when it is lowest cost to them, not when it is most beneficial to the efficient operation of the transmission grid. Retail pricing schemes that do not vary with hourly system conditions result in a virtually inelastic hourly demand for electricity. This can have adverse consequences for system reliability and increase the incentive generators have to withhold capacity from the market to drive up the prices, particularly during peak demand periods.

Final customers purchasing according to real-time prices imply a price-responsive wholesale electricity demand for electricity and ancillary services. These customers benefit retailers and all other consumers because their actions reduce the level and variability of spot market prices. This, in turn, decreases the prices that all customers must pay for fixed-price forward contract for electricity deliveries. The ultimate success of electricity industry restructuring depends on formulating market rules which encourage all customers to manage real-time price risk either through forward contract purchases or by managing their consumption in real-time in response to half-hourly or hourly wholesale prices.

This chapter presents the results of an econometric analysis of the customer-level demand for electricity of large and medium-sized industrial and commercial customers purchasing electricity under half-hourly spot prices and demand charges coincident with system peak in the England and Wales (E&W) electricity market. These customer-level demand estimates can be used to forecast customer-level price responsiveness, and design alternative time-of-use and fixed retail pricing options. This chapter summarizes some of the results developed in Patrick and Wolak (2001b). In particular, we focus on how customers respond to real-time prices and how this response can be used in designing demand side bids. First, we present some background on the E&W electricity market and describe the pool-price contract under which customers purchase electricity according to the half-hourly spot price of electricity.

2. THE ENGLAND AND WALES ELECTRICITY MARKET

March 31, 1990 marked the beginning of an evolving economic restructuring of the electric utility industry in the United Kingdom. This process privatized the government-owned Central Electricity Generating Board and Area Electricity Boards and introduced competition into the generation and supply sectors of the market. In England and Wales, the Central Electricity Generating Board, which prior to restructuring provided generation and bulk transmission, was divided into three-generation companies and the National Grid Company (NGC). National Power and PowerGen took over all fossil fuel generating stations, while nuclear generating plants became the responsibility of Nuclear Electric. The twelve Regional Electricity Companies (RECs) were formed from the Area Electricity Boards, which provided distribution services and retail electricity to final consumers. NGC maintains the transmission network, manages the spot electricity market, and coordinates the transmission and dispatch of electricity generating units.

This basic market structure has served as the template for all electricity industry re-structuring worldwide. That is, competition is generally viewed as feasible for structurally or functionally separated firms generating (wholesale) and/or supplying (retail) electricity, while network services (transmission and distribution) are most efficiently provided by regulated monopolies for a single geographic areas.

RECs are required, with compensation for distribution services provided, to allow competitors to transfer electricity over their systems. The RECs' electricity supply prices for all customers with no alternative choice of suppler (so called franchise customers) were regulated by RPI - X + Y, where Y is an adjustment factor which passes-through unexpected costs the REC incurs, as well as the costs of purchased electricity and transmission and distribution services. The 1 MW limit for a customer to be classified as non-franchise was reduced to 100 KW in 1994. This size restriction on customer peak demand was phased out in1998, when residential customers also had the option to choose a supplier. Subsequently, each of the RECs was mandated to separate into retail (supply) and distribution firms. Retailers are no longer price regulated since all customers now have their choice of retailers. Other changes to the E&W market structure and rules have been implemented, as described in Patrick and Wolak (2001b).

In the spot price determination process, generators submit their willingness to supply electricity as a function of price to the E&W pool during each half-hour of the following day. The National Grid Company (NGC) uses the price and quantity pairs that make up these willingness to supply functions to dispatch generation, reserve capacity, and determine the prices that generators receive for the services they provide.

As part of this dispatch and price-setting process, NGC computes a forecast of half-hourly system demands for the next day. The system marginal price (SMP) for each half-hour of the next day is the bid of the highest-price generation unit required to satisfy NGC's forecast of the system demand for each half-hour of the following day. The SMP is one component of the price paid to generators for each megawatt-hour (MWh) of electricity provided to the pool during each half-hour.

The price paid to generators per MWh in the relevant half-hour is the Pool Purchase Price, defined as $PPP = SMP + CC$. CC is the capacity charge, where $CC = LOLP \times \max\{0, VOLL - SMP\}$, $LOLP$ is the loss of load probability, and $VOLL$ is the value of lost load. $VOLL$ is set for the entire fiscal year to approximate the per MWh willingness of customers to pay to avoid supply interruptions during that year. $VOLL$ was set by the regulator at 2,000 £/MWh for 1990/91 and has increased annually by the growth in the Retail Prices Index (RPI) since that time. The $LOLP$ is determined for each half-hour as the probability of a supply interruption due to the generation capacity being insufficient to meet expected demand. The PPP is known with certainty from the day-ahead perspective.

For each day-ahead price-setting process, the 48 load periods within the day are divided into two distinct pricing-rule regimes, referred to as Table A and Table B periods. The pool selling price (*PSP*) is the price paid by suppliers (retailers) purchasing electricity from the pool to sell to their final commercial, industrial and residential customers. During Table A half-hours the *PSP* is

$$PSP = PPP + UPLIFT = SMP + CC + UPLIFT.$$

UPLIFT is a per MWh charge which covers services related to maintaining the stability and control of the National Electricity System and costs of supplying the difference between NGC's forecast of each half-hourly demand and the actual demands for each load period during that day. UPLIFT is only known at the end of the day in which the electricity is produced and is non-zero during Table A periods. For this reason, the *ex ante* and *ex post* prices paid by suppliers for each MWh are identical for Table B half-hours, i.e., $PSP = PPP$.

By 4 PM each day, the Settlement System Administrator (SSA) provides Pool Members, which includes all of suppliers, with the *SMP*, *CC*, *LOLP*, and identity of the Table A and B pricing periods.

3.   THE POOL PRICE CONTRACT

The price contract (PPC) was first offered at the beginning of the second fiscal year of the E&W market to allow consumers with peak demands greater than 1 MW to assume the risks of pool price volatility and therefore avoid the costs associated with hedging against this price uncertainty. Under the PPC, wholesale electricity costs for both energy and transmission services are directly passed through to the customer. For traditional fixed price contracts, the retailer absorbs all the wholesale price risk associated with wholesale purchases made at the PSP.

The Midlands Electricity Board or MEB had 370 commercial and industrial customers (of approximately 500 customers with demands over 1 MW) purchasing their electricity according to a PPC for the year April 1, 1991 through March 31, 1992. The number of customers on the pool price contract remained stable over the following two years, although approximately one-fourth of the customers changed each year. For the year of April 1, 1994 to March 31, 1995, when the pool price contract was first offered to smaller consumers—those with greater than 100 KW peak demand—a number of commercial customers, as well as smaller industrials, were then given the option to purchase electricity according to pool prices. Approximately 150 customers in this size class signed up for the Pool Price contract for the year 1994-1995. Patrick and Wolak (2001b) describe the types of firms and industries contained in our sample.

The expected PSPs for all 48 half-hourly intervals beginning with the load period ending at 5:30 am the next day until the load period ending at 5:00 am the following day are faxed to all pool price customers immediately following the supplier's receipt of the SMP, CC and the identity of the Table A and Table B periods from NGC. MEB develops forecasts of the UPLIFT component of the PSP and provides these with the 48 half-hourly SMPs and CCs. The actual (*ex post*) PSP paid by electricity consumers on the PPC for Table A periods is known 28 days following the day the electricity is consumed. The actual or *ex post* PSP is equal to the *ex ante* PSP for Table B periods because the UPLIFT is known to be zero in these load periods.

Customers on PPCs also pay a demand charge. A £/MW triad charge is levied on the average capacity used by each PPC customer during the three half-hour load periods ("triads") in which the load on the E&W system is highest, subject to the constraint that each of these three periods is separated from the others by at least ten days. The triad charge is set each year by NGC (subject to their RPI-X price cap regulation) and has ranged from 5,420 to 10,730 £/MW.

There are various mechanisms suppliers use to warn their PPC customers of potential triad periods. *Triad advance warnings* are generally faxed to consumers on Thursday nights and give the load periods during the following week that the supplier feels are more likely to be triad periods. *Triad priority alerts* are issued the night before the day that the supplier considers the probability of a triad period to be particularly high. These alerts also list the half-hours most likely to be triad periods. Actual triad charges have only occurred in the four-month period from November to February.

The actual price for service paid by PPC customers also contains various other factors, not related to the pool prices, which are analogous to charges fixed rate customers pay.

Our data includes the half-hourly consumption of all of MEB's *PPC* customers from April 1, 1991 through March 31, 1995. We also collected the information contained on the faxes sent to each *PPC* customer the day before actual consumption occurred. This fax contains the *ex ante* half-hourly forecasted *PSP* for the sample period—*SMP + CC + Forecasted Uplift Charge*. In addition, we collected information on the actual value of *UPLIFT* for our sample period.

The price determination process in the E&W market does not use the actual market demand to set the two largest components of the PSP, the SMP and CC. These prices are set using a perfectly inelastic (with respect to price) forecast of the market demand for each half-hour of the next day. In addition, few customers pay for electricity during any half-hour at the PSP or even at prices that vary with the PSP throughout the day, month, or year. Consequently, any attempt to estimate a relationship between the PSP for a given half-hour and the total system load for that half-hour will not recover a consistent estimate of the true relationship between final demand and the half-hourly market price for any customer or group of customers. This is because the actual PPP is set by NGC's forecast of total system load and not the actual total system load.

4.   CUSTOMER RESPONSE TO REAL-TIME PRICES

In this section we present the mean price elasticities for some of the firms in the E&W market facing real-time prices under the PPC. Because prices and demands are extremely variable over the course of the year and within the day, there is considerable variability both within the day and across days in these own- and cross-price elasticities. In addition, these elasticities vary significantly across firms within and across industries as well as across days due to changing weather conditions.

Given the amount of price volatility in the PSP and the expected demand charge, even the smallest half-hourly within-load-period own-price elasticities of demand can imply significant load reductions in response to price increases. In addition to this variability, the volatility in the expected demand charge should be taken into account in estimating customer-level demand under the PPC. In particular, it would not be unusual to have values of expected prices across days for the same load period that differ by a factor 20 or 30, which should result in a substantial reduction in the within period demand.

There is a considerable amount of heterogeneity in the degree of price responsiveness observed in the customers we studied. Firms in the water supply industry pump substantial amounts of water into storage and sewage-treatment facilities once or twice a day. These firms generally have the ability to shift this activity to the lowest-priced load periods within the day at very short notice. As expected, there is a considerable amount of heterogeneity in the pattern of within-day price responsiveness, as well as across the firms in this industry. Figure 1 plots the sample mean own-price elasticities as a function of the load period for two firms in the water supply industry. Although during the usual peak total system load periods, 2:30 PM to 6:00 PM (load periods 20 to 26 on Figure 1), we find relatively small mean own-price elasticities for these periods, ranging from .06 to .26 in absolute value. For the load periods immediately preceding and following this time period, the mean own-price elasticities increase rapidly to as large as 0.86, in absolute value. This implies, for example, a 1% increase (decrease) in price during a load period may lead to as much as a 0.86% decrease (increase) in electricity consumed in that period.

Figure 2 plots the sample mean own-price elasticities as a function of the load period for firms in the copper, brass, and other copper alloys manufacturing industry. Again, during the usual peak total system load periods, beginning at 2:30 PM and ending at 6:00 PM, we find a uniform and relatively small mean own-price elasticity. For the load periods immediately preceding and immediately following, the mean own-price elasticity is over 0.2 in absolute value and gets as large in absolute value as 0.3 in load period 4, the period from 6:30 to 7:00 AM. Although these firms are characterized by significantly smaller mean own-price elasticities than the water industry, these elasticities still indicate substantial load response to prices, particularly given the considerable variability in the PSP and expected demand charge across days.

Figure 3 presents analogous information for seven firms in the hand tools and finished metals goods manufacturing industry. Figure 4 presents the own-price elasticities for the food, drink and tobacco manufacturing firms. Figures 1 through 4 are indicative of the range and variability of mean own-price elasticities of firms within and across various industries. Additional information on these firms and industries, as well as others in the E&W market, are reported in Patrick and Wolak (2001b).

By responding to half-hourly price signals, some customers were able to significantly reduce their electricity costs under real-time price contracts. Figures 5-8 are histograms of the customer-level quantity weighted (by that customer's demand each half-hour) average price for all RTP customers for each year of our sample. Comparing the quantity-weighted average prices for RTP customers to fixed price contracts, such as presented in Patrick and Wolak (2001b), demonstrate that many RTP customers paid substantially lower prices per kWh than did fixed price customers. This indicates a substantial premium is earned by E&W retailers supplying fixed price customers and that many customers place a significant premium on the fixed price contract. As can particularly be seen in Figure 8, which represents the year in our sample for which the E&W spot prices were extremely volatile, consumers that can respond to spot prices end up paying even lower average prices when spot prices are more volatile. The corresponding quantity-weighted average prices, weighted by total system load (TSL) for each fiscal year, are £23.74/MWh, £25.08/MWh, £28.36/MWh, and £29.82/MWh, respectively. These total system load weighted prices indicate that many RTP customers were able to benefit from RTP rather than fixed price contracts.

5.   DEMAND SIDE BIDS

Demand side bids effectively introduce elasticity into the demand side in the price determination process, leading to lower SMP for the half-hour than would be set using a perfectly inelastic demand. As emphasized by Wolak and Patrick (2001), the operation of the E&W market illustrates the sort of price volatility that can occur if the demand setting the market price is very price inelastic and only a small fraction of the total electricity consumed in any half-hour is sold to final customers at prices that vary with the half-hourly PSP. Consequently, accurate measurement of the within-day price response of customers is an important necessary ingredient for the design of aggressive demand side bids that build significant price-responsiveness into the price-setting process.

If retailers purchasing from the E&W market are able to accurately predict the response of demand to within-day price changes for their customers on the PPC, this information can be used to formulate a demand side bid function for each supplier. If a supplier is able to entice more customers to face prices for electricity which reflect the current PSP from the E&W market for that half-hour, given accurate estimates of the price-responses of these customers, the supplier can then formulate an aggregate demand side bid function which has a relatively larger price response.

The demand side bid by each retailer is incorporated into their estimated demand function by taking into account PPC customers' responses to the day-ahead market prices. If retailers are provided a means of participating in the spot price determination process and can in fact affect the market price by their demand-side bids, then their actions may significantly lower the magnitude and volatility of market prices.

All of the competitive electricity markets currently operating in the US (the PJM New York, the New England and the California electricity markets) allow demand-side bidding into their day-ahead market and their real-time markets. Consequently, the major barrier associated with implementing significant amounts of demand-side bidding in these markets is the fact that very little final load in these markets faces a retail price that varies with the hourly wholesale electricity price.

All of these markets except the one in New England currently operate a day-ahead market and a real-time market for electricity. In the day-ahead market, generators and loads (retailers or end-use consumers) make forward commitments to supply and consume electricity in a given hour during the following day. If the generator supplies more or less energy in real-time than its day-ahead commitments, it must buy or sell that magnitude from the real-time market at the real-time price. The same logic applies to load-serving entities. For example, in New York, generators and loads bid into the day-ahead market supply and demand schedules at each node in the transmission grid. A nodal-pricing dispatch model computes the least cost dispatch necessary to meet the bid-in demand. This results in a day-ahead schedule for both loads and generators at each node in the grid. Many of the ISOs run within-day markets to allow generators and loads to re-adjust their schedules up until an hour before the power is actually scheduled for delivery. For example, a load-serving entity that believes its demand for energy during a given hour in the day will be higher than the quantity it was able to procure in the day-ahead market will bid a willingness to demand into the within day market. This will allow it to purchase additional energy in the forward market. Alternatively, a generator that believes it has sold too much of its capacity in the day-ahead market, can buy back some of the energy it was committed to sell in a given hour of the day, in the within day market.

Once the hour-ahead market closes, both generators and loads have firm financial commitments that they are expected to honor in real-time. However, both generators and loads will deviate from their hour-ahead schedules in real-time for a variety of reasons. A load-serving entity that consumes more in real-time than its final hour-ahead schedule must purchase this additional energy at the real-time market price, which equates the aggregate real-time supply with the aggregate real-time demand for energy. For example, suppose a generator commits to supply 500 MWh in its hour-ahead schedule. If it happens to actually supply only 400 MWh in that hour, it must purchase the 100 MWh deficit relative to its hour-ahead schedule at the real-time price. A similar logic applies to loads. If a load-serving entity has an hour-ahead schedule for 200 MWh, if it actually consumes 300 MWh during that hour, it must purchase the additional 100 MWh at the real-time price. This multi-settlement mechanism makes it very easy for load-serving entity with no buying power in the spot market to bid optimally into each of these markets. The firm would simply bid in its best estimate of its aggregate demand curve for a given hour conditional on the best information it currently has available concerning the form of this demand. The logic for this bidding strategy can be seen as follows.

A price-responsive demand bid can be treated in the following manner. Suppose a load-serving entity bids in the demand curve $D(p)$. This demand can be translated into an energy supply bid as follows. Let $D(0)$ denote the level of demand at a price of zero dollars. Treat this as the inelastic demand bid of this load serving entity. Then assume this load-serving entity is willing to supply generation (less energy consumption), according to the following supply function $SL(p) = D(0) - D(p)$. Consequently, if all demand side bids are treated in this manner, then it is clear that a price-responsive demand bid is exactly equivalent to a price-responsive supply bid. If $D(p)$ is non-increasing in p, than $SL(p)$ is non-decreasing in p, just like any supply bid function for a generator. Therefore any framework for bidding by a supplier in a competitive electricity market can be used by a demander in a competitive electricity market. As is well-known in standard multi-unit bidding models, when a market participant has no ability to influence the market price through its bidding behavior, its optimal strategy is to bid its marginal cost function into the market. The analogue to the load-serving entity's marginal cost curve is its true supply curve $SL(p)$. Consequently, if a load-serving entity is sufficiently small relative to the market, then it should simply bid in its best estimate of $SL(p)$ for the hour under consideration, given the information it currently has available.

6.   SUMMARY

Several important lessons emerge from our analyses in Patrick and Wolak (2001b). These include:

- Day-ahead price elasticity estimates vary substantially by time-of-day, industry, and firms within industries; with sample mean averages ranging from essentially zero to 0.86 in absolute value.
- Significant across-period price responses are also found, indicating substitution possibilities exist across some pricing periods within the day, while complementarities exist across other pricing periods.
- Real-time pricing contract customers provide a significant price response in restructured electricity markets.
- A significant price response in the market price determination process will reduce the magnitude and volatility of spot prices in restructured electricity markets, the extent depends upon the price responsiveness of real-time pricing customers and how demand side bids are considered in the market price determination process.
- Even if there is no demand-side bidding in spot market, by assuming the spot price risk and responding to it, customers can significantly reduce their electricity costs relative to fixed price contracts.
- Retailer incentives for demand side management and conservation result during peak system demand periods, when the spot prices are greater than the fixed prices at which most retail power is sold. Conversely, any time the spot price is below the fixed price, the retailer has the incentive to encourage electricity consumption. Substitution of consumption from the former to the latter pricing periods can increase retailer profits.

The incentives for retailers to demand side bid are also relevant for pricing structures in addition to spot pricing contracts. For example, interruptible rate structures and other load management programs, may allow retailers to credibly increase their demand side bids. Eakin, Glyer, and Faruqui (2000) discuss a range of possible interruptible pricing options and their potential impact on reducing demand during high price demand periods.

The difficulty in incorporating a demand side into the price determination process in restructured electricity markets has been that few retail consumers face prices that vary with the half-hourly wholesale price of electricity. Customers have little incentive to shift their electricity consumption away from peak periods if they do not face half-hourly prices that reflect the real-time cost of purchasing wholesale electricity. This can have adverse consequences for system reliability, enhance the ability of generators to exercise market power in the spot electricity market, and significantly reduce the likelihood that consumers will benefit from electricity industry re-structuring.

REFERENCES

Eakin, K., D. Glyer, and A. Faruqui. 2000. "The Next Generation of Curtailable Pricing Products." EPRI Technical Brief, Electric Power Research Institute, Palo Alto, CA.

Patrick, R.H., and F.A. Wolak. 1997. *Customer Load Response to Spot Prices in England: Implications for Retail Service Design.* Electric Power Research Institute TR-109143 (November), Palo Alto, CA.

Patrick, R.H., and F.A. Wolak. 2000. "Using Customer-Level Response to Spot Prices to Design Pricing Options and Demand side Bids." In *Pricing in Competitive Electricity Markets*, edited by A. Faruqui and K. Eakin. Kluwer Academic Publishers: Boston, MA.

Patrick, R.H., and F.A. Wolak. 2001a. "Estimating the Customer-Level Demand for Electricity Under Real-Time Market Pricing." NBER Working Paper 8213, National Bureau of Economic Research, Cambridge MA, April.

Patrick, R.H., and F.A. Wolak. 2001b. *Using Customer Demands Under Spot Market Prices for Service Design and Analysis.* EPRI WO 2801-11, Electric Power Research Institute, Palo Alto, CA (available from the EPRI Retail Commodity Service Design site www.epri.com/Target 24.2).

Wolak, F.A., and R.H. Patrick. 2001. "The Impact of Market Rules and Market Structure on the Price Determination Process in the England and Wales Electricity Market," NBER Working Paper 8248, National Bureau of Economic Research, Cambridge MA, April.

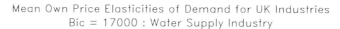

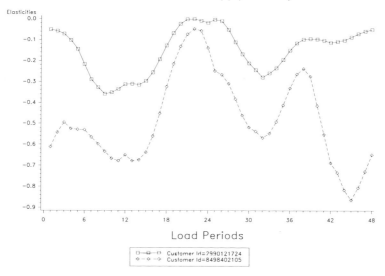

Figure 1. Mean own price elasticities for water supply firms.

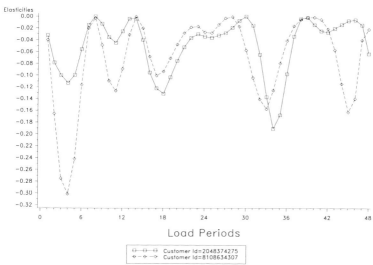

Figure 2. Mean own price elasticities for copper, brass, and other copper alloys manufacturing firms.

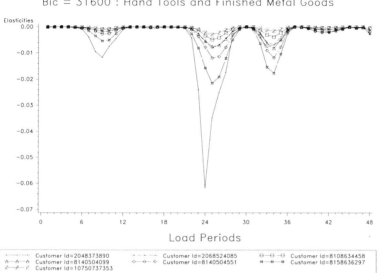

Figure 3. Mean own price elasticities for hand tools and finished metal goods manufacturers.

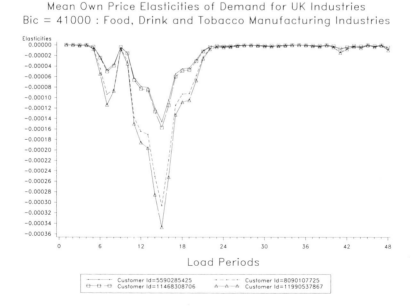

Figure 4. Mean own price elasticities for food, drink, and tobacco manufacturing firms.

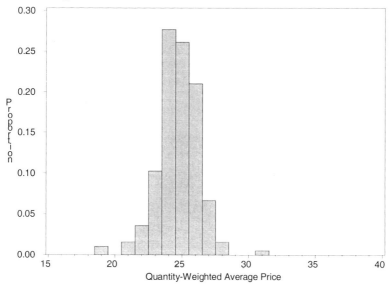

Figure 5.

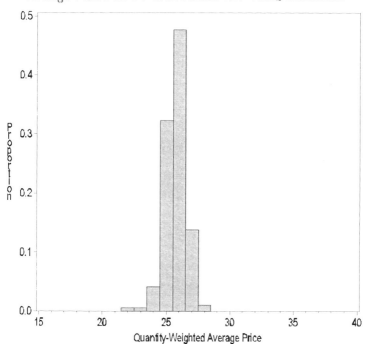

Figure 6

Average Prices for FY 1993 Across RTP Retail Customers

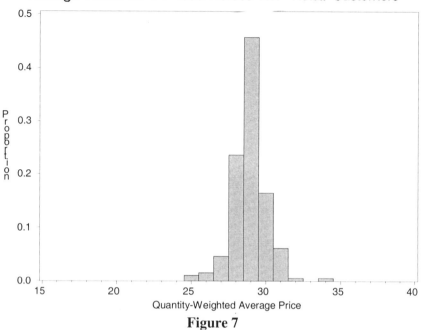

Figure 7

Average Prices for FY 1994 Across RTP Retail Customers

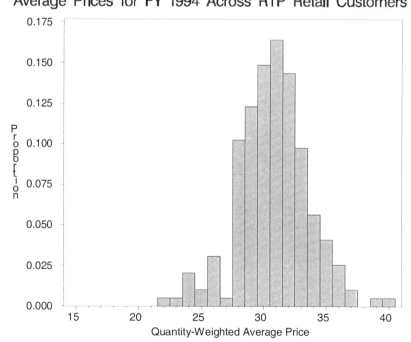

Figure 8.